ALS

Advances in Life Sciences

Metallothionein III

Biological Roles and Medical Implications

Edited by
K.T. Suzuki
N. Imura
M. Kimura

Birkhäuser Verlag
Basel · Boston · Berlin

Editors

Prof. K.T. Suzuki
Faculty of Pharmaceutical Sciences
Chiba University
1-33 Yayoi
Inage, Chiba 263
Japan

Prof. N. Imura
Kitasato University
School of Pharmaceutical Sciences
9-1, Shirokane 5-chome, Minato-ku
Tokyo 108
Japan

Prof. M. Kimura
Keio University
School of Medicine
35 Shinano-machi
Shinjuku-ku
Tokyo 160
Japan

A CIP catalogue record for this book is available from the Library of Congress, Washington D.C., USA

Deutsche Bibliothek Cataloging-in-Publication Data
Metallothionein III : biological roles and medical implications / [Third International Conference on Metallothionein]. Ed. by K. T. Suzuki ... – Basel ; Boston ; Berlin : Birkhäuser, 1993
(Advances in life sciences)
ISBN 3-7643-2769-3 (Basel ...)
ISBN 0-8176-2769-3 (Boston)
NE: Suzuki, Kazuo T. [Hrsg.]; International Conference on Metallothionein <03, 1992, Tsukuba>

Camera-ready copy prepared by the authors
Printed on acid-free paper produced from chlorine-free pulp
Printed in Germany
ISBN 3-7643-2769-3
ISBN 0-8176-2769-3

9 8 7 6 5 4 3 2 1

Table of Contents

Molecular and Cellular Biology

Preface

More than 35 years have passed since the first publication on metallothionein appeared. For a long time, following the biochemical characterization as a low molecular weight metal-binding protein rich in cysteinyl residue and heavy metals, metallothionein was believed to be related primarily with toxicity of heavy metals. Following several studies related to cadmium toxicity, the roles of metallothionein in the metabolism, nutritional and stress-related aspects were investigated. In parallel to these biological interests, metallothionein was studied in coordination and protein chemistry because of its unique structure. Then the promoter region of metallothionein gene found a use as a fusion gene for researchers not only in metal-related fields but also in many new fields of gene technology.

The present book comprises the most interesting and new aspects of metallothionein research presented at the Third International Conference on Metallothionein which took place in Tsukuba, Japan, in December 1992, following the First and Second Meetings, held in Zürich, in 1978 and 1985. It includes certain topics in the form of review articles presented by selected speakers as well as introductory overviews by Professors B.L. Vallee, J.H.R. Kägi, J. Otvos, K.H. Summer, M.P. Waalkes, M.G. Cherian and D.H. Hamer which cover respective mini-reviews.

In former years the roles of metallothionein in heavy metal toxicology and the primary structure in protein biochemistry, coordination and gene structure were the centre of attention, whereas this book intends to focus on the biological roles and medical implications of metallothioneins. The roles of metallothionein in various kinds of stresses including exposure to heavy metals, organic and inorganic chemicals, and physical environment have been continuing interests, and the induction mechanisms by cytokines and hormones were discussed. Scavenging ability for active oxygen species, radicals and alkylating reagents have been anticipatory roles for metallothionein in protection of the body, and this property has been related to resistance to anticancer agents and also utilized to prevent adverse effects in cancer chemotherapy.

The discovery of growth inhibitory factor (GIF) in brain and nerve and its characterization as metallothionein-3 has opened a new interest.

GIF was isolated as an inhibitor against overgrowth of brain cells. It has a close resemblance to metallothionein protein structure, including coordination of zinc and copper and is now classified as a new metallothionein, metallothionein-3. However, this protein is not inducible by heavy metals and its binding with heavy metals has not been clarified yet, despite similar binding of zinc and copper to metallothionein. The specific expression of metallothionein-3 in brain and astrocytes is another interesting point to be resolved.

Localization of metallothionein in nucleus in relation to the cell cycle has been mentioned to have some roles in the differentiation and/or proliferation of cells. Site-directed mutagenesis technique has been utilized to make metallothioneins of altered amino acid composition in order to study coordination chemistry and also to determine the biological function. Degradation of metallothionein has been studied in detail recently. Structure of metallothionein by NMR has characterized metal coordination structure in solution. Metallothioneins of known amino acid structures in the classes I and II are phylogenetically compiled and the classification of metallothionein includes phytochelatin as the class III metallothionein.

It is a pleasure to acknowledge the help and assistance of all contributors, and to thank them for the hard work they put into the meeting and the preparation of excellent manuscripts. We are indebted to our many sponsors for their generous financial support which enabled us to organize this meeting; their names are listed in full at the end of this volume. The editors also express their thanks to The Tsukuba EXPO '85 Memorial Foundation for their financial support of the publication of this book.

September, 1993

Kazuo T. Suzuki
Nobumasa Imura
Masami Kimura

List of Corresponding Authors

Akimoto, M.
Nippon Medical School
1-1-5, Sendagi, Bunkyo-ku
Tokyo 113, Japan

Andrews, G.K.
Biochemistry and Molecular Biology
University of Kansas Medical Center
Kansas City, KS 66103, U.S.A.

Bremner, I.
Rowett Research Institute
Greenburn Road
Bucksburn, Aberdeen
AB2 9SB
United Kingdom

Cherian, M. G.
Health Sciences Ceter
Department of Pathology
The University of Western Ontario
London, Ontario
Canada N6A 5C1

Gedamu, L.
Department of Biological Sciences
University of Calgary
22500 University Drive N. W.
Calgary, Alberta T2N 1N4
Canada

Hamer, D. H.
Cell Regulation Section
National Cancer Institute
Bethesda, MD 20892
U.S.A.

Huang, P. C.
The Johns Hopkins University
School of Hygiene and Public Health
Department of Biochemistry
615 North Wolfe Street, Baltimore
Maryland 21205, U. S. A.

Imura, N.
Kitasato University
School of Pharmaceutical Sciences
9-1, Shirokane 5-chome, Minato-ku
Tokyo 108, Japan

Kägi, J. H. R.
Institute of Biochemistry
University of Zürich
Winterthurerstrasse 190
CH-8057 Zürich
Switzerland

Kimura, M.
Keio University, School of Medicine
35 Shinano-machi, Shinjuku-ku
Tokyo 160
Japan

Klaassen, C. D.
Department of Pharmacology,
Toxicology and Therapeutics
University of Kansas Medical Center
Kansas City, Kansas 66103, U. S. A.

Koizumi, S.
National Institute of Industrial Health
6-21-1, Nagao, Tama-ku
Kawasaki 214
Japan

Lazo, J. S.
Department of Pharmacology
University of Pittsburgh
School of Medicine
E1340 Biomedical Science Tower
Pittsburgh, PA 15261, U.S.A.

Naganuma, A.
Kitasato University
School of Pharmaceutical Sciences
9-1, Shirokane 5-chome, Minato-ku
Tokyo 108
Japan

Nakajima, K.
Japan Immunoresearch Laboratories
17-5 Sakae, Takasaki
Gunma 370, Japan

Otvos, J.
Department of Biochemistry
North Carolina State University
Box 7622 Raleigh
NC 27695, U.S.A.

Palmiter, R. D.
Howard Hughes Medical Institute
University of Washington
HHMI SL-15
Seattle, WA 98195
U.S.A.

Petering, D. H.
Department of Chemistry
University of Wisconsin-Milwaukee
P. O. Box 413
Milwaukee, WI 53201
U.S.A.

Rauser, W. E.
Dept. of Botany
University of Guelph
Ontario N1G 2W1
Canada

Roesijadi, G.
The University of Maryland
Chesapeake Biological Laboratory
Box 38, Solomons
MD 20688, U.S.A.

Saijo, N.
Pharmacology Division
National Cancer Center Research Institute
5-1-1 Tsukiji, Chuo-ku
Tokyo 104, Japan

Sato, M.
Central Research Laboratory
Fukushima Medical College
1, Hikarigaoka
Fukushima 960-12, Japan

Summer, K. H.
Gesellshaft für Strahlen und
Umweltforschung mbH
Institut für Toxiocogie München
Ingolstadter Landstr. 1
D-8042, Neuherberg, Germany

Suzuki, K. T.
Faculty of Pharmaceutical Sciences
Chiba University
1-33 Yayoi
Inage, Chiba 263
Japan

Tanaka, K.
Faculty of Pharmaceutical Sciences
Osaka University
1-6, Senrioka, Suita
Osaka 565, Japan

Tohyama, C.
National Institute for Environmental Studies
16-2 Onogawa
Tsukuba, Ibaraki 305
Japan

Tsujikawa, K.
Faculty of Pharmaceutical Sciences
Osaka University
1-6, Yamadaoka, Suita
Osaka 565, Japan

Uchida, Y.
Tokyo Metropolitan Inst. of Gerontology
35-2, Sakae, Itabashi-ku
Tokyo 173, Japan

Vallee, B. L.
Harvard Medical School
Seeley G. Mudd Buld.
250 Longwood Av.
Boston, MS 02115
U.S.A.

Waalkes, M. P.
National Cancer Institute
Frederick Cancer Research and
Development Center
Bldg. #538, Rm. 205-E
Frederick, MD 21701-1013
U.S.A.

Winge, D. R.
Departments of Medicine
and Biochemistry
Universituy of Utah Medical Center
Salt Lake City, Utah 84132
U.S.A.

Metallothionein III
K. T. Suzuki, N. Imura and M. Kimura (eds)

THE FUNCTIONAL POTENTIAL AND POTENTIAL FUNCTIONS OF METALLOTHIONEINS: A PERSONAL PERSPECTIVE

Bert L. Vallee and Wolfgang Maret

The Center for Biochemical and Biophysical Sciences and Medicine
Harvard Medical School, 250 Longwood Avenue, Boston, MA 02115

SUMMARY: Thirty-five years ago a concerted effort was undertaken to elucidate a possible biological role for cadmium akin to that then becoming known for zinc. This led to the discovery of a low molecular weight, cysteine-rich protein containing several metals, and aptly named metallothionein. After a decade of latency, this protein suddenly became the subject of increasing scrutiny and in recent years has been the central focus of several monographs and two previous international conferences, now a third. This personal view of the coming of age of metallothionein describes the milestones marking progress in this journey. In particular, it considers how the coordination chemistry of zinc ascertained from structural analyses of zinc enzymes and zinc-containing DNA-binding proteins affords new perspectives on the biological function of metallothionein.

I appreciate the singular honor to keynote these proceedings with this prefatory chapter and thank the president, the organizing committee and in particular Profs. Imura and Suzuki, whose help, patience and kindness to me have known no limit. In accord with their suggestion I will try to place into perspective both the scientific progress made since the discovery of metallothionein almost 40 years ago and the scientific, medical, and social circumstances which have had major influences on the course and rate of this progress. It is remarkable that after all this time there is still hardly a clue to the question of what the role of this molecule may be and what potential it may have. I will here recall the milestones that have been reached and attempt to integrate them to the extent possible.[1]

---

[1]In assembling the cardinal events and my recollections, I have had the valuable collaboration of my colleague, Dr. Wolfgang Maret, whom I invited to lend his skills, judgment and objectivity to these efforts.

Metallothionein and the Chemistry and Biochemistry of Cadmium. The discovery of metallothionein (Margoshes and Vallee, 1957) was the result of a search for a biological role of cadmium about which surprisingly, even to this day almost nothing is known. In fact, I have never found a really good discussion of cadmium biochemistry nor, for that matter, any concerted effort to relate this element to biology. Whatever mention of cadmium chemistry there is regarding its presence in metallothionein relates to the methodological aspects of cadmium, e.g., spectroscopy, protein and physical chemistry, kinetics, immunology or toxicology. In regard to metallothionein itself, cadmium chemistry does not seem to have been of much interest either to biologists or inorganic chemists.

In this context a discussion of the chemistry and biochemistry of cadmium, brief as it must be, might serve as a useful backdrop for considering its potential biological significance both in metallothionein and in biology as a whole. Cadmium is chemically and physically quite similar to zinc. The two metals differ most obviously in their boiling points, 907°C for zinc vs 767°C for cadmium. While crude, this is a specific means to differentiate, separate and recognize them. Both form divalent cations with a complete d shell of electrons which accounts for the lack of ligand field stabilization of their coordination complexes. The stereochemistry of these complexes is therefore determined almost entirely by electrostatic considerations, covalent bonding forces and the size of the metal ions. The coordination geometries of both metals are remarkably adaptable and this may be pertinent to their biological roles. Their coordination numbers vary from 2 to 8, but both elements tend to adopt 4, 5 and 6 coordination although Zn(II) (radius = 0.69 Å) distinctly prefers five coordination while Cd(II) (radius = 0.92 Å) prefers six coordination. These features are significant in regard to their interactions with donor atoms. Both favor the non-metals in the order S > Se > P, and replace each other isomorphically in minerals. Zinc occurs in several minerals, principally sphalerite, in which cadmium is typically present at 0.2 to 0.4 %. There are relatively few specific cadmium deposits such as CdS (greenockite).

Finally, zinc ranks 27th to 29th in abundance in the earth's crust, and cadmium is 67th. Knowledge concerning the biochemical interactions of cadmium known up to 1972 were summarized earlier (Vallee and Ulmer, 1972), and will not be extended here.

My search for a biological cadmium complex was based on the inorganic, physico-chemical and geological information just stated. It seemed reasonable to me to ask whether or not the chemical similarities between zinc and cadmium that dictate their geology would produce a similar parallel in biology. It may be appropriate to cite my views at the time (Vallee, 1958). "The similarity between the atomic structure and chemical behavior of cadmium and zinc suggested that cadmium might also occur as an integral part of a natural product. Such a complex had never previously been demonstrated although cadmium has been found in various species." Even as I set out to answer the question there were already strong indications that zinc was biologically very important though at the time only a few zinc proteins were known and, hence, the case for that was far from established. If there was to be a biological corollary of geochemistry then there should be cadmium proteins, cadmium enzymes or other cadmium natural products. Similarly, it seemed valid to ask whether or not cadmium could be an isomorphic substitute for zinc in proteins or other biological ligands much as it does in the inorganic world. Moreover, such a search might uncover novel Zn, Cd, S, Se, P or yet other simple and/or mixed Cd/Zn complexes. It might even be that cadmium would activate or inhibit zinc enzymes on some kind of systematic basis.

Most of these speculative thoughts have never materialized. But they did uncover metallothionein although not its function. I will again recall my comments at the time of its discovery: "The low sedimentation constant and high metal content of [metallothionein] indicate a protein of low molecular weight, probably containing a small number of cadmium atoms per molecule. Thus far, no biological activity has been attributed to this protein. The remarkable aggregation of cadmium would lead to the inference that the element has a significant biological function.

But at present it remains a metalloprotein in search of a biological activity" (Vallee, 1958). Thirty-five years later this statement, while just as valid, continues to be clouded in ignorance; however, there may be a silver lining, which the balance of this discussion will explore. At this juncture I want to reiterate that I am still baffled that overall the occurrence and role of cadmium in biology is as much a mystery today as it was then. When compared with the virtual explosion of our knowledge of zinc biochemistry during this time span the cadmium enigma becomes magnified many fold.

Metallothionein has remained the only protein of its kind so far to be found in animals and microorganisms. Quite the same, the detection of metallothionein isoforms (MT I-III) and perhaps additional structural variations, may be straws in the wind of yet additional variations on this theme. In fact the emergence of the plant phytochelatins (Rauser, 1990; Grill et al., 1991) support this view and suggest that the "discovery period" may not yet be over.

<u>Metallothionein and Toxicology.</u> One of the most important consequences of the discovery of metallothionein was its impact on the study of the toxicology of cadmium and zinc. After lying on fallow biochemical ground for almost a decade, the molecule took root in the hearts and minds of toxicologists who adopted it with unbridled enthusiasm, perhaps - as I often suspected - with the tacit assumption that its very existence provided a raison d'être to their field. After all, metallothionein seemed to be a veritable "deus ex machina" proving that life had to be protected against the toxicity of metals. It mattered little that at least in the case of zinc there was virtually no evidence of toxicity aside from metal fume fever which had long since been recognized to result from inhalation of zinc dust generated during smelting of zinc ores. Acute ingestion of zinc was not known to result in toxic symptoms, a conclusion long since confirmed. If metallothionein was not a detoxifying agent for zinc, that would leave cadmium detoxification as its putative function, most likely

carried out by its metal-free precursor, thionein. Remarkably, evidence for cadmium toxicity has proved to be almost as elusive as that for zinc toxicity, but I suspect it is inappropriate for me to raise that issue under the present circumstances. Suffice it to say that the assertion that the primary role of metallothionein is to detoxify metals rests on evidence which is neither convincing nor persuasive.

Not only is zinc non-toxic, it is the least deserving metal to be on the list of potential pollutants that might endanger human health. In fact, zinc is essential to biological function in all phyla and species. If metallothionein were to play any biological role, and I am certain it does, it seems to me it would most likely be to supply zinc for the chain of reactions and events that accompany metabolism, cell division, differentiation and transmission of the genetic message (see below) rather than to sequester or remove it.

The biological chemistry of zinc has taken giant steps forward since the discovery of metallothionein, but this progress was due exclusively to work on other zinc proteins. I have sometimes thought and perhaps with justification that preoccupation with pollution seriously delayed scientific work on metallothionein for at least a generation. The search for its biological function was hindered by a wave of public concern for ecological problems and presumptions regarding the role of "toxic" metals in human health. These fashionable and seemingly timely speculations were, unfortunately, unrelated to the underlying biological chemistry of zinc and ignored the possible functions of metallothionein in zinc metabolism.

Most discussions of metallothioneins generally forget that its discovery was due entirely to its component metals, cadmium in particular but also zinc. The chemistry of these metals dominates the features of metallothionein which are most unusual when compared to those of other metalloproteins. As I perceive the toxicology line of reasoning, metallothionein was designed by Nature not so much as an adjunct to the chemistry of cadmium and zinc but to anticipate, address and solve such ecological problems as derived from industrial metal pollution. While it may yet turn out, of course, that metallothionein is the answer to these and other societal problems this looks progressively less and less likely. Instead it would seem to me to be a good time to reexamine these earlier premises and problems, to view them from a more scientific perspective, and to discard whatever has proved to be non-productive.

The decade following the discovery of metallothionein witnessed almost complete indifference on the part of the biochemical community as readily confirmed by inspection of the Citation Index (Figure 1). The social preoccupations referred to above, the realities and technical problems of the then contemporary metallobiochemistry, and the competing opportunities emerging in the field of molecular biology combined to direct scientific attention away from this protein and into different directions.

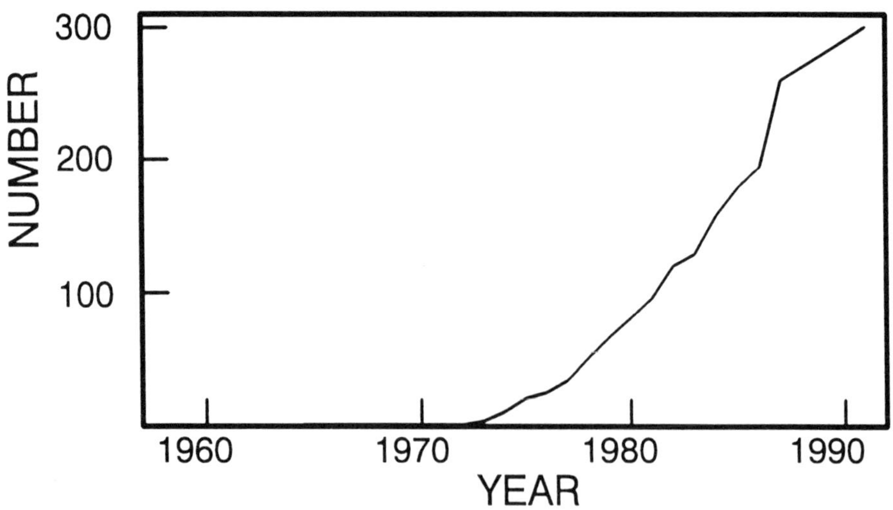

Figure 1. The number of publications per annum give an approximate measure of scientific interest in metallothionein.

Interest began to rise during the 1970s when it was first realized that metals, various kinds of metabolites and yet other organic agents can all induce metallothionein in the liver. Such observations and the prevailing suppositions diverted efforts regarding the occurrence and putative function of metallothionein toward metal scavenging, detoxification and storage. This combined with the coincidental surge in public awareness of the environment resulted in demands for funding of research relevant to the socioecological issues of the time, including pollution. As a consequence, government funds became available and many scientists abandoned their previous scientific interests in favor of problems for which support could be obtained more readily. Whatever the social mechanisms that brought about these circumstances they projected metallothionein to center stage of both scientific and public policy.

With time, the remarkable progress in macromolecular biophysical chemistry and the burgeoning of molecular biology, genetics and bioengineering generated a whole new atmosphere in biology. More specifically, the inducibility of metallothionein called attention to its *promoter* as a tool in biochemical engineering, a completely different and unexpected turn of events identifying previously unknown potentials of this molecule and its genetics. This direction could neither have been predicted nor anticipated based on the structure, chemical properties or occurrence of metallothionein. Moreover, the resultant opportunities generated their own activation energy to find new experimental avenues and approaches to identify, isolate and characterize metallothioneins in different organs of different species, thus greatly enlarging the experimental arena. The evolution of experimental techniques and their impact on important discoveries about metallothionein at the end of this historic perspective are summarized (Figure 2).

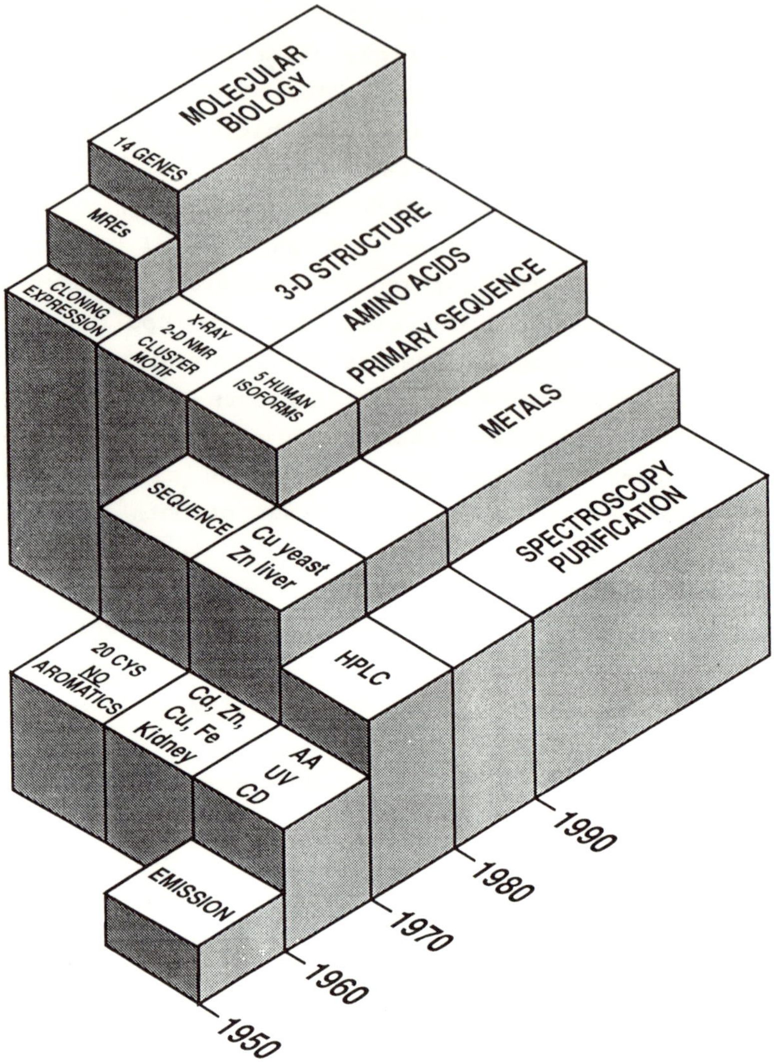

Figure 2. Much work has been expended to delineate the composition, structure and function of metallothionein since its discovery. The approaches have involved multiple disciplines and objectives and have changed as the associated progress in biological and basic sciences has increased in analytical sophistication in the course of the 40 years. This composite graph attempts to integrate the evolution of work on metallothionein in this context from its discovery to the present. It combines both the sequence of events and time table while differentiating them according to techniques employed.

I believe it was essential for me to detail these convoluted and frustrating efforts to establish a relationship between metallothionein and cadmium just as it is necessary now to examine the relationship of the protein to zinc, its other major inorganic component. In the early years of metallothionein the old stand-by zinc was overshadowed by the new-comer, cadmium and, hence, it received almost no attention. Despite this neglect, knowledge of zinc elsewhere in chemistry and biochemistry continued to increase exponentially. In point of fact, the unravelling of zinc biochemistry is perhaps one of the more startling scientific achievements in the course of the last decade and, in retrospect, could not fail to have a critical impact on the metallothionein riddle. I will devote the balance of this memoir to these new developments.

The Chemistry and Biology of Zinc. Among the essential transition and group IIB metals zinc is biologically the most versatile (Vallee and Falchuk, 1993). Progress in the biochemistry of zinc has been intimately connected with the development of instrumental methods for its analysis which have led to a billion-fold lowering of the limit of detection of zinc in a period of just thirty years. As a direct consequence of this analytical capability zinc proteins now constitute the largest group of metalloproteins (Vallee and Galdes, 1984; Vallee and Auld, 1992a). Collectively the physicochemical features of zinc are important for the translation of chemical structure into multiple biological functions. It is remarkable that biology has evolved an entire gamut of zinc-protein coordination structures and conformations that function both in catalysis and in the expression, transcription and translation of the genetic message. Zinc is a constituent of the active site of more than 300 enzymes encompassing all I.U.B.M.B. classes. Moreover, it is indispensable to the function of at least that many DNA-binding proteins. At present it is difficult to identify any cellular process which does not depend on zinc in one way or another, dramatic evidence for the remarkable ability of zinc to interact with proteins engaged in multiple biological functions.

Zinc is inert to oxidoreduction and forms stable complexes with macromolecules. It is amphoteric and, hence, can exist either as a metal hydrate or hydroxide at pH values near neutrality. Its exceptionally flexible coordination allows multiple coordination numbers and geometries, but in proteins zinc accepts four, five and six coordination. This stereochemical adaptability is the basis for a wide range of chemical reactivities and is pivotal to its role in protein, nucleic acid, carbohydrate and lipid metabolism as well as gene transcription, cellular proliferation, differentiation and development. Considering the ubiquity of zinc biomolecules, it is most surprising that virtually nothing is known regarding zinc uptake, intracellular transport, and storage as well as regulation of zinc metabolism.

A great deal was known about zinc physiology and nutrition prior to the recognition of the nature of the underlying zinc biochemistry. The correlation and integration of these disciplines is becoming an exciting and increasingly challenging area of research (Vallee and Falchuk, 1993). Jointly, the ubiquity and biological essentiality of zinc imply the existence of homeostatic mechanisms that regulate its absorption, distribution, cellular uptake and excretion. These mechanisms have been the subject of speculation, but there is no factual knowledge regarding the chemical species and processes involved in any of them.

About 30-40% of all cellular zinc occurs in the nucleus, another 50% is in the cytoplasm and its organelles, and the remainder is in the cell membrane (Vallee and Falchuk, 1993). I have long thought that there should be a regulatory system that stores, transports and distributes zinc among intracellular compartments, and makes provisions for metal transfer from one type of macromolecule to another, thus maintaining the broad range of zinc-dependent biochemical functions. Further, this regulation should be highly coordinated, well-timed and integrated. Indeed, I surmised that the requirement of zinc for fundamental processes is much more stringent than that of many other elements; and that among all zinc-dependent processes, some may be more critically dependent on zinc than others. However, there is no demonstrable molecular hierarchy for the allocation of zinc even when it is in short supply. Thionein and metallothionein have been increasingly mentioned as likely partners in such a distribution system. Since the very existence of the system still remains entirely speculative, I can only cite the biological and chemical properties of metallothionein that might suggest a possible role for it in zinc homeostasis.

Zinc-Binding Sites in Zinc Metalloenzymes. Some perspective on metallothionein can be gained by viewing it in the context of other protein families that participate in zinc metabolism. As I have already noted, metallothionein and zinc enzymes were discovered almost contemporaneously, but while progress in understanding the structure and function of the former remained disproportionately slow, our knowledge of the latter has been truly phenomenal. I believe that it is now possible and, in fact, appropriate to capitalize on that knowledge to try to gain much needed insight for metallothionein. One reason for this is because I am certain that metallothionein plays a role in zinc biology, be it major or minor, and that it may well be at the crossroads of zinc metabolism. Therefore the chemical properties of zinc in metallothionein must be conditioned by those of the other molecules with which it is destined to interact, namely other zinc proteins.

Zinc occurs in a great variety of coordination environments in proteins. High resolution X-ray structural studies of numerous zinc proteins have now given considerable insight regarding the chemical features of these coordination sites. One thing that has emerged from these studies is that, clearly, biology has evolved many forms of zinc chemistry that were not anticipated - at least not by inorganic chemists. Zinc enzymes can have *catalytic*, *structural* and/or *cocatalytic* metal binding sites. A detailed review of these has been given in our recent summaries (Vallee and Auld, 1990a; 1992b; 1993). Briefly, catalytic zinc sites generally consist of three coordinating ligands provided by the protein: histidine nitrogens and/or glutamate or aspartate oxygens and cysteine sulfurs are the usual donors. The frequency of their occurrence is in the order His ≫ Glu > Asp > Cys. In all active sites of enzymes that have a single, catalytic zinc there is a fourth ligand - an activated water molecule (Vallee and Auld, 1990b). We have found the spacing of the three protein ligands in the protein sequence to be remarkably regular: The first two are typically separated by a short spacer of one to three amino acids, whereas the third is separated from the other two by a longer spacer that can range from 5 to 196 amino acids (Vallee and Auld, 1989). So far, the only exception we have found is for the catalytic zinc in the alcohol dehydrogenases. These enzymes have one histidine and two cysteines as ligands to the catalytic zinc atom and the spacer between the first two ligands is relatively long. We think this may be due to the fact that the catalytic zinc site in alcohol dehydrogenase is the only one among the mono-zinc enzymes that undergoes a significant conformational change upon binding the coenzyme, NADH.

Seventeen mono-zinc enzyme structures have been determined by X-ray crystallography and we have used these as standards of reference for the zinc ligands of their respective protein families. Within any family the metal binding site is characteristic. Indeed, it has been possible by sequence alignment to identify seemingly unrelated proteins and even to predict both previously unrecognized catalytic features and putative zinc ligands. Thus, an apparent zinc binding site was discovered by careful comparison of the sequence of leukotriene A_4 hydrolase (EC 3.3.2.6) with those of aminopeptidase M (EC 3.4.11.2) and thermolysin (EC 3.4.24.4). Additional homology considerations led to the inference that the enzyme might have esterase and/or peptidase activity. Subsequent analysis demonstrated that leukotriene A_4 hydrolase contains one atom of zinc *and* has aminopeptidase activity - an unquestionable proof of the heuristic value of this predictive scheme (Haeggström et al., 1990).

The procollagenases and progelatinases are an interesting variation on this theme: these zymogens are activated by converting a structural into a catalytic zinc atom. Their N-terminal activation peptide contains a single cysteine residue, conjectured to be the fourth ligand of the zinc atom in the latent, inactive form of the protein. Apparently, dissociation of the cysteine ligand and concomitant entry of water during autocatalytic cleavage of the prometalloprotease converts this cysteine-blocked zinc site into a catalytic site of the now active enzyme. Thus, cysteine dissociation from the zinc atom of procollagenases in effect constitutes a "new" mechanism of zymogen activation called the "cysteine switch" (Van Wart and Birkedal-Hansen, 1990) and to which we have referred to as the "Velcro" mechanism (Vallee and Auld, 1990a).

The second kind of zinc site containing structural zinc atoms has only been encountered thus far in liver alcohol dehydrogenase and in the regulatory subunit of aspartate transcarbamoylase. In both instances, zinc is coordinated tetrahedrally to four cysteine residues.

Cocatalytic zinc sites are the third type and these are found in multi-zinc enzymes containing two or more zinc and/or magnesium atoms where adjacent metals participate in multi-metal catalysis. Before any X-ray structures were available these zinc atoms were referred to as "modulating" or "regulatory" (Vallee and Auld, 1993). Enough structures have been determined in the past two years to allow systematization of their binding sites. They are the latest motifs to be added to those found in enzymes and are yet another novel aspect of zinc biochemistry.

The most distinctive feature of these cocatalytic sites is a bridging ligand between at least two of the metal atoms; it differentiates them from the catalytic and structural zinc sites of mono-zinc enzymes. Such an arrangement brings the zinc atoms sufficiently close to act as a catalytic unit - hence the name cocatalytic - and to allow multiple interactions between the zinc atoms, on the one hand, and between zinc and substrate, on the other. Such sites also differ in the type of ligands and the existence of 5- or even 6-coordinate zinc complexes in the absence of substrate. The carboxylate and imidazole side chains of aspartate, glutamate and histidine are all suitable to serve as bridging ligands, since they feature two electronegative oxygen or nitrogen donor atoms. The X-ray structures show that aspartate is the bridge amino acid in three enzymes that metabolize phosphate, whereas glutamate is the bridge in lens leucine aminopeptidase. This latter enzyme contains a binuclear site with two zinc atoms whereas, trinuclear metal sites with three zinc atoms or two zinc and one magnesium atom occur in *B. cereus* phospholipase C and *E. coli* alkaline phosphatase, respectively. The unusual proximity of the zinc atoms in these enzymes is reminiscent of the zinc/cadmium clusters of metallothionein.

Zinc-Containing DNA-Binding Proteins. The various families of zinc-containing DNA-binding proteins have three types of zinc binding motifs each of which differs completely from those of enzymatic sites (Vallee et al., 1991). They are referred to respectively as *fingers*, *twists* and *clusters* or collectively as *ftc* proteins. As with structural zinc atoms, and in contrast to catalytic zinc sites, these all have four protein ligands with relatively close spacing, and, there is no open coordination position. These sites help to organize small protein domains for DNA recognition, a corollary of this coordination. Each motif contains multiple zinc atoms and represents a distinctive DNA recognition site consisting of an α-helix - anchored by one or two zinc atoms - that fits into a specific region of the major groove of DNA. The globular peptide domains of zinc *twists* have a DNA recognition site between the two zinc atoms, each of which is coordinated to four cysteine residues. In contrast, two histidines and two cysteines are the zinc ligands of zinc *fingers* whose region of DNA recognition is within the zinc coordination site. The zinc *cluster* structure in DNA-binding proteins consists of six cysteine residues bound to two zinc atoms with four terminal and two bridging cysteine ligands. A short α-helical DNA recognition region originates from the edge of the cluster (Marmorstein et al., 1992). Characteristic zinc coordination to cysteine exists in all three motifs which until about ten years ago, was considered unusual among catalytic zinc atoms with alcohol dehydrogenase the sole example. Now, the predominance and abundance of zinc-sulfur coordination in *ftc* proteins constitutes a new scenario (Vallee et al., 1991) that is at least superficially analogous to metallothionein.

The variety of zinc structures now known (Figure 3) raises the question of whether it is the chemistry of the zinc complexes or an independent biological control mechanism, or both, that is responsible for zinc distribution and also whether or not metallothionein function is in any way connected to these *ftc* proteins. Such a connection should manifest in the dynamics of zinc metabolism which has received attention only recently. The crucial role of zinc in gene transcription and its widespread occurrence in transcription factors and RNA polymerases all suggest that there should be a system to control its availability. Since the concentration of intracellular free zinc is extremely low, zinc

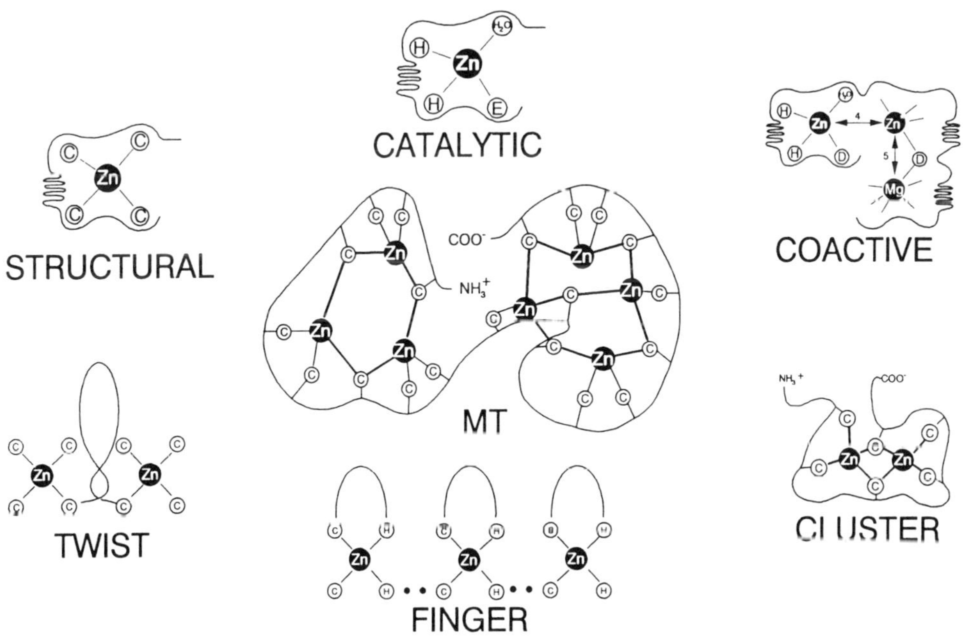

Figure 3. Seven motifs now characterize zinc chemistry in various phases of biology. This composite in which metallothionein forms the center calls attention to its potential and compares its metal binding sites with others currently known.

bound to macromolecules must ultimately be made available to ensure these vital metabolic functions. Metallothionein has many of the attributes that could make it a key player in such a scheme. In fact, the reactivity of the zinc ions in metallothionein and the intimate control of metallothionein gene expression through metals and many other agents could coordinate zinc metabolism and the pivotal roles of zinc.

Metal Reactivity in Metallothionein Clusters. X-ray crystallographic and NMR spectroscopic investigations have revealed two metal clusters in metallothionein with unprecedented structural and functional features (Otvos and Armitage, 1980; Robbins et al., 1991; Wüthrich, 1991). Three metal atoms form a chair arrangement with three bridging and six terminal cysteine ligands in one cluster, while the second cluster has four metal atoms in an adamantane-like coordination with five bridging and also six terminal cysteine ligands. Overall, these cluster structures are unique to biology and have not been observed in inorganic chemistry. The GAL4 transcription factor of yeast, which also turns out to contain a two-metal cluster structure, extends the use of this structural motif to the DNA-binding gene products.

The chemistry of these metal-ligand bonds is both extraordinary and dynamic as evidenced by the selectivity of the clusters for metal atoms, the cooperativity of metal binding and release, and their reactivity with electrophiles (Otvos et al., 1989). It is not inconceivable that this unusual chemistry, including the unique cluster structure itself, underlies and perhaps foreshadows the physiological function of metallothionein.

Metallothionein is a source of labile zinc (Otvos et al., 1989) and therefore could participate in zinc transfer reactions rather than being involved primarily in long-term storage of zinc. Despite the fact that zinc is firmly bound to metallothionein ($K_A' = 2x10^{12}$ M^{-1}, pH 7) it is kinetically very labile as verified *in vitro* and *in vivo*. Hence its release could be controlled and limited to when and where it is needed. Evidence for a possible control mechanism has been seen in zinc deficiency, where the release of zinc from metallothionein occurs on a time scale far shorter than that established for the proteolysis of metallothionein (Krezoski et al., 1988).

The metal dynamics in the metallothionein clusters has been elucidated by Cd-NMR spectroscopy (Nettesheim et al., 1985). Cadmium ions undergo *intramolecular* interchange more rapidly within the 3-metal cluster than within the 4-metal cluster, and exchange between clusters is very slow. *Intermolecular* metal exchange has been observed with different metallothionein isoforms and follows a biphasic reaction. The fast phase ($k = 2\ s^{-1}$) is attributed to the atoms in the 3-metal cluster while the slow phase ($k = 10^{-3}\ s^{-1}$) is thought to involve those in the 4-metal cluster (Kägi, 1991). These exchanges occur only if the two metallothionein molecules are in direct physical contact. Metallothionein dimers are known to exist both in solution and in the crystalline state and could constitute a structural basis for intermolecular metal transfer via ligand-exchange reactions without the intermediate formation of free zinc ions (Palumaa et al., 1992; Palumaa and Vašák, 1992; Robbins et al., 1991). While not demonstrated for interactions of metallothionein with heterologous proteins, such ligand-exchange processes are well-known for metal-sulfur model complexes. The sulfur chemistry of all DNA-binding motifs and metallothionein makes them suitable candidates for such metal transfer-processes through similar ligand-exchange reactions.

Actually two types of zinc transfer reactions have been documented experimentally. In one, the fully metal-complexed metallothionein serves as a metal donor, and in the other metal-free thionein acts as a chelating agent. Apoenzymes prepared from some zinc proteins can abstract zinc from metallothionein *in vitro* (Brady, 1982), though the physiological significance of this observation is unclear since it is difficult to envision how one molecule, e.g. metallothionein, might serve as the universal zinc donor for all apoenzymes and apoproteins.

Inhibition of DNA-binding of TFIIIA and Sp1 by thionein presumably is due to its sequestering their zinc (Zeng et al., 1991a,b), i.e. thionein functions as a chelating agent, as previously anticipated (Vallee, 1979, 1987, 1991). This does not preclude the possibility that metallothionein could also donate zinc to apo versions of the *ftc* protein types or for that matter to any other structural types, under appropriate conditions.

Since each metallothionein molecule can be a reservoir for seven zinc atoms, there would have to be a mechanism for the sequential, simultaneous, partial or complete transfer of these metals to one or more recipients. Circumstantial evidence has implicated a "special reactivity" for two of the seven zinc atoms in metallothionein. The cadmium form that is induced *in vivo* never exceeds a cadmium/zinc stoichiometry of 5/2 and the two zinc atoms that remain occupy specific positions in the 3-metal cluster (Robbins et al., 1991). Presumably these are retained for some special function such as transfer to a designated recipient.

<u>Genetic Polymorphism.</u> Any discussion of the reactivity of metallothionein must also take into consideration the multiplicity of mammalian metallothionein genes and their intricate regulation by metals and other factors. The fourteen different human metallothionein genes located in a gene cluster of about 82 kb on chromosome 16 (West et al., 1990) are ample testimony to the complexity of the metallothionein gene family. Among these, six are functional, two are not and the remainder have not been characterized. The number of genes identified on the basis of nucleotide sequences far exceeds that of known metallothionein proteins, since only three metallothioneins (Ie, IIa, III) have been found to correspond to translated nucleotide sequences. Perhaps more metallothionein forms remain to be discovered. The newly isolated MT-III (Uchida et al., 1991) (see below) is a good example. Perhaps, some of the remaining genes encode for related metallothionein isoforms that have escaped detection so far. Their metal dynamics and functional characteristics could vary widely and it would be sheer speculation to anticipate any possibilities or specificities at this time. The metallothionein genes could encode for closely related isoforms that have not yet been resolved chromatographically.

The variable promoter regions of the metallothionein genes add yet another dimension of complexity and functional potential to the already extensive variation in the metallothionein gene family, and are the most obvious indication that metallothionein has a message to disseminate. These promoter regions differ characteristically in terms of both non-metal and metal inducers of the metallothionein genes. This aspect of the metallothionein gene structure has proved extremely convenient and effective as a workhorse in molecular biology and bioengineering. Numerous applications of gene expression have employed the metallothionein promoter to confer metal-inducibility on a heterologous gene. No doubt the strength of the metallothionein promoter facilitates such experiments, but surprisingly its physiological role in directing the expression of the metallothionein gene itself is still not fully understood.

Non-Metal Inducers. Hormones (cytokines) and cytotoxic agents are inducers of metallothionein implying that it functions in cellular proliferation and differentiation on the one hand and in cellular defense mechanisms on the other. Some inducers are ligands of trans-acting factors and a few act specifically on known response elements of the metallothionein gene. Others like ethanol and some nutritionally significant agents would appear to act through indirect mechanisms. Generally, the inducers of metallothionein gene expression utilize protein kinases A and C, and ras signal transduction pathways. This intriguing use of multiple signals is further evidence that metallothionein is part of a complex network of cellular signal and control mechanisms. However, this does not identify the particular means by which these signals are associated with zinc-dependent processes within the cell. Unfortunately, none of the non-metal inducers of metallothionein have been linked to cellular zinc homeostasis in any way. Since the molecules involved in the regulation process are still unknown, the expectation remains high that sooner or later the metallothionein regulatory system will be integrated into zinc metabolism.

The response of metallothionein to various forms of

physical stress, e.g. UV radiation, exposure to oxygen and viral or bacterial infections, is a second element to be considered in metallothionein gene expression. It cannot be ruled out that metals are involved in these inductions at some point, but perhaps at physiological rather than pharmacological or toxicological concentrations.

A Role for Thionein? What exactly are the possible molecular events that follow the induction of metallothionein by these agents? Let me again remind you that the discovery of metallothionein and many of the subsequent studies of its chemistry were preoccupied with its metal content and characteristics. Little emphasis has been given to thionein, the apoprotein, whose reactivity and potential, while complementary, are quite distinct from those of metallothionein. I have always felt that thionein itself could have a metal-independent function, such as control of essential cysteine/cystine pairs, at a minimum involving the thiol groups. Metals would actually inhibit this property of thionein. Unfortunately, the experimental evidence to support this hypothesis is marginal at best. Perhaps, the cell uses thionein primarily to acquire and sequester zinc, not an unreasonable or novel possibility. How would thionein get the metal? There is no known mechanism by which extracellular zinc could be taken up either by the cell, in general, or by thionein specifically. Further there is no evidence for a cellular zinc depot nor is there sufficient free zinc in the cell to serve as a source for one.

Two potential consequences emerge from this zinc-related function of thionein. For one thionein could act as a specific chelating agent to regulate metal-dependent reactions and thus *inhibit* a physiological process by sequestering metal. It could also, as part of a stress response, act as a scavenger of metals generated during an insult to the cell or even after cell death. Alternatively - or conjointly - thionein could acquire zinc for the express purpose of delivering it to another site/molecule. Thus it could distribute/redistribute zinc in emergency situations to promote such cell functions as the repair of DNA in which case it would *activate* a biological function. While the action of brain MT-III (see below) lends strength to the former view, it is not possible at present to distinguish between these possibilities. The still unknown mechanism of action of MT-III could entail a chain of molecular events involving inhibition as well as activation of biological functions.

<u>Metals as Inducers.</u> Metal induction of metallothionein was an important milestone in the history of metallothionein research. The subsequent advent of molecular biological techniques made it possible to address the molecular mechanism of metal-induced gene expression for which metallothionein has become a paradigm. Gene regulation by metals is no longer an isolated event peculiar to the expression of metallothionein. In fact it has become a "spin-off" subdiscipline of inorganic biochemistry with its own attraction and appeal. This has led to a growing appreciation of the more general importance of metals in gene regulation, and has reinforced a continuing and renewed interest in factors that control metallothionein gene expression. It has even induced us to speculate that the transition of thionein to metallothionein serves as a metal-dependent molecular switch that controls the regulation of its own gene.

Metal induction of metallothionein is mediated by a process that involves the interaction of trans-acting factors, MRE-BFs (metal regulatory element binding factors), with MREs (metal regulatory elements). Up to five different MREs have been identified in some metallothionein promoters. It is unclear whether there is a similar number of MRE-BFs, whether a concerted action of multiple MREs controls promoter strength, and whether different MRE-BFs have different specificities for metal ions or additional recognition sites for other transcription factors. At least two MRE-BFs with molecular masses of 28 and 86 kDa have been identified in human cell lines (Czupryn et al., 1992). MRE-BF1 binds in control cells, while MRE-BF2 binds in metal-treated cells. Both factors are inhibited by 1,10-phenanthroline in a dose-dependent manner, indicating that MRE-BFs are metalloproteins though it is unknown whether their metal binding inhibits or stimulates metallothionein gene expression. In this context, it is noteworthy that the most recently cloned mouse MRE-BF (MTF-1) contains six zinc fingers (Radtke et al., 1993). Furthermore, there is a MRE-BF with a molecular mass of 112 kDa, whose binding to its MRE is inhibited by metals, suggesting a negative regulation of metallothionein gene expression (Koizumi et al., 1992a). Further, a nuclear protein, involved in the zinc-dependent transcriptional activation of the human MT-IIa gene named ZRF (zinc-regulatory factor), has been identified in HeLa cells (Koizumi et al., 1992b).

The role of the MRE-BFs in the context of zinc homeostasis cannot be interpreted at present. The degree of fluctuation in cellular zinc is unknown as are the answers to questions such as how much zinc is too much and how little is enough. Discernment of the strength of zinc binding to MRE-BFs would require knowledge regarding cellular zinc sensors which might control cellular re-distribution of zinc via metallothionein gene expression.

Importantly, MRE-BFs are not the only zinc-dependent trans-acting factors that influence the metallothionein promoter. Evidence now exists that zinc-dependent transcription factors converge at the metallothionein promoter since two additional types of zinc-dependent transcription factors control metallothionein gene expression. These are Sp1 (Lee et al., 1987), a zinc finger protein, and various nuclear hormone receptors of the zinc twist type. Thus, minimally at least three proteins that control metallothionein gene expression depend on zinc. Moreover, thionein inhibits the binding of Sp1 to its cognate DNA (Zeng et al., 1991a). The interaction of Sp1 with the metallothionein promoter therefore suggests a feed-back mechanism in which thionein regulates the function of its own transcription factor. It remains to be shown whether or not thionein can also remove metals from other zinc-dependent transcription factors or if metallothionein can deliver its cargo to their apoforms. This might close another loop in the possible regulatory mechanisms which may not be limited to the specific feedback inhibition/activation of metallothionein gene expression, of course, but would extend to the zinc-dependent transcription of other genes.

Many molecular ratios are known to have important regulatory roles in metabolism. The ATP/AMP, NAD/NADH and GSH/GSSG ratios come readily to mind. The T/MT (thionein/metallothionein) ratio could be thought to be a similarly important regulator, in this case of zinc-dependent gene expression, i.e., by responding to both an excess of and a demand for zinc to bring about homeostasis. While high concentrations of zinc are known to induce metallothionein it is not known that thionein is induced in response to a demand for zinc. In fact, none of the inducers of metallothionein have been linked directly to a cellular demand for zinc, nor is any metal-regulated mechanism known which would induce thionein under conditions of metal deficiency.

<u>The First Metallothionein Physiological Function, MT-III.</u> A growth inhibitory factor from human brain tissue has recently been identified and found to be a metallothionein (Uchida et al., 1991). This discovery must be considered a major breakthrough. First, it elucidates the structure of a new metallothionein (MT-III) that differs from all others in having a glutamic acid-rich, six amino acid insert near its C-terminus. The protein sequence identity between MT-III and the other metallothioneins is only 65%, but all cysteine ligands, presumably the "business end" of the molecule, are strictly conserved. Second, none of the other human metallothionein isoforms are growth inhibitors, indicating a very specific function for MT-III. Third, the expression of MT-III does not appear to be regulated by metals or glucocorticoids (Palmiter et al., 1992). Fourth, this is the first finding to link any metallothionein with a disease process. MT-III is down-regulated in Alzheimer's disease (Tsuji et al., 1992), the very reason that led to its identification. It must now be determined whether or not its mechanism of action as a cellular growth inhibitory factor is a guiding principle for similar actions of other metallothioneins and whether or not the molecular action of MT-III

involves zinc transfer and control of a zinc-dependent process. Nature has been exceptionally reluctant to reveal its secret on the function of metallothionein despite vigorous and persistent assault on this problem over 35 years. It appears that, finally, there is now direct evidence for a physiological function of this particular metallothionein, a function that is not related to detoxification.

In this chapter I have focused on human metallothionein and protein zinc motifs with known structures. I cannot close without some prospective reflections on zinc and metallothionein biochemistry which I have not yet addressed. For one, the term "metallothionein" actually encompasses three classes of molecules with structural variations beyond those of the already complex human system described. Even within class I metallothioneins, there are invertebrate metallothioneins with fewer cysteines, and, accordingly, different cluster structures, and shortened forms such as *Neurospora crassa* metallothionein. When compared to these metallothioneins those of lower eukaryotes, plants, and prokaryotes in class II are even more strikingly different in terms of cysteine content, spacing of cysteine residues, overall length, as well as the presence of histidine and aromatic amino acids. Finally, there are class III metallothioneins, which are γ-glutamylcysteinyl metal-binding peptides, also referred to as cadystins or phytochelatins and bearing resemblance to the metal-binding tripeptide glutathione. If there were only one metallothionein function, then it would have to accommodate all the different structures, metal compositions etc. We do not know yet what the functions of all these molecules are. All that holds them together under the classification metallothionein is their cysteine and metal content, not even structural homology. It remains to be shown whether or not all metallothioneins share common functions, and whether or not the term metallothionein will eventually designate a protein family of similar or related functions. Last but not least, it should not be overlooked that the extensive studies on these metallothioneins of other classes may seed and cross-foster new ideas about possible functions of mammalian metallothioneins.

The second comment I want to make relates to the great variety of cysteine-rich proteins or protein domains that have not been linked to the metallothionein family. These include cysteine-rich intestinal protein (Hempe and Cousins, 1992), cysteine-rich protein (Wang et al., 1992), and several protein families with conserved, yet structurally undefined zinc binding motifs: the LIM-motif (Freyd et al., 1990), the Ring-finger motif (Freemont et al., 1991), and the protein kinase C motif (Hubbard et al., 1991). For cysteine-rich intestinal protein, it has now been postulated that it participates in intestinal zinc uptake and serves as a potential zinc donor to metallothionein (Hempe and Cousins, 1992). It will be exciting to study, test and understand the dynamics of zinc binding and transfer between these various molecules, the possible regulation of their function by zinc, and the participation of metallothionein in these processes. In closing I can only say that it will be equally exciting to be present at this meeting to understand the dynamics of information transfer between scientists, its possible regulation by the organizers, and - again - the participation of metallothionein in these processes.

References

Brady, F.O. (1982) Trends Biochem. Sci. 7, 143.

Czupryn, M., Brown, W.E., and Vallee, B.L. (1992) Proc. Natl. Acad. Sci. U.S.A. 89, 10395.

Freemont, P.S., Hanson, I.M., and Trowsdale, J. (1991) Cell 64, 483.

Freyd, G., Kim, S.K., and Horvitz, H.R. (1990) Nature 344, 876.

Grill, E., Winnacker, E.-L., and Zenk, M.H. (1991) Meth. Enzymol. 205, 333.

Haeggström, J.Z., Wetterholm, A., Shapiro, R., Vallee, B.L., and Samuelsson, B. (1990) Biochem. Biophys. Res. Commun. 173, 431.

Hempe, J.M., and Cousins, R.J. (1992) J. Nutr. 122, 89.

Hubbard, S.R., Bishop, W.R., Kirschmeier, P., George, S.J., Cramer, S.P., and Hendrickson, W.A. (1991) Science 254, 1776.

Kägi, J.H.R. (1991) Meth. Enzymol. 205, 613.

Koizumi, S., Suzuki, K., and Otsuka, F. (1992a) J. Biol. Chem. 267, 18659.

Koizumi, S., Yamada, H., Suzuki, K., and Otsuka, F. (1992b) Eur. J. Biochem. 210, 555.

Krezoski, S.K., Villalobos, J., Shaw III, C.F., and Petering, D.H. (1988) Biochem. J. 255, 481.

Lee, W., Haslinger, A., Karin, M., and Tjian, R. (1987) Nature 325, 368.
Margoshes, M., and Vallee, B.L. (1957) J. Am. Chem. Soc. 79, 4813.
Marmorstein, R., Corey, M., Ptashne, M., and Harrison, S.H. (1992) Nature 356, 408.
Nettesheim, D.G., Engeseth, H.R., and Otvos, J.D. (1985) Biochemistry 24, 6744.
Otvos, J.D., and Armitage, I.M. (1980) Proc. Natl. Acad. Sci. U.S.A. 77, 7094.
Otvos, J.D., Petering, D.H., and Shaw III, C.F. (1989) Comments Inorg. Chem. 9, 1.
Palmiter, R.D., Findley, S.D., Whitmore, T.E., and Durnam, D.M. (1992) Proc. Natl. Acad. Sci. U.S.A. 89, 6333.
Palumaa, P., and Vašák, M. (1992) Eur. J. Biochem. 205, 1131.
Palumaa, P., MacKay, E.A., and Vašák, M. (1992) Biochemistry 31, 2181.
Radtke, F., Heuchel, R., Georgiev, O., Hergersberg, M., Gariglio, M., Dembic, Z., and Schaffner, W. (1993) EMBO J. 12, 1355.
Rauser, W.E. (1990) Ann. Rev. Biochem. 55, 193.
Robbins, A.H., McRee, D.E., Williamson, M., Collett, S.A., Xuong, N.H., Furey, W.F., Wang, B.C., and Stout, C.D. (1991) J. Mol. Biol. 221, 1269.
Tsuji, S., Kobayashi, H., Uchida, Y., Ihara, Y., and Miyatake, T. (1992) EMBO J. 11, 4843.
Uchida, Y., Takio, K., Titani, K., Ihara, Y., and Tomonaga, M. (1991) Neuron 7, 337.
Vallee, B.L. (1958) Proc. 4th Intntl. Cong. Biochem. 8, 138.
Vallee, B.L. (1979) Experientia, Suppl. 34, 19.
Vallee, B.L. (1987) Experientia, Suppl. 52, 5.
Vallee, B.L. (1991) Meth. Enzymol. 205, 3.
Vallee, B.L., and Auld, D.S. (1989) FEBS Lett. 257, 138.
Vallee, B.L., and Auld, D.S. (1990a) Biochemistry 29, 5647.
Vallee, B.L., and Auld, D.S. (1990b) Proc. Natl. Acad. Sci. U.S.A. 87, 220.
Vallee, B.L., and Auld, D.S. (1992a) in Matrix Metalloproteinases and Inhibitors, H. Birkedal-Hansen, Z. Werb, H. Welgus, and H. Van Wart, Eds. Gustav Fischer: Stuttgart, p.5.
Vallee, B.L., and Auld, D.S. (1992b) Faraday Discuss. 93, 47.
Vallee, B.L., and Auld, D.S. (1993) Proc. Natl. Acad. Sci. U.S.A. 90, 2715.
Vallee, B.L., and Galdes, A. (1984) Adv. Enzymol. Rel. Areas Mol. Biol. 56, 283.
Vallee, B.L., and Falchuk, K.F. (1993) Physiol. Rev. 73, 79.
Vallee, B.L., and Ulmer, D.D. (1972) Ann. Rev. Biochem. 41, 91.
Vallee, B.L., Coleman, J.E., and Auld, D.S. (1991) Proc. Natl.Acad. Sci. U.S.A. 88, 999.
Van Wart, H.E., and Birkedal-Hansen, H. (1990) Proc. Natl. Acad. Sci. U.S.A. 87, 5578.
Wang, X., Lee, G., Liebhaber, S.A., and Cooke, N.E. (1992) J. Biol. Chem. 267, 9176.
West, A.K., Stallings, R., Hildebrand, C.E., Chiu, R., Karin, M., and Richards, R.I. (1990) Genomics 8, 513.

Wüthrich, K. (1991) Meth. Enzymol. 205, 502.
Zeng, J., Heuchel, R., Schaffner, W., and Kägi, J.H.R. (1991a) FEBS Lett. 279, 310.
Zeng, J., Vallee, B.L., and Kägi, J.H.R. (1991b) Proc. Natl. Acad. Sci. U.S.A. 88, 9984.

Metallothionein III
K. T. Suzuki, N. Imura and M. Kimura (eds)

EVOLUTION, STRUCTURE AND CHEMICAL ACTIVITY OF CLASS I METALLOTHIONEINS: AN OVERVIEW

Jeremias H.R. Kägi

Institute of Biochemistry, University of Zürich, Winterthurerstrasse 190, CH-8057 Zürich, Switzerland

SUMMARY: Over 80 class I and class II metallothioneins (MTs) have thus far been sequenced. From an alignment of 62 class I amino acid sequences an unrooted evolutionary tree has now been constructed by parsimony analysis yielding plausible organismal relationships from mollusks to mammals and indicating in the latter a molecular differentiation into four structural subgroups, i.e., MT-1, MT-2, MT-3 and MT-4. The most conserved features are the twenty odd cysteine residues which in mammals collectively bind seven bivalent metal ions (M) in two separate oligonuclear metal-thiolate clusters, i.e., M_3Cys_9 and M_4Cys_{11}. The spatial structures of several mammalian MTs now determined by 2D NMR spectroscopy and x-ray crystallography also display a closely conserved molecular architecture with a pseudosymmetric dumbbell-like overall shape and with the clusters buried as "mineral cores" in the interior of the two globular domains formed by the amino- and carboxyl-terminal halves of the polypeptide chain, respectively (Braun et al., Proc. Nat. Acad. Sci. USA 89, 10124, 1992). Notwithstanding their hidden location, the cluster-bound metal ions are remarkably mobile in exchange interactions allowing for a dynamic participation of MT in intracellular metal transfer and equilibration. In mammalian cells, MT is mostly present as a zinc protein with an average apparent binding constant at pH 7.0 of 7×10^{12} M^{-1}. The magnitude of this constant and the variable tissue concentrations of the protein support the suggestion that zinc thionein and its conjugate apoform thionein may function as an adjustable buffering system, regulating cellular free zinc concentration in the subnanomolar range and thereby directing and modulating the flow of this metal to or from some zinc-requiring systems.

Metallothioneins (MTs) are most unusual proteins or polypeptides which have challenged the interest of chemists and life scientists alike for over thirty years. The first MT discovered in search for a potential biochemical function of cadmium was a protein recognized by its containing this metal and accordingly featured as a cadmium protein (Margoshes and Vallee, 1957). It received the designation "metallothionein" on account of its exceptionally high metal and sulfur content (Kägi and Vallee, 1960) and it is still today the only protein known to collect appreciable amounts of cadmium in vivo. However, even the first MTs were shown to include, besides cadmium, also zinc and minor amounts of copper (Kägi and Vallee, 1961; Pulido et al., 1966). In fact, in most mammalian tissues zinc is the most abundant and often sole metallic complement. Thus, in equine and human liver, for instance, MT presents itself mainly as zinc thionein (Zn-MT) and is one of the major zinc proteins of this tissue (Kägi, 1970; Bühler and Kägi, 1974; Onosaka et al., 1986). In fungi MTs are copper proteins (Hamer, 1986).

In spite of decades of extensive research, the exact function of the MTs remains a topic of discussion (Karin, 1985; Bremner, 1991). Suggestive arguments for their playing a central role in metal-related cell-biological processes are their occurrence throughout the entire biosphere and the elaborate regulation and programming of their biosynthesis. Thus, MTs are now known to occur in all animal phyla as well as in fungi and in some plants and in cyanobacteria. Their concentration in tissues is often highly variable and remarkably susceptible to the influence of a large variety of conditions. Thus, as illustrated in Table I, the biosynthesis of the protein thionein is induced not only by many different metal ions - especially cadmium, zinc and copper - but also by hormones, cytokines and a steadily growing number of other endogenous and exogenous agents. Humoral factors are also likely to underlie the programmed emergence of MT in development (Nemer et al., 1984; Nishimura et al., 1989; De et al., 1991) and regeneration (Ohtake et al., 1978) and the increases in biosynthesis observed under a variety of stress conditions also listed (Brady, 1991; Bremner, 1991). The response to so many different stimuli has elicited a strong interest in MT in many diverse fields including metal toxicology, nutritional physiology, pathology and many areas of molecular, cellular and developmental biology and has thus rendered the quest for its biological roles an important and ongoing interdisciplinary effort.

Since the understanding of the biological function of proteins ultimately rests on the knowledge of their molecular features, we have been concentrating our own efforts mainly on the study of the chemical properties of the MTs. Thus, in this overview, too, I shall limit myself essentially to evolutionary, structural, and reactional aspects of these unusual metalloproteins.

Table I. Factors that induce thionein synthesis in cultured cells in vivo

Metal ions	Antibiotics
Cd, Zn, Cu, Hg, Au, Ag, Co, Ni, Bi	Streptozotocin
Hormones and second messengers	Cycloheximide
Glucocorticoids	Mitomycin
Progesterone	Cytotoxic agents
Estrogen	Hydrocarbons
Catecholamines	Ethanol
Glucagon	Isopropanol
Angiotensin II	Formaldehyde
Arg-Vasopressin	Fatty acids
Adenosine	Butyrate
cAMP	Chloroform
Diacylglycerol	Carbon tetrachloride
Ca	Bromobenzene
Growth factors	Iodoacetate
Serum factors	Urethane
Insulin	Ethionine
IGF-1	Di(2-ethylhexyl)phthalate
EGF	α-Mercapto-β-(2-furyl)acrylate
Inflammatory agents and cytokines	6-Mercaptopurine
Lipopolysaccharide (LPS)	Diethyldithiocarbamate
Carrageenan	Penicillamine
Dextran	2,3-Dimercaptopropanol
Endotoxin	2,3-Dimercaptosuccinate
Interleukin-1	EDTA
Interleukin-6	5-Azacytidine
Interferon-α	Acetaminophen
Interferon-γ	Indomethacin
Tumor necrosis factor	Stress-producing conditions
Tumor promoters and oncogenes	Starvation
Phorbol esters	Infection
ras	Inflammation
Vitamins	Laparotomy
Ascorbic acid	Physical stress
Retinoate	X-irradiation
1α,25-Dihydroxyvitamin D_3	High O_2 tension
	Ultraviolet radiation

Adapted from Kägi, 1991, by permission of the publisher. For citations see Kägi, 1991; Templeton and Cherian, 1991; Bremner, 1991; Brady, 1991; Klaassen and Liu, 1991; Schroeder and Cousins, 1991.

General Characteristics and Classification

All MTs known to date can be characterized by the following chemical properties: (1) low molecular weight (<9000 Da), (2) high metal content, (3) high cysteine content, and (4) arrangement of metal ions in oligonuclear complexes (clusters). The mammalian MTs are usually single-chain proteins, containing 61 amino acids, among them 20 cysteines, and binding a total of 7 equivalents of bivalent metal ions. All cysteines occur in the reduced form and are coordinated to the metal ions through mercaptide bonds, giving rise to spectroscopic features characteristic of metal-thiolate clusters.

On structural grounds, the MTs have been subdivided into three classes (Kojima, 1991). Class I MTs include all vertebrate forms and MTs from other phyla with obviously related primary structures. Class II MTs comprise polypeptides displaying no or only a very distant correspondence to the mammalian forms. Class III MTs are no proteins at all but are short atypical non-translationally synthesized metal-thiolate-polypeptides built up from repetitive gamma-glutamyl cysteinyl isopeptide units. In most species one finds expression of multiple isoforms of MT indicative of genetic polymorphism (Kägi and Kojima, 1987). Thus, the human genome is reported to contain 14 MT genes (West et al., 1990) of which seven alone are expressed and identified as protein products in human liver (P. Hunziker, personal communication). In mammals, the multiple isoforms have been subdivided in four structurally distinct subgroups, i.e., MT-1, MT-2, MT-3, and MT-4 (see below).

Amino Acid Sequences

Owing to improvements in protein sequencing techniques and the facile determination of nucleotide sequences, complete primary structure data are now available for 62 class I MTs (Fig. 1), 19 class II MTs (Fig. 2), and various homologous sets of class III MTs (Rauser, 1990). As evident from the first sequence study (Kojima et al., 1976), the key structural motif of all forms is the repeated Cys-X-Cys tripeptide sequence, where X stands for an amino acid residue other than Cys. The multiple sequence alignment of the class I MTs obtained on the basis of the Needleman-Wunsch algorithm (Fig. 1) documents their high degree of homology. The gaps introduced compensate mainly for the unequal lengths of the polypeptide chains. Thus, while most vertebrate MTs contain 60-63 amino acid residues and crustaceans 57 and 58, respectively, there are 68 amino acid residues in the brain-specific mammalian subform MT-3 (Palmiter et al., 1991; Uchida et al., 1991) and 71-75 residues in the ten MTs from bivalve mollusks (Mackay et al., 1993). The cardinal feature of all class I MTs is the very close correspondence in the alignment of the Cys along the chain. In all 39 mammalian forms one finds an almost complete preservation of their arrangement and of nearly all Lys and Arg. Overall, in the 62 class I MTs compared, about 40% of all residues are invariant. As already noticed in early studies (Bühler and Kägi, 1974; Kägi et al., 1974), aromatic amino acid residues are totally absent and there are also only very few bulky hydrophobic residues.

The class II MTs, by definition lacking obvious homology to the mammalian forms and among each other, range in length from 25 - 80 residues (Fig. 2). They are now taken to include the animal MTs from D. melanogaster, the sea urchin S. purpuratus, and the nematode C. elegans, together with fungal and cyanobacterial MTs and the recently identified MT-like proteins from various plants.

Evolution

The appreciable number of class I MT sequences now available (Fig. 1) makes it possible to establish reliable evolutionary relationships, using state of the art methods. Figure 3 shows an unrooted evolutionary tree calculated by the parsimony method of Felsenstein (1993) from the 62 sequences aligned by the use of the Needleman-Wunsch algorithm and minor manual adaptations (P.-A. Binz and T. Hale, personal communication). This unrooted tree reveals the grouping of the MTs with different orders, classes and phyla of the animal kingdom, thereby illustrating

```
                          1        10        20        30          40        50             60
 1  Human    MT-1a       MDP.NCSCAT.GGSCTCTGSCKCKECKCN.SCK.K.S.C.CSCCPMSCAKCA..QGCICKG......ASEKC.S.C.CA.
 2           MT-1b       MDP.NCSCTT.GGSCACAGSCKCKECKCT.SCK.K.C.C.CSCCPVGCAKCA..QGCVCKG......SSEKC.R.C.CA.
 3           MT-1e       MDP.NCSCAT.GGSCTCAGSCKCKECKCT.SCK.K.S.C.CSCCPVGCAKCA..QGCVCKG......ASEKC.S.C.CA.
 4           MT-1f       MDP.NCSCAA.GVSCTCAGSCKCKECKCT.SCK.K.S.C.CSCCPVGCSKCA..QGCVCKG......ASEKC.S.C.CD.
 5           MT-1g       MDP.NCSCAA.GVSCTCASSCKCKECKCT.SCK.K.S.C.CSCCPVGCAKCA..QGCICKG......ASEKC.S.C.CA.
 6           MT-1h       MDP.NCSCAA.GVSCTCAGSCKCKECKCT.SCK.K.S.C.CSCCPLGCAKCA..QGCICKG......ASEKC.S.C.CA.
 7           MT-1i       MDP.NCSCAA.GVSCTCAGSCKCKECKCT.SCK.K.S.C.CSCCPVGCAKCA..QGCICKG......ASEKC.S.C.CA.
 8           MT-1k       MDP.NCSCAAAGVSCTCASSCKCKECKCT.SCK.K.S.C.CSCCPVGCAKCA..QGCICKG......ASEKC.S.C.CA.
 9           MT-1l       MDP.NCSCSP.VGSCACAGSCKCKECKCT.SCK.K.S.C.CSCCPVGCAKCA..QGCICKG......TSDKC.S.C.CA.
10           MT-2        MDP.NCSCAA.GDSCTCAGSCKCKECKCT.SCK.K.S.C.CSCCPVGCAKCA..QGCICKG......ASDKC.S.C.CA.
11           MT-3        MDPETCPCPS.GGSCTCADSCKCEGCKCT.SCK.K.S.C.CSCCPAECEKCA..KDCVCKGGEAAEAEAEKC.S.C.CQ.
12  Monkey   MT-1        MDP.NCSCAT.GVSCTCADSCKCKECKCT.SCK.K.S.C.CSCCPVGCAKCA..QGCVCKG......ASEKC.N.C.CA.
13           MT-2        MDP.NCSCVA.GDSCTCAGSCKCKECKCT.SCK.K.S.C.CSCCPVGCAKCA..QGCICKG......ASDKC.N.C.CA.
14  Horse    MT-1A       MDP.NCSCPT.GGSCTCAGSCKCKECRCT.SCK.K.S.C.CSCCPGGCARCA..QGCVCKG......ASDKC.S.C.CA.
15           MT-1L54     MDP.NCSCPT.GGSCTCAGSCKCKECRCT.SCK.K.S.C.CSCCPGGCARCA..QGCVCKG......ALDKC.S.C.CA.
16           MT-1R39     MDP.NCSCPT.GGSCTCAGSCKCKECRCT.SCK.K.S.C.CSCCPRGCARCA..QGCVCKG......ASDKC.S.C.CA.
17           MT-1B       MDP.NCSCVA.GESCTCAGSCKCKQCRCA.SCK.K.S.C.CSCCPVGCAKCA..QGCVCKG......ASDKC.S.C.CA.
18  Cow      MT-1        MDP.NCSCPT.GGSCTCAGSCKCKACRCP.SCK.K.S.C.CSCCPVGCAKCA..QGCVCKG......ASDKC.S.C.CA.
19           MT-2        MDP.NCSCTA.GESCTCAGSCKCKDCKCA.SCK.K.S.C.CSCCPVGCAKCA..QGCVCKG......ASDKC.S.C.CA.
20  Sheep    MT-1a       MDP.NCSCPT.GGSCSCAGSCTCKACRCP.SCK.K.S.C.CSCCPVGCAKCA..QGCVCKG......ASDKC.S.C.CA.
21           MT-1b       MDP.NCSCPT.SGSCSCAGSCTCKACRCP.SCK.K.S.C.CSCCPVGCAKCA..QGCVCKG......ASDKC.S.C.CA.
22           MT-1c       MDP.NCSCST.GSSCSCAGSCTCKACRCP.SCK.K.S.C.CSCCPVGCAKCA..QGCICKG......ASDKC.S.C.CA.
23           MT-2        MDP.NCSCTA.GESCTCAGSCKCKDCKCA.SCK.K.S.C.CSCCPVGCAKCA..QGCVCKG......ASDKC.S.C.CA.
24  Pig      MT          MDP.NCSCPT.GGSCSCAGSCTCKACRCT.SCK.K.S.C.CSCCPAGCARCA..QGCICKG......ASDKC.S.C.CA.
25  Rabbit   MT-1a       MDP.NCSCAT.GNSCTCASSCKCKECKCT.SCK.K.S.C.CSCCPAGCTKCA..QGCICKG......ASDKC.S.C.CA.
26           MT-2a       MDP.NCSCAAAGDSCTCANSCTCKACKCT.SCK.K.S.C.CSCCPPGCAKCA..QGCICKG......ASDKC.S.C.CA.
27           MT-2b       MDP.NCSCAT.GDSCTCASSCKCKECKCT.SCK.K.S.C.CSCCPAGCTKCA..QGCICKG......ASDKC.S.C.CA.
28           MT-2c       MDP.NCSCATAGDSCTCANSCTCKACKCT.SCK.K.S.C.CSCCPPGCAKCA..QGCICKG......ASDKC.S.C.CA.
29           MT-2d       MDP.NCSCAT.RDSCACASSCKCKECKCT.SCK.K.S.C.CSCCPAGCTKCA..QGCICKG......ASDKC.S.C.CA.
30           MT-2e       MDP.NCSCAT.RDSCACASSCKCKECKCT.SCK.K.S.C.CSCCPAGCTKCA..QGCICKG......ALDKC.S.C.CA.
31  Mouse    MT-1        MDP.NCSCST.GGSCTCTSSCACKNCKCT.SCK.K.S.C.CSCCPVGCSKCA..QGCVCKG......AADKC.T.C.CA.
32           MT-2        MDP.NCSCAS.DGSCSCAGACKCKQCKCT.SCK.K.S.C.CSCCPVGCAKCS..QGCICKQ......ASDKC.S.C.CA.
33           MT-3        MDPETCPCPT.GGSCTCSDKCKCKGCKCT.NCK.K.S.C.CSCCPAGCEKCA..KDCVCKGEEGAKAEAEKC.S.C.CQ.
34           MT-4        MDPGECTCMS.GGICICGDNCKCTTCNCK.TCR.K.S.C.CPCCPPGCAKCA..RGCICKG......GSDKC.S.C.CP.
```

			1 10 20 30 40 50 60
35	Hamster	MT-1	MDP.NCSCST.GSTCTCSSSCGCKDCKCT.SCK.K.S.C.CSCCPVGCSKCA..QGCVCKG......ASDKC.T.C.CA.
36		MT-2	MDP.NCSCAT.DGSCSCAGSCKCKECKCT.TCK.K.S.C.CSCCPVGCAKCS..QGCVCKE......ASDKC.S.C.CA.
37	Rat	MT-1	MDP.NCSCST.GGSCTCSSSCGCKNCKCT.SCK.K.S.C.CSCCPVGCSKCA..QGCVCKG......ASDKC.T.C.CA.
38		MT-2	MDP.NCSCAT.DGSCSCAGSCKCKQCKCT.SCK.K.S.C.CSCCPVGCAKCS..QGCICKE......ASDKC.S.C.CA.
39	Dolphin	MT-II	MDP.NCSCTA.GGSCACPGSCKCKECKCT.SCK.K.S.C.CSCCPVGCAKCA..QGCICKG......ASDKC.S.C.CA.
40	Chicken	MT	MDPQDCTCAA.GDSCSCAGSCKCKNCRCR.SCR.K.S.C.CSCCPAGCNNCA..KGCVCKE....PASSKC.S.C.CH.
41	Pigeon	MT-1	MDSQDCPCAA.GGTCTCGDNCKCKNCKCT.SCK.K.G.C.CSCCPAGCAKCA..QGCVCKG....PSSAKC.S.C.CH.
42		MT-2	MDPQDCTCAA.GDSCSCAGSCKCKNCRCQ.SCR.K.S.C.CSCCPASCSNCA..KGCVCKE....PSSSKC.S.C.CH.
43	Duck	MT	MDPQDCTCAA.GDSCSCAGSCKCKNCRCR.SCR.K.S.C.CSCCPAGCNNCA..KGCVCKE....PASSKC.S.C.CH.
44	Flounder	MT	MDP..CECSK.TGTCNCGGSCTCKNCSCT.TCN.K.S.C.CPCCPSGCPKCA..SGCVCKG......KTCDT.T.C.CQ.
45	Plaice	MT	MDP..CECSK.TGTCNCGGSCTCKNCSCT.TCN.K.S.C.CPCCPSGCPKCA..SGCVCKG......KTCDT.S.C.CQ.
46	Trout	MT-A	MDP..CECSK.TGSCNCGGSCKCSNCACT.SCK.KAS.C.CDCCPSGCSKCA..SGCVCKG......KTCDT.S.C.CQ.
47		MT-B	MDP..CECSK.TGSCNCGGSCKCSNCACT.SCK.K.S.C.CPCCPSDCSKCA..SGCVCKG......KTCDT.S.C.CQ.
48	Loach	MT	MDP..CECSK.TGTCNCGATCKCTNCQCT.TCK.K.S.C.CSCCPSGCSKCA..SGCVCKG......NSCDS.S.C.CQ.
49	Pike	MT	MDP..CECSK.TGTCNCGATCKCTNCQCT.TCK.K.S.C.CSCCPSGCSKCA..SGCICKG......NSCDS.S.C.CQ.
50	Crab	MT-1	.PG.PC.CND.K..CVCKEGGCKEGCQCT.SCRCSP....CEKCSSGC.KCA..NKEECSKT.CSKA....C.S.C.CPT
51		MT-2	.PD.PC.CND.K..CDCKEGECKTGCKCT.SCRCPP....CEQCSSGC.KCA..NKEDCRKT.CSKP....C.S.C.CP.
52	Lobster	MT	.PG.PC.CKD.K..CECAEGGCKTGCKCT.SCRCAP....CEKCTSGC.KCP..SKDECAKT.CSKP....C.X.C.CXX
53	Oyster	MT	MSD.PCNCIE.TGTCACSDSCPATGCKCGPGCKCGD.DCKCAGCKVKC.SCTSEGGCKCGEK.CTGPATCKCGSGCSCKK
54	Mussel	CdBP10-I	.PA.PCNCIE.TNVCICDTGCSGEGCRCGDACKCSGADCKCSGCKVVC.KCS..GRCECGKG.CTGPSTCKCAPGCSCK.
55		CdBP10-II	.PA.PCNCIE.TNVCICDTGCSGEGCRCGDACKCSGADCKCSGCKVVC.KCS..GSCECGKG.CTGPSTCKCAPGCSCK.
56		CdBP10-III	.PA.PCNCIE.SNVCICGTGCSGEGCRCGDACKCSGADCKCSGCKVVC.KCS..GSCACEAG.CTGPSTCRCAPGCSCK.
57		CdBP10-IV	.PA.PCNCIE.TNVCICDTGCSGEGCRCGDACKCSGADCKCSGCKVVC.KCS..GSCACEGG.CTGPSTCKCAPGCSCK.
58		CdBP20-Ia	.PG.PCNCIE.TNVCICGTGCSGKCCRCGDACKCA.SGCGCSGCKVVC.KCS..GTCKCGCD.CTGPTNCKCESGCSCK.
59		CdBP20-Ib	.PG.PCNCIE.TNVCICGTGCSGKCCRCGDACKCA.SGCGCSGCKVVC.KCS..GTCKCGCD.CTGPTNCKCEAGCSCK.
60		CdBP20-II	.PG.PCNCIE.TNVCICGTGCSGKCCRCGDACKCA.SGCGCSGCKVVC.KCS..GTCACGCD.CTGPTNCKCESGCSCK.
61		CdBP20-IIIa	.PG.PCNCIE.TNVCICGTGCSGKCCQCGDACKCA.SGCGCSGCKVVC.RCS..GTCACGCD.CTGPINCKCESGCSCK.
62		CdBP20-IIIb	.PG.PCNCIE.TNVCICGTGCSGKCCQCGDACKCA.SGCGCSGCKVVC.RCS..GTCACGCD.CTGPTNCKCDSGCSCK.

Fig. 1. Amino acid sequences and best alignment of 62 class-I metallothioneins. For details of alignment see text. The numeration of residues relates to the standard sequence of human MT-2. The classification of the major mammalian subgroups MT-1 and MT-2 is based on the respective absence and presence of an acidic amino acid residue in position 10 or 11 of the standard sequence (Kojima, 1991; Kojima and Kägi, 1987). By this structural criterion equine MT-1B and dolphin MT-II are to be reclassified as belonging to the subgroups MT-2 and MT-1, respectively. For references see following page.

References to sequences of Fig. 1

1. Richards, R.I., Heguy, A. and Karin, M. (1984) Cell 37, 263-272, *b*
2. Heguy, A., West, A., Richards, R.I. and Karin, M. Mol. Cell. Biol. 6, 2149-2157, *b*
3,4. Schmidt, C.J., Jubier, M.F. and Hamer, D.H. (1985) J. Biol. Chem. 260, 7731-7737, *b*
5. Foster, R., Jahroudi, N., Varshney, U. and Gedamu, L. (1988) J. Biol. Chem. 263, 11528-11535, *b*
6-9,26,27,29,30. Hunziker, P. (1993) Personal Communication, *a*
10. Karin, M. and Richards, R.I. (1982) Nature 299, 797-802, *b*
11. Uchida, Y., Takio, K., Titani, K., Ihara, Y. and Tomonaga, M. (1991) Neuron 7, 337-347, *c*
12,13. Schmidt, C.J. and Hamer, D.H. (1983) Gene 24, 137-146, *c*
14-17. Kägi, J.H.R., Kojima, Y. (1987) Experientia Supplementum 52, 25-61, *a*
18,19. Münger, K., German, U.A., Beltramini, M., Niedermann, D., Baitella-Eberle, G., Kägi, J.H.R. and Lerch, K. (1985) Biol. Chem. 260, 10032-10038, *a*
20. Peterson, M.G., Lazdins, I., Danks, D.M. and Mercer, J.F.B (1984) Eur. J. Biochem. 143, 507-511, *c*
21-23. Peterson, M.G., Hannan, F. and Mercer, J.F.B. (1988) Eur. J. Biochem. 174, 417-424, *b*
24. Buchman, T.G., Cabin, D.E., Porter, J.M. and Bulkley, G.B. (1989) Surgery 106, 283-291, *b*
25. Tam, Y.C., Chopra, A., Hassan, M. and Thirion, J.-P. (1988) Biochem. Biophys. Res. Commun. 153, 209-216, *b*
28. Wagner, G., Neuhaus, D., Wörgötter, E., Vasak, M., Kägi, J.H.R. and Wüthrich, K. (1986) Eur. J. Biochem. 157, 275-289, *a*
31. Durnam, D.M., Perrin, F., Gannon, F. and Palmiter, R.D. (1980) Proc. Natl. Acad. Sci. USA 77, 6511-6515, *b*
32. Huang, I., Kimura, M., Hata, A., Tsundoo, H. and Yoshida, A. (1981) J. Biochem. 89, 1839-1845,*a*
33. Palmiter, R.D., Findley, S.D., Whitmore, T.E. and Durnam, D.M. (1992) Proc. Natl. Acad. Sci. USA 89, 6333-6337, *c*
34. Palmiter, R.D. (1992) Personal Communication, *c*
35,36. Griffith, B.B., Walters, R.A., Enger, M.D., Hildebrand, C.E. and Griffith, J.K. (1983) Nucleic Acids Res. 11, 901-910, *c*
37. Andersen, R.D., Birren, B.W., Ganz, T., Piletz, J.E. and Herschman, H.R (1983) DNA 2, 15-22, *c*
38. Winge, D.R., Nielson, K.B., Zeikus, R.D. and Gray, W.R. (1984) J. Biol. Chem. 259, 11419-11425, *a*
39. Kwohn, Y.T., Okubo, A., Hirano, H., Kagawa, H., Yamazaki, S. and Toda, S. (1988) Agric. Biol. Chem. 52, 837-841, *a*
40. Wei, D. and Andrews, G.K. (1988) Nucleic Acids Res. 16, 537-553, *c*
41,42. Lin, L., Lin, W.C. and Huang, P.C.(1990) Biochem. Biophys. Acta 1037, 248-255, *a*
44. Chan, K.M., Davidson, W.S., Hew, C.L., and Fletcher, G.L. (1989) Can. J. Zool. 67, 2520-2527, *c*
45. Leaver, M.J. and George S.C., EMBL databank accession no. X56743
46,47. Bonham, K., Zafarullah, M. and Gedamu, L. (1987) DNA 6, 519-528, *c*
48,49. Kille, P., Stephens, P.E. and Kay, J. (1991) Bioch. Biophys. Acta 1089, 407-410, *c*
50,51. Lerch, K., Ammer, D. and Olafson, R.W. (1982) J. Biol. Chem. 257, 2420-2426, *a*
52. Brouwer, M., Winge, D.R. and Gray, W.R. (1989) J. Inorg. Biochem. 35, 289-303,
53. Roesijadi, G., Kielland, S. and Klerks, P. (1989) Arch. Biochem. Bioph. 273, 403-413, *c*
54-62. Mackay, E.A., Overnell, J.,Dunbar, B., Davison, I., Hunziker, P.E., Kägi, J.H.R. and Fothergill, J.E. (1993) Eur. J. Biochem. (in press) *a*

Sources of sequence information : *a*, protein; *b*, DNA; *c*, cDNA

1	D.melanogaster	Ntn	MPCPCGSGCKCASQATKGSCNCGSDCKCGGDKKSACGCSE
2		Mto	MVCKGCGTNCQCSAQKCGDNCACNKDCQCVCKNGPKD
3	N.crassa	Mt	MGDCGCSGASSCNCGSGCSCSNCGSK
4	A.bisporus	MT	GDCGCSGASSCTCASGQCTCSGCGK
5	Sea urchin	MTa	MPDVKCVCCTEGKECACFGQDCCVTGECCKDGTCCGICTNAACKCANGCKCGSGCSCTEGNCAC
6		MTb	MPDVKCVCCKEGNECACTGQDCCTIGKCCKDGTCCGKCSNAACKTCADGCTCGGCSCTEGNCPC
7	C. elegans	CeMT-I	ACKCDCKNKQCKCGDKCECSGDKCCEKYCCEEASEKKCCPAGCKGDCKCANCHCAEQKQCGDKTHQHQGTAAAH
8		CeMT-II	MVCKCDCKNQNCSCNTGTKDCDCSDAKCCEQYCCPTASEKKCCKSGCAGGCKCACECAQAAH
9	S. cerevisiae		MFSELINFQNEGHECQCQCGSCKNNEQCQKSCSCPTGCNSDDKCPCGNKSEETKKSCCSGK
10	C. glabrata	MT-I	MANDCKCPNGCSCPNCANGGCQCGDKCECKKQSCHGCGEQCKCGSHGSSCHGSGCGDKCDCK
11		MT-II	MPEQVNCQYDCHCSNCACENTCNCCAKPACACTNSASNECSCQTCKCQTCKC
12	Cynobacteria	MT	TSTTLVKCACEPCLCNVDPSKAIDRNGLYYCCEACAHGHTGGSKGCGHTGCNC
13	Mouse-ear cress	MT	MSCCGGNCGCGSGCKCGNGCGGCKMYPDLGFSGETTTTETFVLGVAPAMKNQYESGESNNAESDACKCGSDCKCDPCTCK
14	Pea	MT	MSGCGCGSSCNCGDSCKCNKRSSGLSYSEMETTETVILGVGPAKIQFEGAEMSAASEDGGCKCGDNCTCDPCNC
15	Mimulus guttatus	MT	MSSGCSCGSGCKCGDNCSCSMYPDMETNTTVTMIEGVAPLKMYSEGSEKSFGAEGGNGCKCGSNCKCDPCNC
16	Wheat germ	Ecprot.	GCNDKCGCAVPCPGGTGCRCTSARSGAAAGEHTTCGCGEHCGGNPCACGGEGTPSGCAN...
17	Maize	MT like	MSCSCGSSCGCGSSCKCGKKYPDLEETSTAAQPTVVLGVAPEKKAAPEFVEAAAESGEAAHGCSCGSGCKCDPCNC
18	Barley	MT	MSCSCGSSCGCGSNCNCGKMYPDLEEKSGATMQVTVIVLGVGSAKVQFEEAAEFGEAAHGCSCGANCKCNPCN
19	Ricin	MT-like	MSCCGGNCGCGSGCKCGNGCGGCKMYPDMSFSEKTTTETLVLGVGAEKAHFEGGEMGVVGAEEGGCKCGDNCTCNPCTCK

Fig. 2. Amino acid sequences of class-II metallothioneins. For references see following page.

References to sequences of Fig. 2

1. Lastowski-Perry, D., Otto, E. and Maroni, G.(1985) J. Biol. Chem. 260, 1527-1530, *c*
2. Mokdad, R., Debec, A. and Wegnez, M. (1987) Proc. Natl. Acad. Sci. USA 84, 2658-2662 *b*
3. Lerch, K. (1979) In: Metallotionein. J.H.R. Kägi and M. Nordberg, eds., pp. 173-179. Birkhäuser Verlag Basel, Boston, Stuttgart, *a*
4. Münger, K. and Lerch, K. (1985) Biochemistry 24, 6751-5756, *a*
5,6. Wilkinson, D.G. and Nemer, M. (1987) Mol. Cell. Biol. 7, 48-58, *b*
7. Slice, L.W., Freedman, J.H. and Rubin, C.S. (1990) J. Biol. Chem. 265, 256-263, *c*
8. Imagawa, M., Onozawa, T., Okumura, K., Osada, S., Nishihara, T. and Kondo, M.: (1990) Biochem. J. 268, 237-240, *c*
9. Butt, T.R., Sternberg, E.J., Gorman, J.A., Clark, P., Hamer, D., Rosenberg, M. and Crooke, S.T. (1984) Proc. Natl. Acad. Sci. USA 81, 3332-3336, *b*
10,11. Mehra, R.K., Garey, J.R., Butt, T.R., Gray, W.R. and Winge, D.R. (1989) J. Biol. Chem. 264, 19747-19753, *b*
12. Olafson, R.W., McCubbin, W.D. and Kay, C.M. (1988) Biochem. J. 251, 691-699, *a*
13. Takahashi K. (1991) EMBL Databank Accession Nr. S18069, *b*
14. Evans, K.M., Gatehouse, L.N., Gatehouse, J.A., Robinson, N.J. and Croy, R.R.D. (1990) FEBS Lett. 262, 29-32, *b*
15. De Miranda, J.R., Thomas, M.A., Thurman, D.A. and Tomsett, A.B. (1990) FEBS Lett. 260, 277-280, *b*
16. Hoffmann, T., Kells, D.I.C. and Lane, B.G. (1984) Can. J. Biochem. Cell Biol. 62, 908-913, *a*
17. De Framont, A.J. (1991) FEBS Lett. 290, 103-106, *b*
18. Okumura, N., Nashizawa, N.K., Umehara, Y., Mori, S. (1991) Plant Mol. Biol. 17, 531-533, *c*
19. Weig, A. and Komor, E. (1992) EMBL Databank Accession Nr. L02306, *c*

Sources of sequence information: *a*, protein; *b*, DNA; *c*, cDNA

their extensive evolutionary conservation. The separation of the subgroups MT-1 and MT-2, evident in three mammalian orders, validates them as definite evolutionary entities and suggests that the gene duplication leading to the separation of the two subgroups has preceded the divergence of the mammalian orders. The tree also indicates that the two recently discovered and tissue-specifically expressed mammalian subforms MT-3 (Uchida et al., 1991; Palmiter et al., 1992) and MT-4 (R. Palmiter, personal communication) are structurally more remote and must thus have arisen earlier in evolution. Analogous divergences in isoforms are also seen in crustaceans and mollusks. The significance of the very extensive genetic polymorphism seen in some mammalian orders and in mussels is unclear. One explanation for this multiplicity might be the need for differential expression of these genes in different tissues, in development, and in regeneration and in the selective adaptation to challenges by metals, hormones and cytotoxic agents (Sadhu and Gedamu, 1988; Jahroudi et al., 1990).

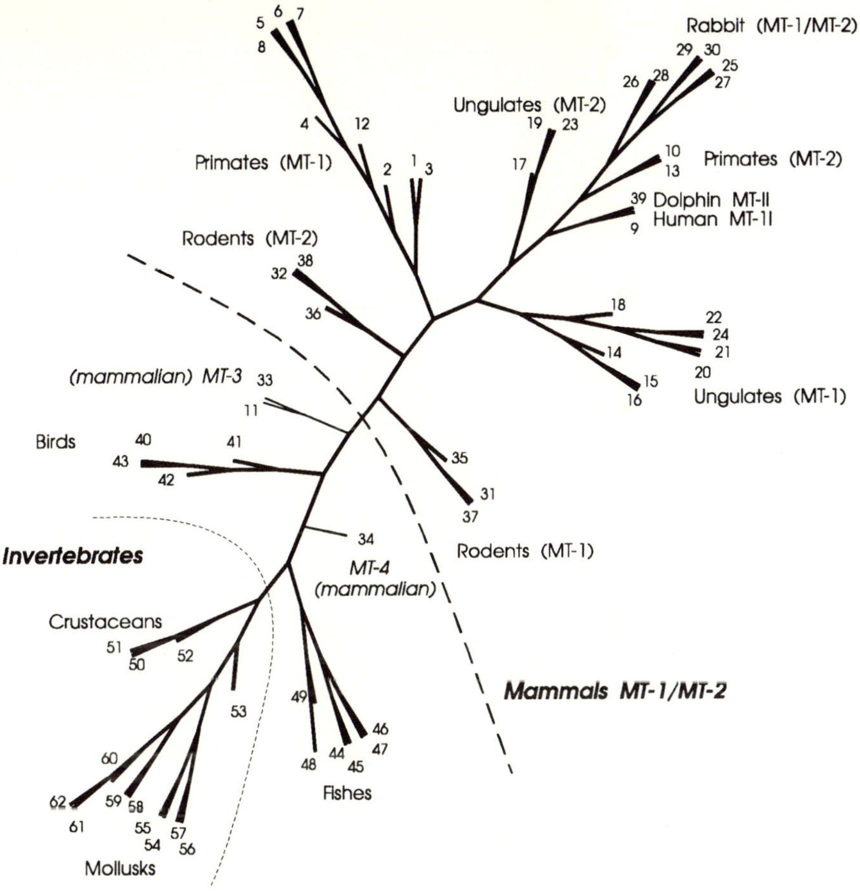

Fig. 3. Unrooted evolutionary tree of class I metallothioneins calculated by parsimony analysis (Felsenstein, 1993) from data of Fig. 1.

Metal Binding and Spectroscopy

The abundance of cysteine in the polypeptide chain and their arrangement in chelating Cys-Cys, Cys-X-Cys and Cys-X-Y-Cys sequences, where X and Y are residues other than Cys, is the basis for the high affinity of the apoprotein thionein for "soft" transition and d^{10} metal ions. In the mammalian forms, all 20 Cys participate as deprotonated thiolate ligands in the binding of the seven bivalent metal ions.

The coordinational properties of the organization of the metal-binding sites have been elucidated by the application of a large variety of spectroscopic methods. As indicated in Table II, for many of these studies derivatives of MT had to be prepared, in which the naturally bound metal ions were replaced by spectroscopically "active" bivalent metal ions such as Co or special isotopes suitable for a particular spectroscopic method, such as ^{113}Cd, ^{111}Cd, ^{111m}Cd or ^{57}Fe. The data provided by these studies have allowed the unambiguous and direct identification of the thiolate nature of the ligands, have established the tetrahedral coordination of each metal ion by four sulfur atoms of Cys, and have revealed the collective arrangement of these structures in metal-thiolate clusters.

The accommodation of the metal ions in metal-thiolate clusters is the cardinal structural feature of all MTs classifying them as prototypes of a growing set of oligonuclear metal-thiolate proteins (Vallee et al., 1991). This arrangement in which some of the thiolate ligands are shared by adjacent metal ions (bridging ligands) reconciles the spectroscopically established tetrahedral tetrathiolate coordination of each metal ion with the measured overall stoichiometry of less than three thiolate ligands per metal ion bound (see above). The most direct spectroscopic evidence for the existence of discrete clustered structures and the relative positions of the metal ions has come from ^{113}Cd nuclear magnetic resonance (NMR) studies of exclusively ^{113}Cd-containing forms of mammalian (Otvos and Armitage, 1980) and crab (Otvos et al., 1982) MTs. These forms which in their ^{113}Cd NMR spectra display J-coupled ^{113}Cd resonances, each attributable to one of the seven and six metal sites, respectively, allowed the unambiguous identification of the topological relationships among the ^{113}Cd nuclei. The results implied that the metals are partitioned in two completely separate clusters, i.e., a three-metal cluster, $^{113}Cd_3Cys_9$, having a cyclohexane-like structure, and a four-

Table II. Spectroscopic methods applied to the study of metal-binding sites in metallothionein

Objective of study	Spectroscopic method	Metal derivative examined
Ligand identification	Electronic absorption spectroscopy	Cd, Zn, Hg, Co
	X-ray photoelectron spectroscopy	Cd, Zn, Hg, Cu
Elucidation of coordination geometry	Electronic absorption spectroscopy	Co, Fe
	Electron paramagnetic resonance (EPR)	Co
	Magnetic circular dichroism	Co
	^{113}Cd nuclear magnetic resonance (NMR)	^{113}Cd
	Perturbed angular correlation of γ ray (PAC)	^{111m}Cd
	Mössbauer spectroscopy	^{57}Fe
	Extended X-ray absorption fine structure (EXAFS)	Zn, Cu
Demonstration of metal-thiolate clusters	^{113}Cd nuclear magnetic resonance (NMR)	^{113}Cd
	Electron paramagnetic resonance (EPR)	Co
	Mössbauer spectroscopy	^{57}Fe
	Electronic absorption spectroscopy	Zn, Cd, Hg, Co, Fe
	Circular dichroism	Zn, Cd, Hg, Co, Fe
	Magnetic circular dichroism	Zn, Cd, Hg, Co, Fe
	Luminescence	Cu, Ag, Au

Reproduced from Kägi, 1991, by permission of the publisher. For citations see Vašák and Kägi, 1983; Kägi and Kojima, 1987; and appropriate chapters in Riordan and Vallee, eds., 1991.

metal cluster, $^{113}Cd_4Cys_{11}$, having a bicyclo (3.3.1.) nonane-like structure in the mammalian MTs, and two three-metal clusters, $^{113}Cd_3Cys_9$, in crab MTs, respectively. The inferred existence of two independent metal-thiolate clusters in the mammalian MTs found its support and correspondence in the independently established partitioning of the folded protein in two about equally sized half-molecules of which the amino-terminal beta-domain harbours the three-metal cluster and the carboxyl-terminal alpha-domain the four-metal cluster (Winge and Miklossy, 1982).

The organization of the metal ions in mineral-like, oligonuclear structures manifests itself also in anomalous magnetic, electron paramagnetic resonance (EPR) and Mössbauer properties of Co(II)- and Fe(II)-substituted forms of MT (Vašák and Kägi, 1983; Ding et al., 1987). It is also the basis of the unusual electronic absorption and emission features of the various metal forms (Stillman, 1992) and of the very conspicuous multiphasic chiroptical properties arising from excitonic coupling of dissymmetrically oriented thiolate transition moment dipoles within each cluster (Willner et al., 1987; Willner et al., 1992).

Spatial Structure

The most important exploit of the application of spectroscopic methods was the elucidation of the spatial structure of mammalian MTs in aqueous solution by two-dimensional NMR techniques (Wüthrich, 1991). An integral part of this analysis was the topological allocation of the 28 coordination bonds between the seven ^{113}Cd ions and the 20 Cys by a heteronuclear ^{1}H ^{113}Cd multiple quantum coherence transfer (HMQC) procedure specially developed for the study of $^{113}Cd_7$-MT (Frey et al., 1985). This method allows the direct observation of the scalar connectivities of the seven ^{113}Cd resonances differentiated in the ^{113}Cd NMR spectrum with the ^{1}H side chain resonances of the appropriate Cys, the chemical shift positions of which beforehand had been assigned in the ^{1}H NMR spectrum by "classical" homonuclear two-dimensional $^{1}H^{1}H$ NMR measurements. The resulting network of ^{113}Cd-Cys connectivities representing Cd-Cys bonds as determined in ^{113}Cd-substituted rat liver MT-2 is displayed in Figure 4. It confirms the complete separation of the clusters (see above) and unambiguously identifies the eight metal-connecting, twofold coordinated, bridging ligands. Exactly the same organization of ^{113}Cd-Cys connectivities pertains to rabbit liver MT-2a (Arseniev et al., 1988) and to human liver MT-2 (Messerle et al., 1990a). This agreement between three mammalian MTs which differ in more than 25% of their noncysteine amino acid residues is strong evidence that the cluster structures are dictated by the conserved arrangement of Cys in the chains.

From the geometric constraints set by the network of 42 intramolecular Cys-Cd-Cys crosslinks and from long-range through-space proton-proton distance constraints (<5 Å), deter-

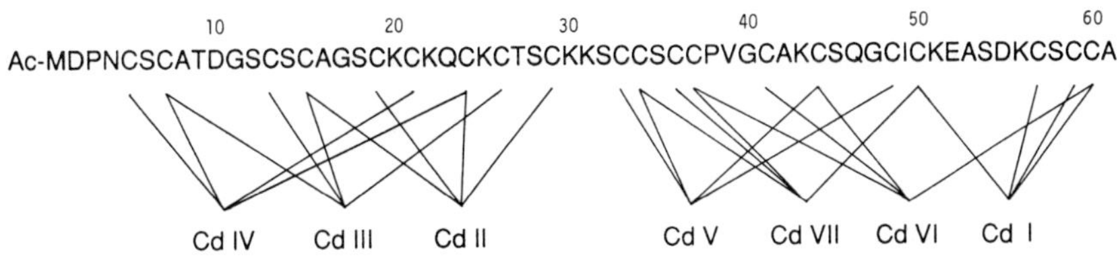

Fig. 4. Cadmium-cysteine connectivities of rat liver $^{113}Cd_7$-thionein, as established by two-dimensional NMR spectroscopy (adapted from Vašák et al., 1987). The numbering of Cd refers to the ^{113}Cd resonances assigned by Otvos and Armitage (1980). Reproduced from Kägi and Schäffer (1988) by permission of the publisher.

mined by "classical" homonuclear, two-dimensional $^1H^1H$ NMR measurements, the most probable spatial polypeptide fold and cluster puckering was computed, using a distance geometry algorithm developed for the calculation of tertiary protein structures. A simplified model depicting the chainfold of the two domains of rat liver-MT-2 and the Cd-thiolate clusters, as evaluated from NMR measurements, is shown in Figure 5A. The structures reveal that the clusters are located as "mineral cores" in the interior of the about equally sized globular domains each formed by two large helical turns of the respective halves of the polypeptide chain. In the amino-terminal beta-domain the chainfold is righthanded, in the carboxyl-terminal alpha-domain it is lefthanded (Schultze et al., 1988).

The same chainfolds and cluster organizations were also found for the independently determined solution structures of rabbit MT-2a and human MT-2 (Messerle et al., 1992), disclosing the strong preservation of the molecular architecture in the mammalian MTs and evincing the reliability of the two-dimensional NMR method. In each case the domains are connected through the invariant Lys 30 - Lys 31 segment. However, owing to interdomain flexibility resulting in a lack of interdomain contacts recognizable by NMR spectroscopy, the mutual orientation of the domains cannot be specified in the NMR-derived models.

Reliable structural information has now been obtained also by x-ray diffraction analysis of crystallized native (Cd_5Zn_2)-MT-2 isolated from the liver of rats preinjected with $CdCl_2$ (Robbins et al., 1991). The disposition of the polypeptide chain and the two metal-thiolate clusters of the 2 Å resolution crystal structure (Fig. 5B) indicate a very close similarity to the NMR-derived structure of $^{113}Cd_7$-MT-2. This correspondence is also documented quantitatively by a recently completed detailed comparison of the positions of the backbone atoms and the dihedral angles Φ and Ψ of the superimposed NMR- and x-ray structures (Braun et al., 1992). The only noticeable minor deviations in

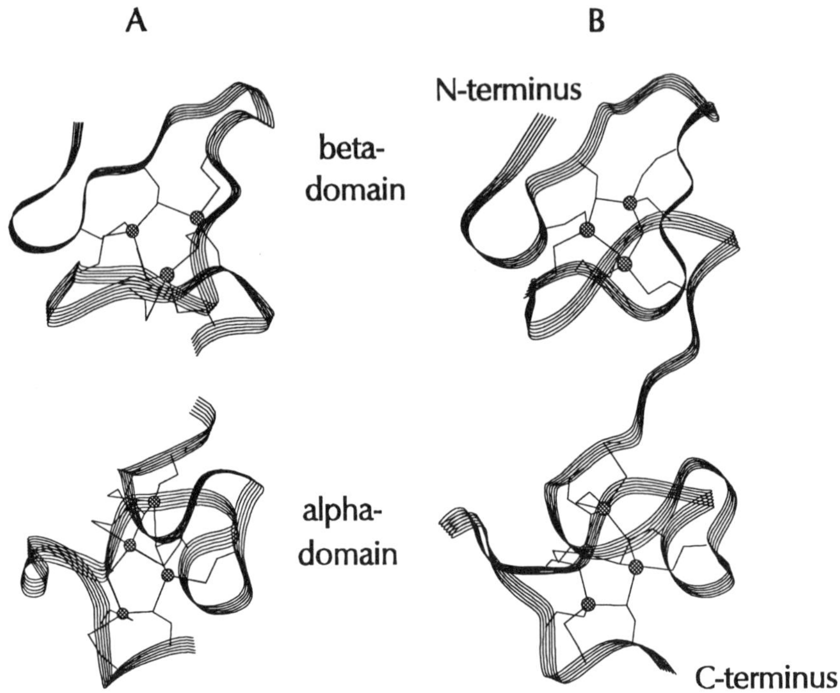

Fig. 5. Spatial structure models of rat metallothionein-2. Ribbon representation of main chain fold with enclosed Cys metal clusters (indicated by thin lines and filled circles, respectively). The alpha-domain enfolds the M_4Cys_{11} cluster, the beta-domain the M_3Cys_9 cluster. A: Model derived from coordinates of 2D NMR solution structure (courtesy of Prof. K. Wüthrich, Zürich). The disconnection of the domains in the latter model reflects the lack of definition of the mutual orientation of the two domains due to flexibility in the hinge region in aqueous solution. B: Model derived from coordinates of crystal structure (courtesy of Prof. D. Stout, San Diego).

the global polypeptide fold between the structures were seen in the amino-terminal beta-domain. They could be caused by crystal packing effects, by the difference in the spatial requirement of the $(Cd,Zn_2)Cys_9$ cluster in the beta-domain of the native protein as opposed to that of the $^{113}Cd_3Cys_9$ cluster in the derivative used for the NMR analysis, or perhaps most likely by the greater dynamic flexibility of this domain (see below) and the resulting poorer definition of the NMR data.

In its overall structure, mammalian MT has a dumbbell-like pseudosymmetric shape in which, except for the connecting Lys 30 - Lys 31 segment, there are no contacts between the two domains (Robbins et al., 1991). This is in keeping with the independence of the formation of the folded domains upon titration of the apoprotein with Cd^{2+} (Nielson and Winge, 1983; Stillman and Zelazowski, 1988) and the flexibility of the molecule in solution as inferred from the absence of interdomain "crosstalk" in the two-dimensional $^1H^1H$ spectra (see above). This extended and flexible structure also provides the most likely explanation for the large Stokes' radius typical of mammalian MTs (Vašák et al., 1984).

The comparison of the solution and crystal structures also shows almost completely identical cluster structures. In both models the metal-sulfur rings display the same puckered conformations which are best described as distorted boat forms (Braun et al., 1992). In Figure 6 this is illustrated by a set of stereoviews of the crystallographically derived cluster models (Robbins et al., 1991).

The preservation of the same cluster geometry in the native both Cd- and Zn-containing form and in the reconstituted solely ^{113}Cd-containing form of rat MT-2 reveals a remarkable ability of the apoprotein to accommodate metals of differing size. This faculty has now also been explored by a direct comparison of the solution structures of native human Zn_7-MT-2 with reconstituted human Cd_7-MT-2 using homonuclear two-dimensional $^1H^1H$ NMR spectroscopy (Messerle et al., 1992). The results showed that aside from a few localized differences the secondary and tertiary structures with their hydrogen bonding pattern are closely similar in the two forms, indicating the same overall molecular architecture and the same metal coordination topology. This ability to undergo without appreciable structural distortion on about 20% cluster volume increase upon substituting Cd (covalent atomic radius, 1.48 Å) for Zn (covalent atomic radius 1.31 Å) was attributed to the adaptability of the peptide loop structures interspaced between the metal-bound cysteine residues. These loops which make up the major portion of the protein allow the polypeptide chains to follow isomorphous contractions or expansions of the metal-thiolate cores with only small changes in bond angles and side

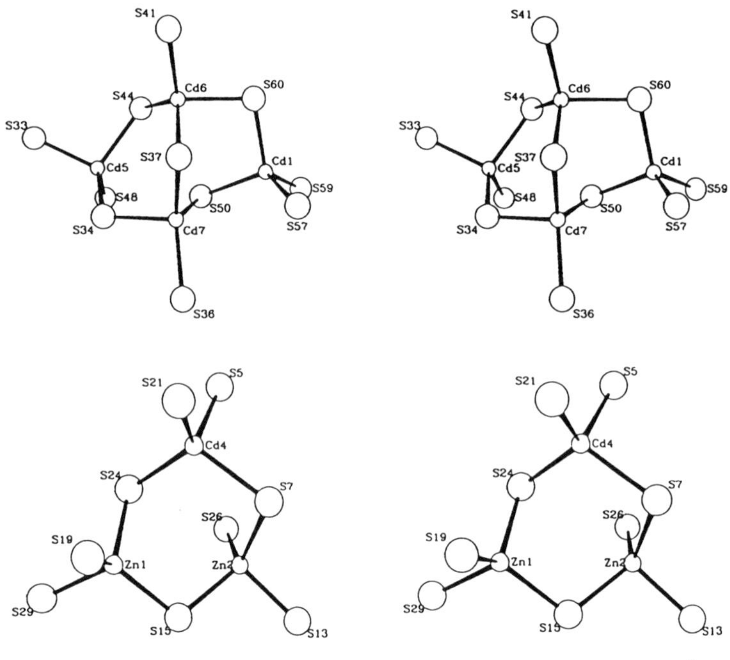

Fig. 6. Stereoviews of the metal-thiolate clusters in native rat metallothionein-2. Stick-and-ball representation of the arrangement of metal atoms (M) and of sulfur atoms (S) in the M_4Cys_{11} cluster (top) and the M_3Cys_9 cluster (bottom) derived from x-ray crystallographic analysis. The numbering of the sulfur atoms relates to the position of the corresponding Cys in the sequence. The numbering of the metal atoms relates to the NMR assignments (Otvos and Armitage, 1980; Wüthrich, 1991). Adapted from Figure 5 of Robbins et al. (1991), by permission of the publisher.

chain orientations. This flexibility is also signalled by the fact that both in the crystal and in the solution structure of rat MT-2 these polypeptide loops are less precisely defined and thereby indicate a greater degree of dynamical disorder than what is generally seen in proteins (Braun et al., 1992). This structural plasticity conditions the astonishing property of MT to accommodate metals of widely differing size, ranging from Co (covalent atomic radius 1.16 Å) to Hg (covalent atomic radius 1.49 Å), in the interior of the globular domains without causing principal changes in coordination geometry (Vašák and Kägi, 1993) and in the conformation of the enfolding protein. It is also the basis of the frequent occurrence of mixed metal forms (see above) and of the protective function apoMT is believed to have against a wide spectrum of toxic metals (Templeton and Cherian, 1991).

Dynamic Aspects

Notwithstanding the stable equilibrium structures now established by NMR-spectroscopic and crystallographic means, there is abundant evidence for dynamic fluctuations within the clusters and the polypeptide chain, allowing for ready chemical reactivity and for functional interactions with the cellular environment. The lack of rigidity of the clusters arises from the flexibility of the polypeptide fold (see above) and from the inherently high kinetic lability of all complexes of the group-12 metal ions, i.e., Zn^{2+}, Cd^{2+}, Hg^{2+} (Eigen, 1962) which, irrespective of the moderate to very high thermodynamic stability of these complexes, allows for continuous fluxional breaking and re-forming of the coordination bonds. This fluxional lability is reflected in the quality of the ^{113}Cd NMR resonances (Vašák et al, 1985; Otvos et al., 1989) and in a variety of other spectroscopic features (Vašák, 1986). The motility of the protein moiety is documented by the ready exchangeability of almost all peptides hydrogens with the solvent (Messerle et al., 1990b) and by the reactivity of the coordinated Cys side chains with alkylating (Petering et al., 1992) and oxidizing (Shaw et al., 1991) agents.

A readily measurable and functionally relevant manifestation of this motile state is the facile exchange of the bound metal ions with metal ions in solution, with other ligands, and with other MT molecules (Nettesheim et al., 1985; Petering et al., 1992). A rapid turnover in the metal occupation of the various thiolate binding sites was first demonstrated by Cd NMR saturation transfer experiments in which the redistribution of selectively saturation-labeled ^{113}Cd nuclei among the different binding sites in the clusters was followed with time (Otvos et al., 1987). These and subsequent studies with $^{111}Cd_7$-MT showed that the metal ions of the three-metal cluster of MT are scrambled by an interprotein exchange reaction with a half-life of about 0.008 min, while the rate for the exchange of the four-metal cluster was found to be orders of magnitude lower (Otvos et al., 1989). In a study in which the radioisotope redistribution between the charge-separable isoform fractions $^{109}Cd_7$-MT-I and Cd_7-MT-II was followed with time, we have now confirmed this fast intermolecular exchange and have determined that the four-metal clusters undergo at 20° isotopic exchange in a first order reaction with a half-life of about 16 min. Thus, as displayed by Figure 7, there is a nearly 3000fold difference in exchange rates of the metals in the two domains, supporting the much lower chemical reactivity of the carboxyl-terminal domain and its enclosed four-metal cluster established previously by chemical modification (Bernhard et al., 1986; Shaw et al., 1991),ligand substitution (Petering et al., 1992), and peptide hydrogen exchange (Messerle et al., 1990b) studies.

These exchange reactions depend on contact between the MT molecules since no isotopic redistribution to speak of was observed when the $^{109}Cd_7$-MT-I and Cd_7-MT-II fractions were separated by a dialysis membrane impermeable to protein (Schmid, 1991). We submit, therefore, that the exchange does not proceed through an intermediate dissociation and hydration step of the metal but occurs by direct transfer between the clusters, mediated by intermolecular ligand swapping during a transient

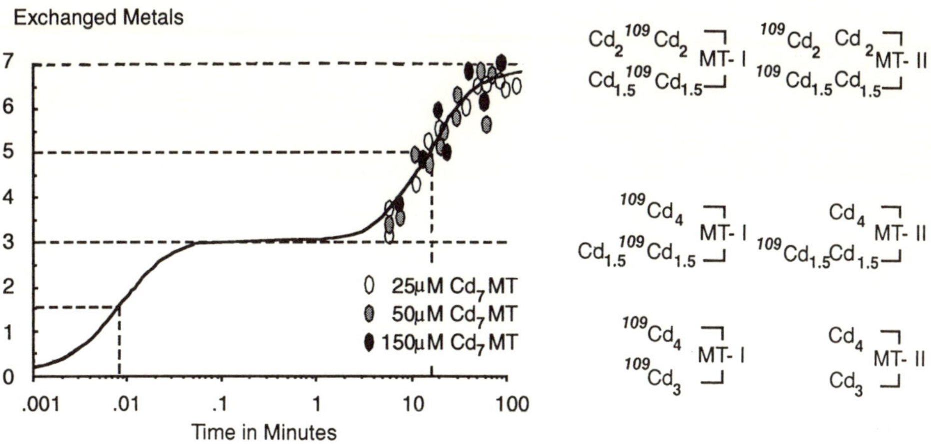

Fig. 7. Metal exchange between charge-separable $^{109}Cd_7$-MT-I and Cd_7-MT-II in function of time. MT-I and MT-II represent ion exchange chromatographic fractions from rabbit liver, comprising isoforms of charge -2 and -3, respectively. Left: Progress of isotopic equilibration initiated by mixing of equal amounts of the appropriately reconstituted metal derivatives in 33 mM Tris-HCl, pH 8.6, 20°, and monitored by sampling and separating them at different times by native PAGE (pH 8.6) and measuring their specific radioactivity. The results indicate partitioning into three sites (presumably the three-metal cluster) undergoing exchange before the first measurements were made and four sites (presumably the four-metal cluster) undergoing exchange with an average half-life of about 16 min. The exchange rate is unaffected by changes in protein concentration. The fast exchange step is pictured to proceed by first order with a half-life of 0.008 min, as determined earlier by ^{111}Cd NMR saturation transfer measurements (Otvos et al., 1989). Right: Schematic view of exchange process.

association of the MT molecules. The occurrence of such associations is supported by the observation that in the crystal pairs of MT molecules are intimately associated (Robbins et al., 1991) and that in aqueous solution, in the presence of a slight excess of Cd^{2+}, stable dimers of MT are formed (Palumaa et al., 1992). Interestingly, however, we found no dependency of the isotope exchange reaction on protein concentration. We take this as evidence for a pseudo-first order process in which the metal exchange is rate-limited by an intramolecular conformational transition of the protein required to establish the productive chemical contact between the clusters. This interpretation is supported by our observation that the measured exchange rate of the four-metal cluster is also the same with other metal species, i.e., $^{65}Zn_7$-MT-I vs. Zn_7-MT-II, $^{203}HG_7$-MT-I vs. HG_7-MT-II (Schmid, 1991) and that rates of the same order of magnitude are also seen in chemical modification reactions (Savas et al., 1991) and in the release of Zn from Zn-MT in cells incubated in Zn-deficient medium (Krezoski et al., 1988). While the biological significance of intermolecular metal exchange between MT isoforms is unclear, it highlights a reactivity of the domains that is likely to be fundamental for understanding metal transfer and equilibration functions MT may have in the cellular environment.

Thermodynamic Aspects

As for any chemical change, the direction of metal transfer to or from MT is set by the thermodynamic constraints of the system. By adapting a simple model of equivalent binding sites for all seven bivalent metal ions we have previously determined by spectrophotometric pH-titration of Cd_7-MT and Zn_7-MT from rabbit dissociation constants of approximately 10^{-22} M and 10^{-18} M, respectively. Extrapolating these values to pH 7.0 yields the apparent average dissociation constants $K'_{Cd} = 5 \times 10^{-17}$ M and $K'_{Zn} = 5 \times 10^{-13}$ M (Vašák and Kägi, 1983). For human Zn_7-MT an average value of the same order of magnitude, i.e., 1.4×10^{-13} M, has now also been measured by a direct method at pH 7.0, using competition with the chromogenic chelator 8-hydroxyquinoline-5-sulfonic acid (Bienz, 1992). These constants representing the metal ion concentrations at which 50% of the binding sites are occupied by the metal equally apply to the more realistic cooperative metal binding model (Nielson and Winge, 1983) where they characterize the all-or-none release of the metal ions from the clusters. An isotherm describing the dependency of such a cooperative dissociation/association equilibrium on free Zn^{2+} concentration is displayed in Figure 8. The idealized curve which is based on an overall cooperativity coefficient of 4 but which makes no allowance for the differences in the affinity of the two domains for the metal (for

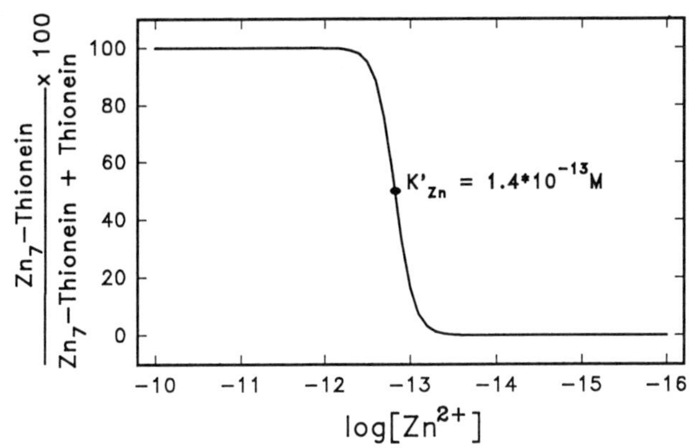

Fig. 8. Idealized isotherm for cooperative binding of Zn^{2+} to apometallothionein (thionein) at pH 7.0, 20°, in 1 M HEPES.

references see Kägi and Kojima, 1987), illustrates the strong Zn^{2+} buffering features of the apoprotein thionein and its ability to form fully saturated complexes with Zn^{2+} at picomolar concentrations of the ion. The average dissociation constant of Zn_7-MT lies in the same range as those of several well-characterized zinc metalloenzymes (Coleman, 1971; Petering et al., 1987) and is thus compatible with observations that under appropriately chosen conditions, Zn-containing forms of MT can serve as Zn donator to Zn-depleted apoenzymes, thereby restoring their activity (Udom and Brady, 1980; Li et al., 1980; Churchich et al., 1989). On the other hand, the high affinity of thionein for Zn^{2+} makes it also feasible that this chelator, which at pH 7 is nearly equal to EDTA in binding power, abstracts Zn^{2+} from less stable zinc protein complexes and thus interferes with their functions. Thus, we have recently shown that thionein effectively removes Zn^{2+} from mammalian, amphibian and viral zinc finger transcription factors and thereby suppresses their functionality in gene expression (Zeng et al., 1991a, 1991b; Fraefel et al., in preparation). Yet, there is no firm experimental evidence that any of these in vitro observed direct Zn transfer reactions, either from Zn_7-MT or to thionein, are of actual importance under in vivo conditions. On principal grounds it seems in fact far more likely that the major function of the conjugate pair Zn_7-MT and thionein lies in controlling and regulating the intracellular free Zn^{2+} concentration and thereby directing the movements of Zn in the cell and also safeguarding the maintenance and functioning of the many Zn-dependent systems. Since MT is present in most mammalian tissues in at least micromolar concentration (Onosaka et al., 1974), and this concentration is believed to be by orders of magnitude larger than the steady-state concentration of thionein, one can conjecture that the corresponding equilibrium concentration of Zn^{2+} lies in the subnanomolar region to the left of the titration step in the isotherm (Fig. 8). At such conditions, the concentration of free Zn^{2+} will be modulated mainly, in an inversed manner, by changes in the concentration of thionein. Thus, an increase in the production of thionein, elicitable by many regulatory stimuli (Table I) would be expected to produce a lowering of the concentration of Zn^{2+} and, hence, a down-regulation of Zn^{2+}-sensitive processes in metabolism, gene expression and signal transduction. Since Zn^{2+} by itself is one of the most potent inducers of thionein synthesis, the same mechanism also provides for efficient autoregulation of its own concentration.

Acknowledgements

The authors research in the field of this chapter was supported by grants from the Swiss National Science Foundation, the Bonizzi-Theler Foundation, and the Kanton Zürich. Special thanks are due to T. Hale and P.-A. Binz for carrying out the parsimony analysis and to Dr. P. Hunziker for permission to include hitherto unpublished sequences.

REFERENCES

Arseniev, A., Schultze, P., Wörgötter, E., Braun, W., Wagner, G. Vašák, M., Kägi, J.H.R., and Wüthrich, K. (1988) J. Mol. Biol. Biol. 201, 637-657.

Bernhard, W.R., Vašák, M., and Kägi, J.H.R. (1986) Biochemistry 25, 1975-1980.

Bienz, A. (1992) Diploma thesis, University of Zürich.

Brady, F.A. (1991) Meth. Enzymol. 205, 559-567.

Braun, W., Vašák, M., Robbins, A.H., Stout, C.D., Wagner, G., Kägi, J.H.R., and Wüthrich, K. (1992) Proc. Natl. Acad. Sci. USA 89, 10124-10128.

Bremner, I. (1991) Meth. Enzymol. 205, 25-35.

Bühler, R.H.O. and Kägi, J.H.R. (1974) FEBS Lett. 39, 229-234.
Churchich, J.E., Scholz, G., and Kwok, F. (1989) Biochim. Biophys. Acta 996, 181-186.
Coleman, J.E., (1971) In: Progress in Bioorganic Chemistry, Vol. 1 (E.T. Kaiser and F.J. Kézdy, eds.), Wiley-Interscience, New York, pp. 159-344.
De, S.K., Enders, G.C., and Andrews, G.K. (1991) Mol. Endocrinol. 5, 628-636.
Ding, X., Bill, E., Good, M., Trautwein, A.X., and Vašák, M. (1988) Eur. J. Biochem. 171, 711-714.
Eigen, M. (1962) Proc. Intern. Congr. Coord. Chem., 7th, Stockholm, pp. 97-115.
Felsenstein, J. (1993) PHYLIP (Phylogeny Inference Package) Version 3.5 (Computer software package and manual distributed by the author, Dept. Genetics, University of Washington, Seattle, WA, 1993).
Fraefel, C., Zeng, J., Engels, M., Schwyzer, M., and Ackermann, M. (1993) J. Virol., submitted for publication.
Frey, M.H., Wagner, G., Vašák, M., Sørensen, O.W., Neuhaus, D., Wörgötter, E., Kägi, J.H.R., Ernst, R.R., and Wüthrich, K. (1985) J. Am. Chem. Soc. 107, 6847-6851.
Hamer, D.H. (1986) Ann. Rev. Biochem. 55, 913-951.
Jahroudi, N., Foster, R., Price-Haughey, J., Beitel, G., and Gedamu, L. (1990) J. Biol. Chem. 265, 6506-6511.
Kägi, J.H.R. and Vallee, B.L. (1960) J. Biol. Chem. 235, 3460-3465.
Kägi, J.H.R. and Vallee, B.L. (1961) J. Biol. Chem. 236, 2435-2442.
Kägi, J.H.R. (1970) 8th Int. Congr. Biochem., Switzerland, Abstracts, pp. 130-131.
Kägi, J.H.R., Himmelhoch, S.R., Whanger, P.D., Bethune, J.L., and Vallee, B.L. (1974) J. Biol. Chem. 249, 3537-3542.
Kägi, J.H.R. and Kojima, Y. (1987) Experientia Suppl. 52, 25-61.
Kägi, J.H.R. and Schäffer, A. (1988) Biochemistry 27, 8509-8515.
Kägi, J.H.R. (1991) Meth. Enzymol. 205, 613-626.
Karin, M. (1985) Cell 41, 9-10.
Klaassen, C.D. and Liu, J. (1991) Meth. Enzymol. 205, 567-574.
Kojima, Y., Berger, C., Vallee, B.L., and Kägi, J.H.R. (1976) Proc. Natl. Acad. Sci. USA 73, 3413-3417.
Kojima, Y. (1991) Meth. Enzymol. 205, 8-10.
Krezoski, S.K., Villalobos, J., Shaw III, C.F., and Petering, D.H. (1988) 255, 483-491.
Li, T.-Y., Kraker, A.J., Shaw III, C.F., and Petering, D.H. (1980) Proc. Natl. Acad. Sci. USA 77, 6334-6338.
Mackay, E.A., Overnell, J., Dunbar, B., Davidson, I., Hunziker, P.E., Kägi, J.H.R., and Fothergill, J.E. (1993) Eur. J. Biochem., in press.
Margoshes, M. and Vallee, B.L. (1957) J. Am. Chem. Soc. 79, 4813-4814.
Messerle, B.A., Schäffer, A., Vašák, M., Kägi, J.H.R., and Wüthrich, K. (1990a) J. Mol. Biol. 214, 765-779.
Messerle, B.A., Bos, M., Schäffer, A., Vašák, M., Kägi, J.H.R., and Wüthrich, K. (1990b) J. Mol. Biol. 214, 781-786.

Messerle, B.A., Schäffer, A., Vašák, M., Kägi, J.H.R., and Wüthrich, K. (1992) J. Mol. Biol. 225, 433-443.
Nemer, M., Travaglini, E.C., Rondinell, E., and D'Alonzo, J. (1984) Dev. Biol. 102, 471-482.
Nettesheim, D.G., Engeseth, H.R., and Otvos, J.D. (1985) Biochemistry 24, 6744-6751.
Nielson, K.B. and Winge, D.R. (1983) J. Biol. Chem. 258, 13063-13069.
Nishimura, H., Nishimura, N., and Tohyama, C. (1989) J. Histochem. Cytochem. 37, 715-722.
Ohtake, H., Hasegawa, K., and Koga, M. (1978) Biochem. J. 174, 999-1005.
Onosaka, S., Tanaka, K., and Cherien, M.G. (1984) Environ. Health Perspect. 65, 67-72.
Onosaka, S., Min, K.-S., Fukuhara, C., Tanaka, K., Tashiro, S.-I., Shimizu, I., Furuta, M., Yasutomi, T., Kobashi, K., and Yamamoto, K. (1986) Toxicology 38, 261-268.
Otvos, J.D. and Armitage, I.M. (1980) Proc. Natl. Acad. Sci. USA 77, 7094-7098.
Otvos, J.D., Olafson, R.W., and Armitage, I.M. (1982) J. Biol. Chem. 257, 2427-2431.
Otvos, J.D., Engeseth, H.R., Nettesheim, D.G., and Hilt, C.R. (1987) Experientia Suppl. 52, 171-178.
Otvos, J., Chen, S.-m., and Liu, X. (1989) In: Metal Ion Homeostasis: Molecular Biology and Chemistry (D.H. Hamer and D.R. Winge, eds.) Alan R. Liss, Inc., New York, pp. 197-206.
Palmiter, R.D., Findley, S.D., Whitmore, T.E., and Durnam, D.M. (1992) Proc. Natl. Acad. Sci. USA 89, 6333-6337.
Palumaa, P., Mackay, E.A., and Vašák, M. (1992) Biochemistry 31, 2181-2186.
Petering, D.H., Krezoski, S., Villalobos, J., Shaw III, C.F., and Otvos, J.D. (1987) Experientia Suppl. 52, 573-580.
Petering, D.H., Krezoski, S., Chen, P., Pattanaik, A., and Shaw III, C.F. (1992) In: Metallothioneins (M.J. Stillman, C.F. Shaw III, and K.T. Suzuki, eds.) VCH Publishers, Inc., New York, pp. 164-185.
Pulido, P., Kägi, J.H.R., and Vallee, B.L. (1966) Biochemistry 5, 1768-1777.
Rauser, W.E. (1990) Annu. Rev. Biochem. 59, 61-86.
Riordan, J.F. and Vallee, B.L., eds. (1991) Metallobiochemistry, Part B, Meth. Enzymol. 205.
Robbins, A.H., McRee, D.E., Williamson, M., Collett, S.A., Xuong, N.H., Furey, W.F., Wang, B.C., and Stout, C.D. (1991) J. Mol. Biol. 221, 1269-1293.
Sadhu, C. and Gedamu, L. (1988) J. Biol. Chem. 263, 2679-2684.
Savas, M.M., Petering, D.H., and Shaw III, C.F. (1991) Inorg. Chem. 30, 581-583.
Schmid, R. (1991) Diploma thesis, University of Zürich.
Schroeder, J.J. and Cousins, R.J. (1991) Meth. Enzymol. 205, 575-584.
Schultze, P., Wörgötter, E., Braun, W., Wagner, G., Vašák, M., Kägi, J.H.R., and Wüthrich, K. (1988) J. Mol. Biol. 203, 251-268.

Shaw III, C.F., Sawas, M.M., and Petering, D.H. (1991) Meth. Enzymol. 205, 401-414.
Stillman, M.J. and Zelazowski, A.J. (1988) J. Biol. Chem. 263, 6128-6133.
Stillman, M.J. (1992) In: Metallothioneins (M.J. Stillman, C.F. Shaw III, and K.T. Suzuki, eds.) VCH Publishers, Inc., New York, pp. 55-127.
Templeton, D.M. and Cherian, M.G. (1991) Meth. Enzymol. 205, 11-24.
Uchida, Y., Takio, K., Titani, K., Ihara, Y., and Tomonaga, M. (1991) Neuron 7, 337-347.
Udom, A.O. and Brady, F.O. (1980) Biochem. J. 187, 329-335.
Vallee, B.L., Coleman, J.E., and Auld, D.S. (1991) Proc. Natl. Acad. Sci. USA 88, 999-1003.
Vašák, M. and Kägi, J.H.R. (1983) In: Metal Ions in Biological Systems (H. Sigel, ed.), Vol. 15, Marcel Dekker, Inc., New York, pp. 213-273.
Vašák, M., Berger, C., and Kägi, J.H.R. (1984) FEBS Lett. 168, 174-178.
Vašák, M., Hawkes, G.E., Nicholson, J.K., and Sadler, P.J. (1985) Biochemistry 24, 740-747.
Vašák, M. (1986) Environ. Health Perspect. 65, 193-197.
Vašák, M., Wörgötter, E., Wagner, G., Kägi, J.H.R., and Wüthrich, K. (1987) J. Mol. Biol. 196, 711-719.
Vašák, M. and Kägi, J.H.R. (1993) In: Encyclopedia of Inorganic Chemistry (R.B. King, ed.), John Wiley & Sons, Ltd., in press.
West, A.K., Stallings, R., Hildebrand, C.E., Chiu, R., Karin, M., and Richards, R.I. (1990) Genomics 8, 513-518.
Willner, H., Vašák, M., and Kägi, J.H.R. (1987) Biochemistry 26, 6287-6292.
Willner, H., Bernhard, W.R., and Kägi, J.H.R. (1992) In: Metallothioneins (M.J. Stillman, C.F. Shaw III, and K.T. Suzuki, eds.) VCH Publishers, Inc., New York, pp. 128-143.
Winge, D.R. and Miklossy, K.-A. (1982) J. Biol. Chem. 257, 3471-3476.
Wüthrich, K. (1991) Meth. Enzymol. 205, 502-520.
Zeng, J., Heuchel, R., Schaffner, W., and Kägi, J.H.R. (1991a) FEBS Lett. 279, 310-312.
Zeng, J., Vallee, B.L., and Kägi, J.H.R. (1991b) Proc. Natl. Acad. Sci. USA 88, 9984-9988.

Metallothionein III
K.T. Suzuki, N. Imura and M. Kimura (eds)

DYNAMIC ASPECTS OF METALLOTHIONEIN STRUCTURE

J.D. Otvos, X. Liu, H. Li, G. Shen and M. Basti

Department of Biochemistry, North Carolina State University, Raleigh, NC, USA

SUMMARY: The dynamics of metal ion interaction with the two structural domains of metallothionein (MT) were investigated to provide insight into how the protein might carry out its proposed dual functions of "active" Zn transfer agent and "passive" repository of Cd, Cu, and other heavy metals. Using mainly NMR methods, facile interprotein metal exchange reactions were shown to provide a metal redistribution mechanism for conversion of mixtures of single-metal MT species into thermodynamically more stable mixed-metal species. The data suggest that the α- and β-domains of MT are structurally optimized for interaction with divalent (Zn, Cd) and monovalent (Cu, Ag) metal ions, respectively, and that the metal bound to one domain does not influence the reactivity of the metal in the other domain. Support for a mechanism of metal exchange involving transient formation of intermolecular metal-thiolate bonds was obtained for Cd-MT by the direct demonstration by gel filtration HPLC and 2D NMR of the presence of dimeric MT in solution.

INTRODUCTION

During the past decade, great progress has been made in defining the structural details of divalent metal ion binding to mammalian metallothionein (MT). Owing mainly to two-dimensional NMR studies of Cd_7-MT from several sources (Schultze et al., 1988; Arseniev et al., 1988; Messerle et al.,1990) and X-ray crystallographic analysis of rat liver Cd,Zn-MT (Robbins et al., 1991), we now know precisely how the polypeptide chain enfolds the 3-metal and 4-metal clusters of the amino-terminal β-domain and carboxyl-terminal α-domain, respectively. While this information has been useful, particularly in reinforcing the structural independence of the two domains and the relatively high degree of solvent exposure of both metal clusters, it has not provided many new insights into how the structure of MT specifically relates to its biological function(s). It is important to recognize, however, that we still know very little about how MT interacts with copper. Since MT is believed to play central roles in both copper and zinc metabolism, and the two metals bind very differently to form clusters of different stoichiometry (Nielson et al., 1985), there may be aspects of the structure that relate specifically to Cu(I) coordination that are not apparent from the existing X-ray and NMR structures.

The difficulty in defining classical structure-function relationships for MT stems from the likelihood that the protein performs multiple cellular functions, though none are yet firmly established (Bremner, 1987; 1991; Kägi & Schäffer, 1988; Templeton & Cherian, 1991; Zeng et al., 1991a). In some of its proposed roles, such as providing protection against the deleterious effects of excessive intracellular concentrations of both toxic and essential heavy metal ions, MT performs the relatively passive function of providing binding sites of high thermodynamic stability for metal sequestration. While coordination to multiple cysteine thiolate ligands adequately fulfills this function, there is no obvious need for the protein to have two discrete domains containing metal clusters of different structure.

In contrast, the more active role proposed for MT as a metal transfer agent, donating zinc to and/or pulling zinc from zinc-dependent proteins (Li et al.,1980; Udom & Brady, 1980, Zeng et al., 1991a; 1991b), requires a protein structure that confers on its thiolate ligands and bound metal ions a high degree of kinetic reactivity. Consistent with these requirements are both the small size of the protein, which gives potential reactants relatively unhindered access to the metal centers, and the arrangement of bound metal ions in polynuclear clusters that display unusually facile metal-thiolate ligand exchange (Hagen et al., 1982; Otvos et al., 1989a). Under many physiological conditions, it is likely that MT is called on to simultaneously perform both the "dynamic" role of zinc donor/acceptor and the "passive" homeostatic role of providing a stable repository for metals such as copper and cadmium. Such functional versatility would almost certainly be impossible to achieve were it not for the presence of two non-interacting domains with different numbers of cysteine ligands capable of forming metal clusters of differing thermodynamic stability and kinetic lability.

For reasons alluded to above, the focus of our work has been less on defining the "static" structures of the metal clusters in MT than on understanding the dynamics of Cd(II), Zn(II), Cu(I), and Ag(I) interaction with the two domains of the protein. Using a variety of experimental methods, we have uncovered evidence that the metals bound to MT are anything but inert. Rather, they are in a constant state of flux in which extremely facile metal-ligand exchange promotes a continual intermolecular transfer of metal between clusters (Otvos et al., 1987; 1989a). While this reactivity may serve a functional need, such as providing a pathway for metals to relocate to their preferred binding sites in mixed-metal systems (Nettesheim et al, 1985), it is probably an unavoidable consequence of the structural features of MT that evolved to facilitate rapid metal transfer to and from other metalloproteins.

The reactions we have investigated that are responsible for metal ion movement into and out of the binding sites of thionein (Th) are listed below:

(1)	Me[1]-Th	→	Me[1] + Th	(SLOW)
(2)	Me[1]-Th + Me[2]	→	Me[2]-Th + Me[1]	(FAST)
(3)	Me[1]-Th + Me[2]-Th*	→	Me[2]-Th + Me[1]-Th*	(FAST)
(4)	Me[1]-Th + Ligand	→	Me[1]-Ligand + Th	(FAST)

Eq.(1) describes metal dissociation from MT; the slow rates are consistent with the high thermodynamic stability of the metal clusters. The apparent association constant for Zn(II) at pH 7 is about 10^{11}-10^{12} M^{-1} (Otvos et al., 1989b; Vašàk & Kägi, 1983) and Cd(II), Cu(I), and Ag(I) bind several orders of magnitude more tightly. There is good evidence that the two domains form clusters of unequal stability, with Zn(II) and Cd(II) binding more strongly to the α-domain than the β-domain and Cu(I) and Ag(I) exhibiting the opposite preference (Briggs & Armitage, 1982; Nielson & Winge, 1983; 1984). Eq.(2) indicates that despite binding very tightly, metal ions are displaced very rapidly from their binding sites (typically within seconds) by metal ions from solution (Nettesheim et al., 1985). The free metal ions apparently actively facilitate dechelation of the bound metals by reacting with the solvent-exposed cysteine ligands of the metal clusters. Depicted in Eq.(3) is the facile intermolecular metal exchange reaction between MT molecules. First demonstrated between Cd_7-MT and Zn_7-MT as the route to formation of the mixed-metal clusters produced in animals in response to Cd(II) administration (Nettesheim et al., 1985), the reaction was subsequently found to be responsible for rapid Cd self-exchange among the 3 sites of the β-domain cluster of Cd_7-MT (Otvos et al, 1987; 1989a). Recently we have shown that hybrid MTs containing exclusively Cu or Ag in the β-domain and Zn or Cd in the α-domain are products of the <u>in vitro</u> reactions of Cu-MT or Ag-MT with Zn-MT or Cd-MT (Li, 1991). Eq.(4) depicts the ligand substitution reaction by which metal is transferred from MT to a competing ligand (L). The unexpectedly rapid rate of this reaction was first demonstrated for Zn transfer to several apo-Zn metalloenzymes (Li et al., 1980; Udom & Brady, 1980), and similar results have been obtained for Zn donation to the "zinc twist" DNA-binding domain of the glucocorticoid receptor (Liu, 1989). The question of whether the two domains exhibit differences in the kinetics of Zn transfer to a competing ligand (nitrilotriacetate, NTA) will be addressed later in the paper; the answer turns out to be that the α-domain cluster is much more reactive despite binding Zn more tightly than the β-domain cluster. The result serves as an excellent reminder that kinetic lability and thermodynamic stability are not necessarily related and that structural features that favor one may not have the same effect on the other.

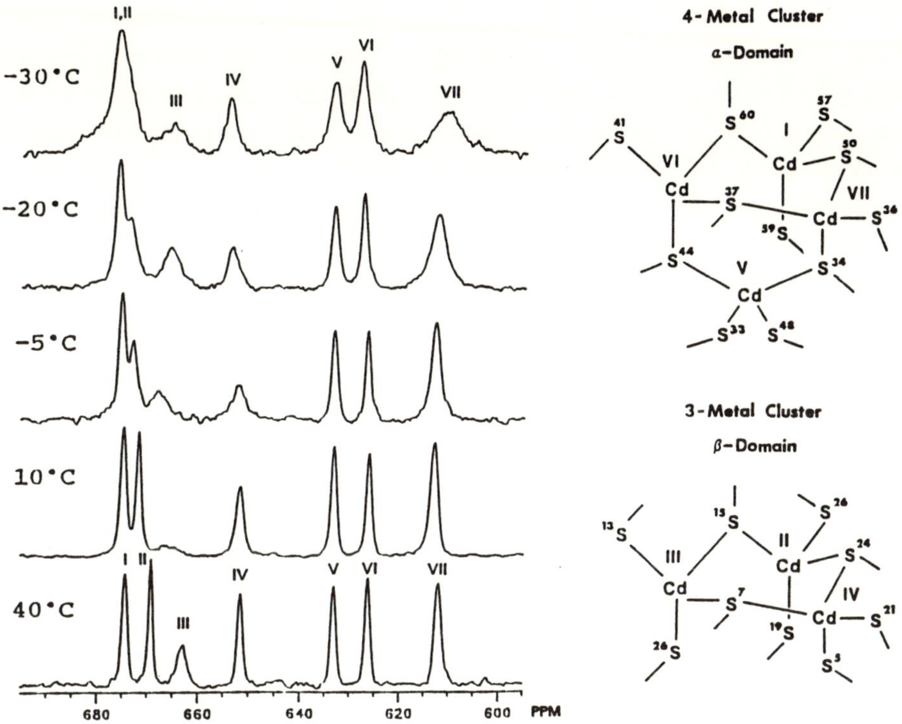

Fig.1. (Left) ^{111}Cd NMR spectra (106 Mhz) of rabbit liver Cd-MT-2 in 35% (v/v) MeOH acquired at different temperatures. (Right) Schematic depiction of the protein's metal clusters, with numbers indicating the sequence positions of the cysteine ligands.

DYNAMICS OF METAL AND LIGAND EXCHANGE IN CADMIUM-MT

Cd NMR spectroscopy is uniquely suited to studying the dynamics of Cd interaction with MT owing to the responsiveness of Cd chemical shifts to very small differences in coordination environment (Armitage & Otvos, 1982). Thus, the slight structural differences among the 7 binding sites create enough spectral dispersion to allow separate signals to be detected from Cd in each site, as shown in the bottom spectrum in Fig. 1 acquired at 40°C. The only hint in this spectrum of any dynamic modulation in the environments of any of the metals is the greater linewidth of peak III, which originates from one of the Cd ions in the 3-metal cluster of the β-domain. As the temperature is lowered, it is easy to see that the chemical shifts and/or linewidths of the signals from all 3 metals in the β-domain (peaks II-IV) undergo significant changes, while those of the α-domain metals do not. The only exception is peak VII, which begins to exhibit differential broadening at very low temperatures. Since 3 of the 4 ligands to Cd^{VII} (Cys 34, 36, and 37) are from residues near the junction between the 2 domains, it is possible that the broadening of peak VII is an indirect manifestation of dynamic events centered in the β-domain.

Though we have been unable to identify the dynamic events by going to low enough temperatures to "freeze out" the species involved, the data indicate that MT at physiological temperature undergoes continual rapid interconversion between two or more "states" of the 3-metal cluster that have different local coordination environments. We consider the most likely identity of these "states" to be 1) configurational isomers that interconvert rapidly via intramolecular metal-thiolate ligand exchange, and 2) oligomeric forms of MT that are in rapid equilibrium in solution. The former possibility is suggested by analogy to the facile exchange of terminal and bridging thiolate ligands observed in model inorganic cage complexes (Hagen et al, 1982) and the latter by experimental evidence to be presented below. A different dynamic process that is surprisingly rapid, but too slow to influence the appearance of a standard Cd NMR spectrum, is that of mutual Cd interchange among the 3 sites of the β-domain cluster. The phenomenon was discovered by saturation transfer experiments such as shown in Fig. 2 in which the resonance of each Cd ion is selectively perturbed by a presaturation pulse (Otvos et al., 1987;

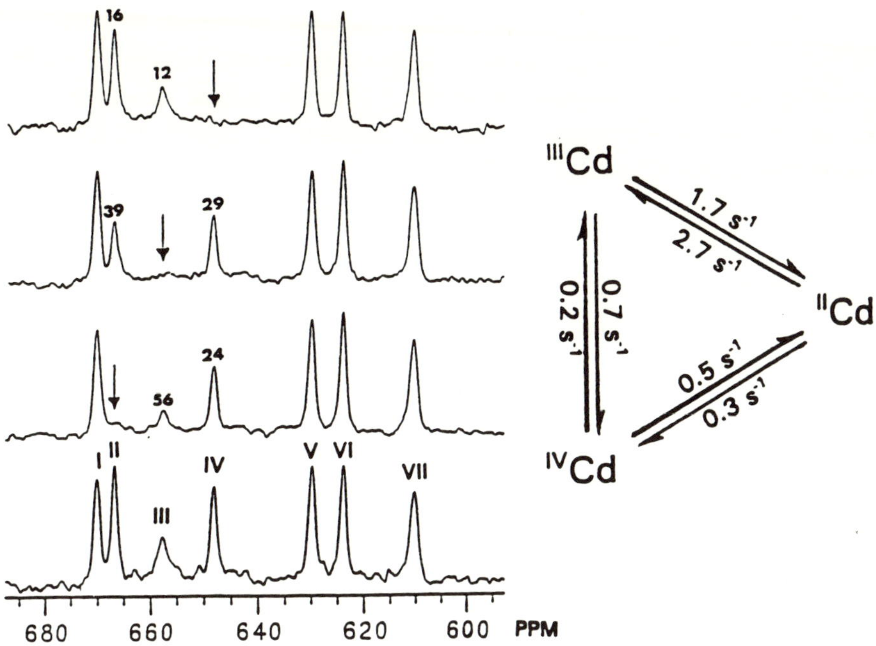

Fig. 2. (left) ^{111}Cd saturation transfer spectra of Cd-MT-2. The arrows indicate the frequencies at which the saturating pulse was applied, and the numbers give the percent reductions in intensity of the other resonances. (Right) Calculated unidirectional rate constants for metal exchange among the 3-metal cluster sites at 35°C.

1989a). The extent to which this perturbation is transferred to the resonances of Cd located at other binding sites provides a measure of the rates of intersite metal exchange. A complete analysis of such data yields the 6 unimolecular rate constants characterizing metal exchange into and out of each 3-metal cluster site of MT-2 at 35°C, which are given in Fig. 2 (Liu, 1989). Exchange between 4-metal cluster sites and exchange between clusters is several orders of magnitude slower and cannot be measured by saturation transfer (Otvos et al., 1987). It can be seen in Fig. 2 that the rates of exchange between 3-metal cluster sites are not equal; metal migrates significantly faster between sites II and III than between the others, which causes the lifetimes of Cd in these two sites to be about four times shorter than in site IV (0.3-0.4 vs 1.4 sec).

By demonstrating transfer of saturation between MT-1 and MT-2 isoforms, we have shown conclusively that the route by which Cd moves so rapidly between 3-metal cluster sites is intermolecular metal exchange (Otvos et al., 1989a; Liu, 1989). What apparently takes place is bimolecular association of MT molecules to form

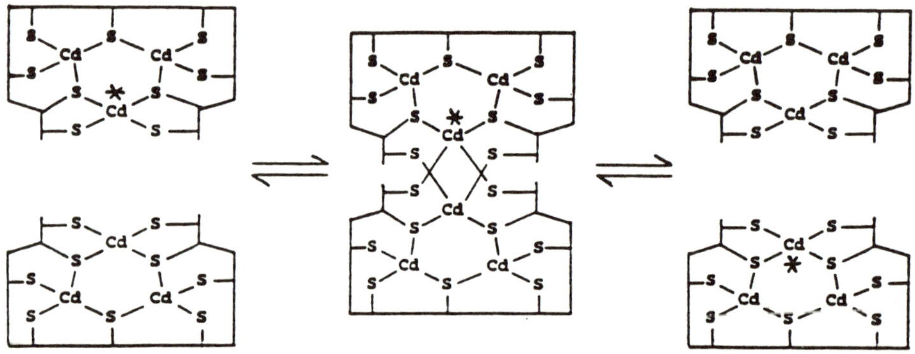

Fig. 3. Schematic depiction of the process by which two β-domain clusters interact via shared metal-thiolate coordination to accomplish the transfer of Cd from one MT molecule to the other.

dimers, followed by a series of intermolecular metal-ligand exchange events that result in Cd being transferred between 3-metal clusters across the dimer interface. A greatly oversimplified depiction of this process is given in Fig. 3, which shows only one of a manifold of possible intermediates in which the β-domain metal ions of one MT molecule become transiently coordinated to the thiolate ligands of the other. The dynamics predicted by this model, with many individual metal-thiolate bond-breaking and bond-making events accompanying the transfer of each metal, are consistent with the temperature-dependent modulation of the β-domain Cd NMR signals seen in Fig. 1.

If Cd-MT indeed exists for much of the time in a self-associated state stabilized by the novel type of intermolecular interaction depicted in Fig. 3, one would expect to be able to find experimental evidence for MT dimers in solution. In fact, several reports describing dimeric MTs produced by oxidation or exposure to excess metal ion have appeared (Otvos et al., 1985; Palumaa et al., 1992), but these species are not the same as those proposed to be in continuous reversible equilibrium with MT monomers. Our group is currently reinvestigating the hydrodynamic properties of Cd-MT and Zn-MT by gel filtration, sedimentation equilibrium, and other methods and has obtained convincing support for the proposed self-association reaction. One technique that clearly demonstrates the presence in solution of MT monomers and dimers and the expected dependence of their relative concentrations on protein concentration is gel filtration HPLC. Fig. 4 contains some representative data showing that under appropriate high resolution conditions, monomeric and dimeric species can be separated. Although it has been known for many years that MT exhibits gel filtration properties of a protein roughly twice its size (Kägi et al., 1974; Vašàk et al., 1984), the reason was assumed to be its elongated shape. This conclusion should now be reevaluated. Ongoing sedimentation equilibrium studies confirm the existence of MT dimers and possibly larger oligomers; a preliminary estimate of the self-association constant for Cd-MT at pH 6.8 is about 3×10^4 M^{-1}.

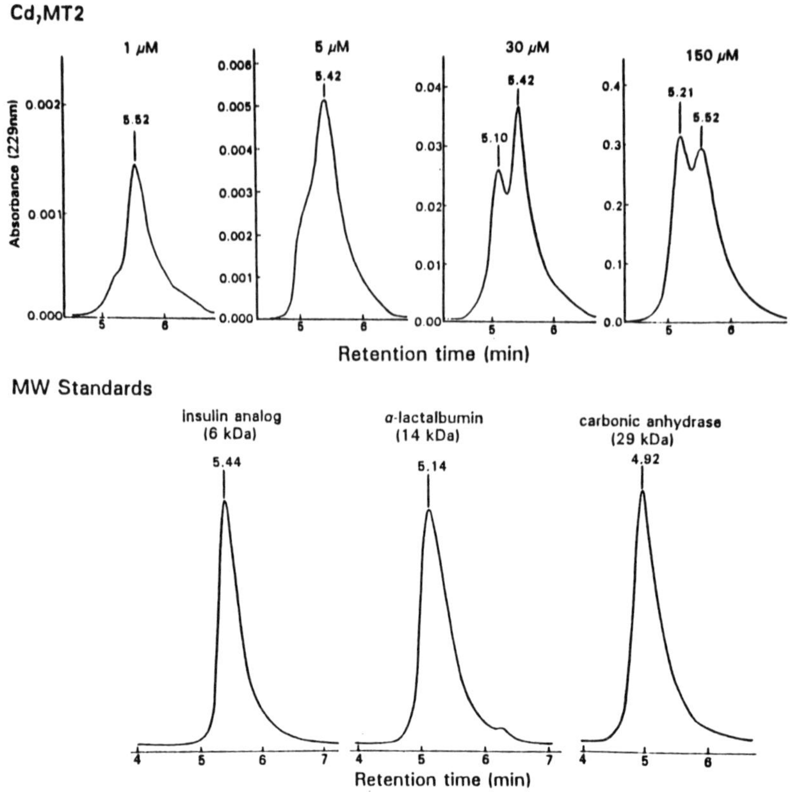

Fig. 4. (Top) Protein concentration dependence of the gel filtration HPLC profiles of Cd-MT-2 in 10 mM phosphate, pH 7.0 at 25°C. 10 μl samples prepared by dilution of a 3mM stock Cd-MT solution were applied to a GPC-100 column (SynChrom) and eluted at a flow rate of 0.5 ml/min. (Bottom) Gel filtration HPLC profiles of 3 molecular weight standards run under identical conditions.

A method that could potentially provide not only confirmation of the existence of MT dimers in solution but also give their structure(s) is 2D ^{1}H NMR spectroscopy (Wüthrich, 1991). The needed information would come from identification of protons at the dimer interface that exhibit NOEs to protons on the other molecule. Despite the prediction that the great majority of MT is in a

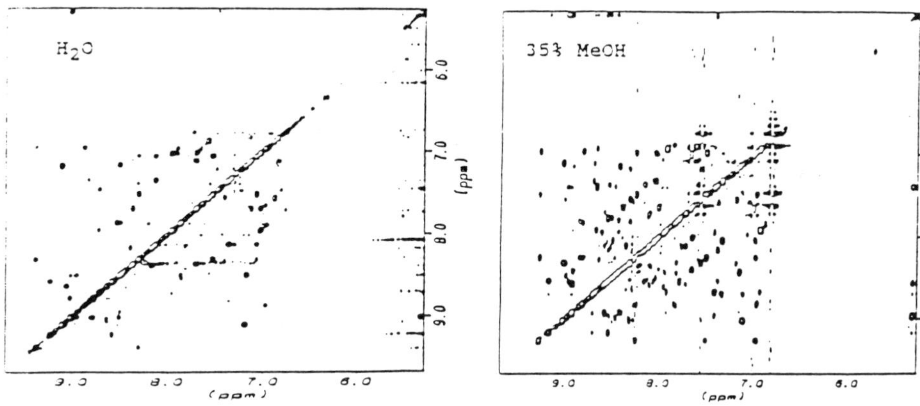

Fig. 5. The "fingerprint" region of 500 MHz NOESY spectra of rabbit liver Cd-MT-2 in 90% H_2O/10% D_2O (left) and 35% MeOH-d_3/65% H_2O (right) acquired at 10°C.

dimeric state at NMR concentrations (ca. 10 mM), no intermolecular NOEs have ever been reported. The reason almost certainly is that these NOEs are rendered undetectable by the complicated dynamics at the dimer interface associated with interprotein metal exchange. Our approach to this problem has been to conduct low temperature 2D NMR studies in MeOH/H_2O solutions with the aim of sufficiently slowing down the unfavorable dynamics to allow detection of intermolecular NOEs. Unexpectedly, as illustrated in Fig. 5, we found that 35% MeOH solutions of Cd-MT at normal temperatures yield NOESY spectra containing a far greater number of cross-peaks than seen in H_2O. Analysis of these and related spectra indicate that most of the new crosspeaks are not intermolecular in origin, nor are they a result of structural changes induced by the change of solvent. Rather, most arise from short-range NOEs between β-domain protons that, based on the 3-dimensional structures, would have been expected to be observed in H_2O were it not for dynamic structural fluctuations that apparently occur on different time-scales in MeOH. Four NOE cross-peaks, by contrast, have been identified from pairs of β-domain protons that are too far apart in the 3D structure to be intramolecular in origin. They are therefore interpreted as arising from interactions between MT molecules. This finding not only confirms the existence of dimeric MT under these conditions, but supports the proposed model (Fig. 3) of a β-domain interface across which metals are continually being transferred.

DYNAMICS OF METAL EXCHANGE IN MIXED-METAL METALLOTHIONEINS

The facile interprotein metal exchange reaction described above makes it impossible for Cd_7-MT and Zn_7-MT to coexist in solution without producing species containing mixed-metal clusters in both the α- and β-domains (Nettesheim et al., 1985). Formation of such clusters is possible because of the similar preferences of Zn(II) and Cd(II) for tetrahedral coordination. Cu(I) and its air-stable analog, Ag(I), on the other hand, appear to form 6-metal complexes in each domain with trigonal coordination (Winge, 1991), which might make sharing a cluster with Zn(II) or Cd(II) unfavorable. Since copper is invariably found in vivo in mixed-metal Cu,Zn-MTs, we have been interested in determining how the two metal ions are distributed among the two domains and whether that distribution can be achieved by intermolecular metal exchange between Cu-MT and Zn-MT.

One experimental approach has been to use ^{111}Cd and ^{109}Ag as structural analogs of Zn and Cu, respectively, so that ^{111}Cd and ^{109}Ag NMR can be employed to monitor metal ion distribution between the clusters. Fig. 6 shows the results of one such study in which increasing amount of Ag(I) were added to displace Cd(II) from Cd-MT. The progressive reduction seen in the intensities of the 3-metal cluster signals (II-IV) brought about by addition of the first 6 Ag ions confirms the β-domain as the preferred site of Ag (and Cu) binding and indicates that no mixed-metal clusters are produced. The only intermediate formed along the route to Ag_{12}-MT is $Ag_6^{\beta}Cd_4^{\alpha}$-MT. This species has been found by other Cd NMR studies

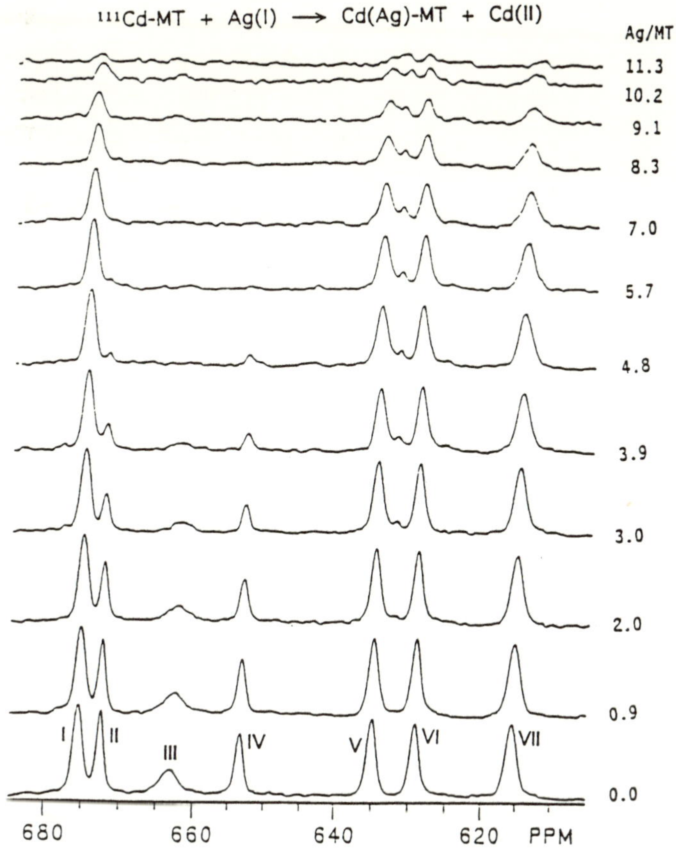

Fig. 6. ^{111}Cd NMR titration of Cd_7-MT-2 with Ag(I). Protein (1.6 mM) was incubated with successive aliquots of $AgNO_3$ in the presence of Chelex-100. After 30 minutes, the Chelex was removed by centrifugation and the ^{111}Cd NMR spectrum acquired (4 hr. each).

(not shown) to be the only mixed-metal MT formed by intermolecular metal exchange between Ag_{12}-MT and Cd_7-MT.

Evidence of a different type for direct metal transfer between Ag_{12}-MT and Zn_7-MT leading to production of $Ag_6{}^{\beta}Zn_4{}^{\alpha}$-MT comes from kinetic studies of the products formed by Ag(I) reaction with Zn-MT

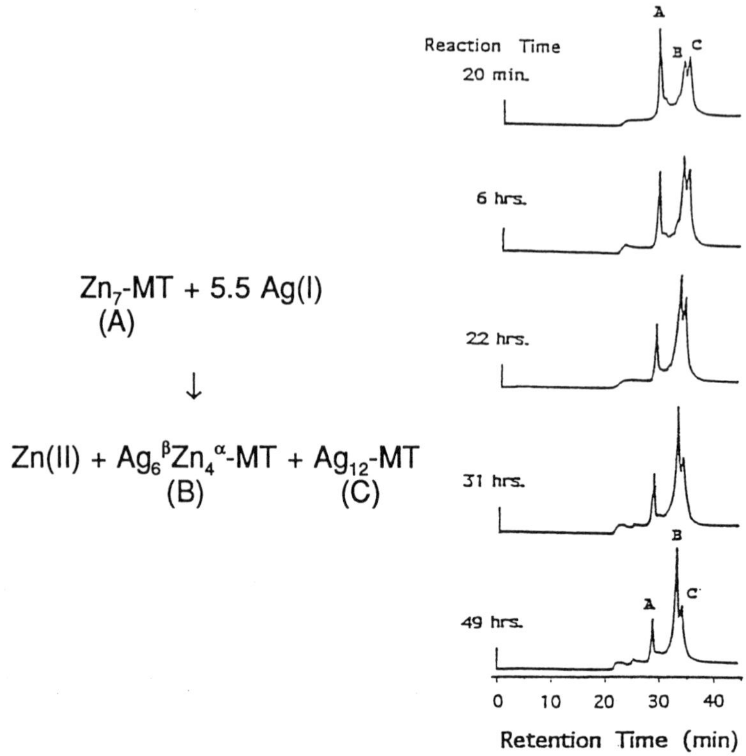

Fig. 7. Time dependence of the formation of $Ag_6^{\beta}Zn_4^{\alpha}$-MT-2 following reaction of Zn-MT-2 with Ag(I) monitored by ion exchange (DEAE) HPLC. A 100 uM Ag,Zn-MT sample was prepared from Zn-MT by addition of 5.5 equivalents of $AgNO_3$ followed by addition of Chelex-100. The sample was maintained at 23°C and 100 μl aliquots were withdrawn at the indicated times and applied to a Bio-Gel DEAE-5-PW HPLC column (Bio-Rad). Elution was with a 30 min. linear gradient of 5-300mM Tris-HCl, pH 8.5 at a flow rate of 1 ml/min.

monitored by ion exchange HPLC. The experiment shown in Fig. 7 indicates that the rapid displacement of Zn(II) by 5.5 equivalents of Ag(I) proceeds with little or no kinetic selectivity to produce both $Ag_6^{\beta}Zn_4^{\alpha}$-MT (peak B) and Ag_{12}-MT (peak C) as well as unreacted Zn_7-MT (peak A). After free metal ions are removed by Chelex, the product distribution is seen to undergo continual change as Zn-MT and Ag-MT exchange metal to produce the more thermodynamically stable $Ag_6^{\beta}Zn_4^{\alpha}$-MT. As expected, the rate of this bimolecular metal

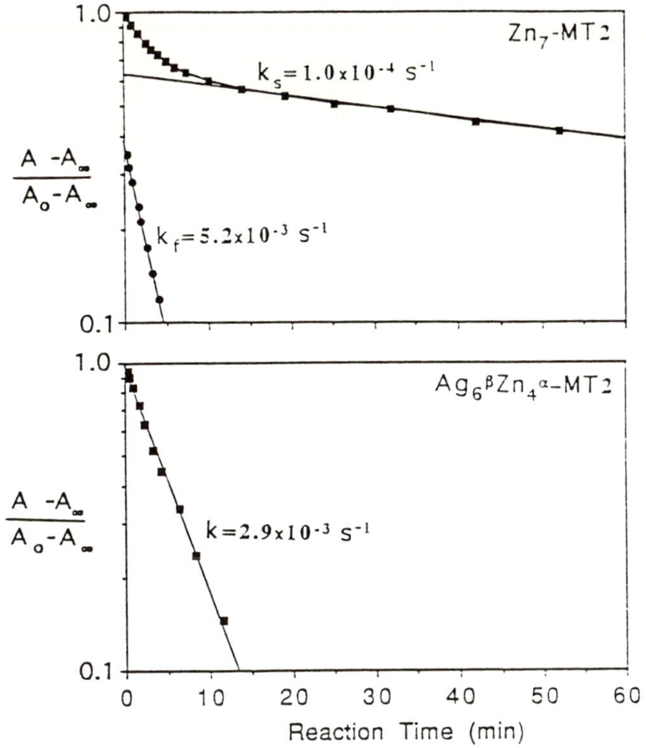

Fig. 8. Kinetics of Zn(II) removal from 7.7 μM Zn_7-MT-2 (top) and 13.3 μM $Ag_6{}^{\beta}Zn_4{}^{\alpha}$-MT-2 (bottom) by 50 mM nitrilotriacetate (NTA). The reactions were conducted in 20 mM phosphate, pH 7.5, plus 0.1 M KF at 25°C and monitored by loss of UV absorbance at 230 nm. Experimental data are given by the filled squares and the constants for the fast and slow steps of the Zn-MT reaction and for the monophasic reaction of Ag,Zn-MT are as indicated.

exchange reaction is highly dependent on protein concentration, which explains why only thermodynamically-controlled product mixtures are likely to be observed in NMR experiments such as shown in Fig. 6 which use high concentrations of MT and require hours to obtain each spectrum.

DYNAMICS OF METAL TRANSFER FROM MT TO COMPETING LIGANDS

If an essential biological function of MT is to serve as a kinetically-reactive donor of Zn to Zn-requiring proteins, it makes teleological sense that a mechanism exists to prevent more strongly-binding metals such as Cu from interfering with this function. In the preceding section it was demonstrated that both a thermodynamic driving force and accessible kinetic pathway exist for interprotein metal exchange between Zn-MT and Cu-MT that accomplishes a redistribution of Cu to the β-domain and Zn to the α-domain. If this redistribution is the means by which MT preserves its dual functions in Zn and Cu metabolism, it might be predicted that the α-domain cluster is the most reactive in Zn transfer reactions with competing ligands.

Evidence supporting this prediction is presented in Fig. 8, which shows the time-courses for Zn transfer from Zn_7-MT and $Ag_6^{\beta}Zn_4^{\alpha}$-MT to the competing ligand, NTA. In common with several other small ligands, NTA displays biphasic kinetics of Zn removal from Zn_7-MT under pseudo first-order conditions, with about 50% of the metal reacting more than an order of magnitude faster than the rest (Otvos et al., 1989b). In distinct contrast, the Zn in Ag_6Zn_4-MT reacts with monophasic kinetics at a rate that is virtually identical to that of the fast step of the reaction with Zn_7-MT. This result not only proves that the thermodynamically more stable Zn_4 α-domain cluster is unexpectedly more reactive in Zn transfer reactions (at least with NTA) than the Zn_3 β-domain cluster, but also indicates that Ag (and presumably Cu) bound to the β-domain does not significantly alter this reactivity. Such characteristics are in accord with the proposition that the structures of the two domains of MT evolved to 1) bind Cu and Zn with differential affinities to assure that each metal is bound to a different domain in the presence of both metals, and 2) assure that Cu binding does not compromise the structural characteristics that allow Zn to be rapidly transferred from MT to essential cellular ligands.

Acknowledgements

We thank Qun Zhou and Xiaowen Ma for able technical assistance. This research was supported by NIH Grant ES04036.

References

Armitage, I.M. and Otvos, J.D. (1982) in "Biological Magnetic Resonance" (Berliner, L.J. and Reuben, J., eds.) Vol. 4, pp. 79-144. Plenum, New York.
Arseniev, A., Schultze, P., Wörgötter, E., Braun, W., Wagner, G., Vašàk, M., Kägi, J.H.R., and Wüthrich, K. (1988) J. Mol. Biol. 201, 637-657.
Bremner, I. (1987) Experentia, Suppl. 52, 81-107.
Bremner, I. (1991) Meth. Enzymol. 205, 25-35.
Briggs, R.W. and Armitage, I.M. (1982) J. Biol. Chem. 257, 1259-1262.
Hagen, K.S., Stephan, D.W., and Holm, R.H. (1982) Inorg. Chem. 21, 3928-3936.
Kägi, J.H.R., Himmelhoch, S.R., Whanger, P.O., Bethune, J.L., and Uallee, B.L. (1974) J. Biol. Chem. 249, 3537-3542.
Kägi, J.H.R. and Schäffer, A. (1988) Biochemistry 27, 8509-8515.
Li, H. (1991) Ph.D. thesis, University of Wisconsin-Milwaukee.
Li, T.-Y., Kraker, A.J., Shaw, C.F., and Petering, D.H. (1980) Proc. Natl. Acad. Sci. USA 77, 6334-6338.
Liu, X. (1989) Ph.D. thesis, University of Wisconsin-Milwaukee.
Messerle, B.A., Schäffer, A., Vašàk, M., Kägi, J.H.R., and Wüthrich, K. (1988) J. Mol. Biol. 214, 765-779.
Nettesheim, D.G., Engeseth, H.R., and Otvos, J.D. (1985) Biochemistry 24, 6744-6756.
Nielson, K.B. and Winge, D.R. (1983) J. Biol. Chem. 258,13063-13069.
Nielson, K.B. and Winge, D.R. (1984) J. Biol. Chem. 259, 4941-4946.
Nielson, K.B., Atkin, C.L., and Winge, D.R. (1985) J. Biol. Chem., 260, 5342 5350.
Otvos, J.D., Engeseth, H.R., and Wehrli, S. (1985) Biochemistry 24, 6735-6740.
Otvos, J.D., Engeseth, H.R., Nettesheim, D.G., and Hilt, C.R. (1987) Experentia, Suppl. 52, 171-178.
Otvos, J.D., Chen, S-m., and Liu, X. (1989a) in "Metal Ion Homeostasis: Molecular Biology and Chemistry" (Hamer, D.H. and Winge, D.R., eds.), pp. 197-206. Alan R. Liss, New York.
Otvos, J.D., Petering, D.H., and Shaw, C.F. (1989b) Comments Inorg. Chem. 9, 1-35.
Palumaa, P., Mackay, E.A., and Vašàk, M. (1992) Biochemistry 31, 2181-2186.
Robbins, A.H., McRee, D.E., Williamson, M., Collett, S.A., Xuong, N.H., Furey, W.F., Wang, B.C., and Stout, C.D. (1991) 221, 1269-1293.
Schultze, P., Wörgötter, E., Braun, W., Wagner, G., Vašàk, M., Kägi, J.H.R., and Wüthrich, K. (1988) J. Mol. Biol. 203, 251-268.

Templeton, D.M. and Cherian, M.G. (1991) Meth. Enzymol. 205, 11-24.
Udom, A. and Brady, F.O. (1980) Biochem. J. 187, 329-335.
Vašàk, M. and Kägi, J.H.R. (1983) in "Metal Ions in Biological Systems", Vol. 15 (Sigel, H., ed.), p. 213-273. Marcel Dekker, New York.
Vašàk, M., Berger, C., and Kägi, J.H.R. (1984) FEBS Lett. 168, 174-178.
Winge, D. (1991) Meth. Enzymol. 205, 458-469.
Wüthrich, K. (1991) Meth. Enzymol. 205, 502-520.
Zeng, J., Vallee, B.L., and Kägi, J.H.R. (1991a) Proc. Natl. Acad. Sci. USA 88, 9984-9988.
Zeng, J., Heuchel, R., Schaffner, W., and Kägi, J.H.R. (1991b) FEBS Lett. 279, 310-312.

Metallothionein III
K. T. Suzuki, N. Imura and M. Kimura (eds)

QUANTIFICATION OF METALLOTHIONEIN IN BIOLOGICAL MATERIALS

Karl H. Summer and Dominik Klein

Institute of Toxicology, GSF-Research Center, W-8042 Neuherberg, Germany

SUMMARY: Although several methods are available, the accurate quantification of metallothionein (MT) in complex biological materials still represents a problem which is far from being solved. This originates, among others, by the fact that with most commonly used methods MT is determined indirectly via its thiol or metal content. The direct immunological methods, on the other hand, may lead to erroneous results due to often weakly defined antibodies and standard MT. Since so far there is no universal method for the quantification of MT in biological materials and only few methods give additional informations e.g. on MT isoforms, metal composition or oxidation state, the parallel use of different methods is considered to be most appropriate.
In the following the major advantages and problems of the various methods for MT determination will be summarized.

INTRODUCTION

To quantitate metallothionein (MT) in complex biological materials several methods requiring largely different expenses in time, laboratory skills and instrumentation have been developed. The majority of these methods takes advantage of the unique characteristics of MT namely its binding capacity for metals, high content of thiol groups and antigenic properties. However, in order to measure MT, one has to consider that each of these methods has its own inherent as well as experimental limitations.

METHODS FOR METALLOTHIONEIN DETERMINATION

Due to the relatively low concentration of MT as compared to other proteins and its predominant localization in the soluble fraction of cells, tissue homogenates or body fluids, the determination of MT in biological materials usually requires special

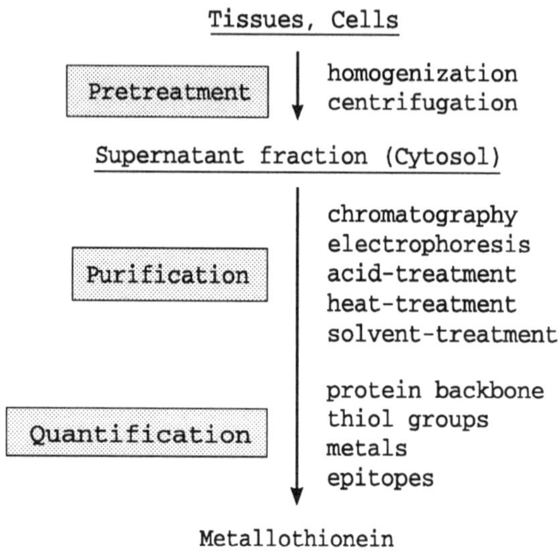

Fig. Schematic Outline of Metallothionein Assays

treatment of the sample. This is performed by centrifugating tissue homogenates or body fluids at 100,000 *g*. In the resulting supernatant fraction, MT normally is determined after separation from high molecular weight (HMW) proteins by chromatographic or electrophoretic methods or by acid, heat or solvent treatment (Figure). Dependent on the efficiency of the separation, MT subsequently is quantified indirectly via protein staining or its content of metals and thiols, or directly via immunological methods (Table).

Table: Principles for MT Quantification in Biological Materials

indirect	direct
protein (dye staining)	protein epitopes (RIA, ELISA)
thiol groups (spectroscopy, polarography)	
metal content (Hg-, Cd-, Ag-saturation)	

Since separation procedures and analytical methods may be combined in many ways, in the following, current protocols for MT quantification are classified to five basic principles, namely chromatography, electrophoresis, metal saturation, immunology and electrochemistry. At present, Cd saturation assays and radioimmunoassays (RIAs) are mostly used to quantitate MT in biological materials.

CHROMATOGRAPHY

MT determination involving chromatographic methods consists of a separation step followed by quantification of either the protein thiol or metal content in the MT specific elution volume. By measuring metals useful informations on the metal composition of the protein are obtained. Such determinations most favourably can be done on-line with the separation step (Suzuki 1980, Suzuki 1991).

Previously, chromatographic separation was performed by conven-

tional gel filtration with Sephadex G-75® (Kägi and Nordberg, 1979). This approach is increasingly replaced by the faster and more specific high-performance liquid chromatography (HPLC) principles of similar or even better specificity such as ion exchange (Suzuki et al. 1983; Lehman and Klaassen 1986, Klaassen and Lehman-McKeeman 1991) or hydrophobic interactions (Klauser et al. 1983; Richards 1989). Additionally, these techniques permit the quantification of MT isoforms.
The specificity of a particular chromatographic method largely depends on the resolution, being best in reverse phase HPLC. In general, possibly due to unspecific interaction with the column matrix, chromatographic methods seem to underestimate MT levels (Dieter et al. 1986).
The detection limits of chromatographic methods by analyzing SH groups via colorimetric methods or MT-bound metals via AAS are approximately 200 and 50 ng MT/ml, respectively.

ELECTROPHORESIS

MT determination by using electrophoretic methods consists of a separation step which in this case is based on the mobility of either the native (PAGE or isoelectric focusing, PAGIF) or denatured (sodium dodecyl sulfate(SDS)-PAGE) protein in polyacrylamide gels. Nondenaturing PAGE and PAGIF (Felley-Bosco et al. 1990) as well as a modified SDS-PAGE procedure (Aoki et al. 1991) proved to be also valuable in separating MT isoforms.
To enhance the specificity of the electrophoretic separation, common interferences are usually diminished by repeated heat or acid treatment of the supernatant fraction of tissue homogenates. Specificity has been additionally improved by carboxylating MT prior to SDS-PAGE (Otsuka et al. 1988; Kimura et al. 1991) or by the combination of an initial Coomassie blue stain followed by silver stain enhancement in nondenaturing PAGE (McCormick and Lin 1991). The various sample pretreatments may diminish MT recovery from the gels, therefore, appropriate standards have to be included.

The minimal detection limit of the individual method again largely depends on the parameter measured such as dye or silver staining, immunochemical detection or measuring of Cd after saturation of the protein with ^{109}Cd. The most sensitive procedure detecting 1.5 - 50 ng MT was reported for SDS-PAGE combined with silver or Coomassie blue staining (Otsuka et al. 1988; Kimura et al. 1991). Immunochemical staining using protein-A/colloidal gold conjugate after SDS-PAGE with a Western blotting procedure was able to quantify 60-160 ng (Aoki at al. 1991; Aoki and Suzuki 1991) and quantification of the protein by Cd saturation after SDS-PAGE/Western blotting had a detection limit of 400 ng (Aoki at al. 1991; Aoki and Suzuki 1991).

METAL SATURATION

Metal saturation methods take advantage of the high affinity of MT to certain metals and the known and constant stoichiometry between the protein and the metal. In choosing a particular metal saturation method one has to take into account that, due to their binding affinities declining in the order of Ag,Hg > Cu > Cd > Zn, metals bound to MT are only displaced by metals of higher affinity. Quantification of MT is performed by determining metals via atomic absorption spectrometry (AAS) or their radioactivity and finally by calculating the MT content on the basis of the stoichiometric ratio of metal to MT. As an inherent problem of using metal saturation assays with biological materials, the affinity of the saturating metal to other proteins or peptides requires an adequate pretreatment of the sample. The earlier method of separating high molecular weight proteins by gel filtration with Sephadex G-75 (Chen and Ganther 1975) is time consuming and has lost importance. More recent approaches take advantage of the stability of MT at low pH, the heat stability of Zn- or Cd-MT and its solubility in organic solvents.

Hg saturation

The analytical principle of saturating MT with metals at first

has been introduced by Piotrowski et al. (1973; Zelazowski and Piotrowski 1977) by adding ^{203}Hg followed by trichloroacetic acid (TCA) to tissue homogenates. The amount of ^{203}Hg in the TCA supernatant fraction was used to calculate the content of MT. In spite of some earlier controversy about the correct molar ratio of Hg:MT (Vasak et al. 1981; Day et al. 1984; Holt et al. 1980; Patierno et al. 1983; Cai and Stillman 1988), the ratio of 7 seems to be reliable (Vasak 1991). The major pitfall of the Hg saturation method consists in its moderate specificity resulting in an overestimation of the protein (Webb 1979; Waalkes et al. 1985; Dieter et al. 1986). To resolve this problem further modifications such as gel filtration to remove Hg bound to low molecular weight non-MT proteins/peptides (Kotsonis and Klaassen 1977,1979; Paternio et al. 1983) have been proposed.
While the detection limit of the Hg saturation method in one publication has been reported to be 0.7 ng (Patierno et al. 1983), others, in contrast, emphasize its rather low sensitivity (Eaton and Toal 1982; Waalkes et al. 1985) and a detection limit of 700 ng (Dieter et al. 1986).

Cd saturation

Titration with Cd is the most frequently used metal saturation method with the major advantage of a well established ratio of Cd:MT = 7 (Otvos and Armitage 1980; Winge and Miklossy 1982). In agreement with the original procedure (Onosaka et al. 1978), subsequent modifications still include red blood cell lysates or crude bovine hemoglobin to bind excess of Cd and several heat treatment steps to remove HMW proteins and Cd bound to hemoglobin (Onosaka and Cherian 1982; Eaton and Toal 1982). Recently, a more cautious procedure using acetonitril to separate HMW proteins and Chelex® to remove excess Cd has been introduced (Bartsch et al. 1990). Furthermore, HMW proteins and excess Cd have been separated from MT by adsorbing MT to polystyrol (Payan et al. 1988). With special modifications, such as the removal of Cu bound to MT at low pH (Payan et al. 1988) or with ammonium-tetrathiomolybdate (Klein et al. 1990), Cd saturation procedures even can be valuable in determining Cu containing MT.

According to recovery tests and as confirmed by chromatography, specificity of Cd saturation assays is reasonably well (Eaton and Toal 1982; Kotsonis and Klaassen 1977; Dieter et al. 1986; Bartsch et al. 1990; Klein et al. 1990). Overestimation of MT due to endogenous glutathione (GSH) which binds Cd and at higher concentrations may prevent the denaturation of hemoglobin (Eaton and Toal 1982; Bartsch et al. 1990) is unlikely to occur at physiological GSH concentrations. Underestimation of MT, however, may take place when Cd is removed from MT by extensively repeated heat precipitation steps with hemoglobin (Eaton and Toal 1982; Eaton and Cherian 1991).
Whereas with one test protocol an exceptionally low detection limit of 100 pg is obtained (Payan et al. 1988), most Cd saturation procedures correspondingly are able to detect between 10 and 100 ng MT.

Ag saturation

The high affinity of Ag towards MT has been used to develop a MT determination procedure which should be fairly independent of the metal composition of the protein (Scheuhammer and Cherian 1986, 1991). Similar to some Cd saturation methods, the assay protocol includes addition of red blood cell hemolysate followed by heat precipitation of excess Ag bound to hemoglobin and other proteins except MT. The stoichiometry for Ag:MT was assumed to be 17 gatoms Ag per mol MT, a value which is supported by recent spectroscopic studies revealing the formation of an Ag_{18}-MT complex following addition of Ag to rabbit liver apoMT-2 (Zelazowski et al. 1989). With appropriate experimental precautions to avoid the formation of AgCl, the Ag saturation assay is rather specific for MT as demonstrated by gel chromatography. Similar results were obtained with both the Ag saturation and a Cd heme assay (Scheuhammer and Cherian 1986), but consistently lower results were obtained comparing the Ag saturation assay to a RIA (Leibbrandt et al. 1991).
The detection limit of the Ag saturation assay when determining

Ag by AAS was 100 ng (Scheuhammer and Cherian 1986). By using radioactive Ag, sensitivity yet might be substantially higher (Scheuhammer and Cherian 1986).

IMMUNOLOGICAL METHODS

Immunological methods are unique in directly measuring the protein moiety of MT. Immunological procedures for MT quantification have been introduced in the late 70's (Tohyama and Shaikh 1978; Vander Mallie and Garvey 1979) and repeatedly modified according to the samples studied (Chang et al. 1980; Mehra and Bremner 1983; Mulder et al. 1990). More recently, several ELISA techniques with colorimetric or fluorometric detection have been developed (Thomas et al. 1986; Ghaffar et al. 1989; Grider et al. 1989). Although usually less sensitive than RIAs, ELISA procedures have the advantage to be less time consuming and not to require permanent preparation of radioactive tracers. Antibodies against MT, moreover, are used in immunodiffusion, immunoblotting, and at least semiquantitatively, in immunohistochemistry (Aoki and Suzuki 1991; Cherian and Banerjee 1991; Jasani and Elmes 1991; Garvey 1991a).
Specificity of an immunological method primarily depends on the particular antibody. Whereas the use of monoclonal antibodies for MT determination still is quite limited (Garvey 1991a,1991b) polyclonal antibodies seem to be valuable under most conditions. However, the various modifications in the antigen preparation techniques and the different species used for antibody production often result in inadequately defined antibodies with scarcely documented specificities (Vander Mallie and Garvey 1978; Mehra and Bremner 1983, Leibbrandt et al. 1991, Nakajima et al. 1991). Accordingly, antibodies raised against MT in one species might poorly react with MT from other species (Mehra and Bremner 1983, Mulder et al. 1990; Bremner et al. 1991). The reliable quantification of MT in different samples, therefore, requires proper characterization of the antibodies, particularly with respect to their cross-reactivity towards MT from different

species and isoforms as well as towards MT binding different metals. Errors also might be associated with the use of inappropriate standard MT. For maximal specificity, standard MT should have as many as possible similarities to the MT to be determined. Immunological assays, therefore, might not be considered to be of general use particularly for screening MT in various biological systems. Nevertheless, only these methods so far allow the measurement of MT in the picogram range and thus are sufficiently sensitive to quantify MT in body fluids.

ELECTROCHEMISTRY

Among the voltametric methods, the differential pulse polarography as introduced by Olafson and Sim (1979) seems to be appropriate for MT quantification. The method is based on the detection of the changes in current that occurs when a compound is either oxidized or reduced. When MT is quantified, the reduction of hydrogen in the sulfhydryl groups of the protein is measured. The major advantage of the method is, that the analysis is not influenced by the metal composition of MT. Specificity of the method seems to be independent of species (Hogstrand and Haux 1992), which, however, is only guaranteed if HMW proteins with sulfur groups of similar wave potential are eliminated after heat or ethanol treatment. Remaining small molecular weight thiol compounds such as GSH apparently seem to be a minor problem since in most cases they have little polarographic activity (Olafson and Olsson 1991). Dependent on the analytical equipment used, the detection limits of electrochemical methods vary between 100 and 600 ng MT (Olafson 1981; Olafson and Olsson 1991).

COMMENT

The basic restrictions and the sensitivity of a particular test protocol have to be considered, when deciding which is the most suitable method to quantify MT in biological materials. Basic restrictions, for example, are the impossibility to measure MT containing Cu with a traditional Cd saturation assay or to analyze MT of different species with immunological methods without proper knowledge of the interspecies cross-reactivity of the antibodies. With regard to the minimal detection limits of a particular method one has to consider that the overall sensitivity of a particular test protocol not necessarily equals the detection limit of the corresponding parameter measured. Therefore, the accuracy of a particular method always has to be assured by investigating the linearity of the assay with several appropriate dilutions of the sample.

Finally, an universal method for the quantification of MT, particularly concerning additional informations such as MT isoforms, metal composition and oxidation state of the protein does not exist. Therefore, in order to obtain the necessary informations, the parallel use of different methods might be required.

REFERENCES

Aoki, Y., and Suzuki, K.T. (1991) Methods Enzymol. 205, 108-114

Aoki, Y., Tohyama, C., and Suzuki, K.T. (1991) J. Biochem. Biophys. Methods 23, 207-216

Bartsch, R., Klein, D., and Summer, K.H. (1990) Arch. Toxicol. 64, 177-180

Bremner, I., Wood, A.M., Noble, N.A., Robertson, A. (1991) in: Metallothionein in Biology and Medicine (C.D. Klaassen, and K.T. Suzuki, eds.) pp 323-338, CRC Press, Boca Raton

Cai, W., and Stillman, M.J. (1988) J. Am. Chem. Soc. 110, 7872-7873

Chang, C.C., Vander Mallie, R.J., and Garvey, J.S. (1980) Toxicol. Appl. Pharmacol. 55, 94-102

Chen, R.W., and Ganther, A.E. (1975) Environ. Physiol. Biochem. 5, 378-388

Cherian, M.G., and Banerjee, D. (1991) Methods Enzymol. 205, 88-95

Day, F.A., Funk, A.E., and Brady, F.O. (1984) Chem.-Biol. Interactions 50, 159-174

Dieter, H.H., Müller, L., Abel, J., and Summer, K.H. (1986) Toxicol. Appl. Pharmacol. 85, 380-388

Eaton, D.L., and Cherian, M.G. (1991) Methods Enzymol. 205, 83-88

Eaton, D.L., and Toal, B.F. (1982) Toxicol. Appl. Pharmacol. 66, 134-142
Felley-Bosco, E., Hunziker, P.E., and Savolainen, H. (1990) Scand. J. Clin. Lab. Invest. 50, 343-346
Garvey, J.S. (1991a) Meth. Enzymol. 205, 141-174
Garvey, J.S. (1991b) in: Metallothionein in Biology and Medicine (C.D. Klaassen, and K.T. Suzuki, eds.) pp 25-48, CRC Press, Boca Raton
Ghaffar, A., Aggett, P.J., and Bremner, I. (1989) in: Nutrient Availability: Chemical and Biological Aspects (Southgate, D.A.T., Johnson, J.T., and Fenwick, G.R., eds.) pp 74-76, Royal Soc. of Chemistry, London
Grider, A., Kao, K.J., Klein, P.A., and Cousins, R.J. (1989) J. Lab. Clin. Med. 113, 221-228
Hogstrand, C., and Haux, C. (1992) Anal. Biochem 200, 388-392
Holt, D., Magos, L., and Webb, M. (1980) Chem.-Biol. Interactions 32, 125-135
Jasani, B., and Elmes, M.E. (1991) Methods Enzymol. 205, 95-107
Kägi, J.H.R., and Nordberg, M. (1979) (eds.) Experientia, Suppl. 34
Kimura, M., Koizumi, S., and Otsuka, F. (1991) Methods Enzymol. 205, 114-119
Klaassen, C.D., and Lehman-McKeenman, L.D. (1991) Methods Enzymol. 205, 190-198
Klauser, S., Kägi, J.H.R., and Wilson, K.J. (1983) Biochem. J. 209, 71-80
Klein, D., Bartsch, R., and Summer, K.H. (1990) Anal. Biochem. 189, 35-39
Kotsonis, F.N., and Klaassen, C.D. (1977) Toxicol. Appl. Pharmacol. 42, 583-588
Kotsonis, F.N., and Klaassen, C.D. (1979) Toxicol. Appl. Pharmacol. 51, 19-27
Lehman, L.D., and Klaassen, C.D. (1986) Anal. Biochem. 153, 305-314
Leibbrandt, M.E.I., Koropatnick, J., Harris, J.F., and Cherian, M.G. (1991) Biol. Trace Elem. Res. 30, 245-256
McCormick, C.C., and Lin, L.Y. Methods Enzymol. 205, 71-78
Mehra, R.K., and Bremner, I. (1983) Biochem. J. 213, 459-465
Mulder, T.P.J., Janssens, A.R., Verspaget, H.W., and Lamers, C.B.H.W. (1990) J. Immunol. Meth. 130, 157-161
Nakajima, K., Suzuki, K., Otaki, N., and Kimura, M. (1991) Methods Enzymol. 205, 174-189
Olafson, R.W. (1981) J. Biol. Chem. 256, 1263-1268
Olafson, R.W., and Olsson, P.E. (1991) Methods Enzymol. 205, 205-213
Olafson, R.W., and Sim, R.G. (1979) Anal. Biochem 100, 343-351
Onosaka, S., and Cherian, M.G. (1982) Toxicol. Appl. Pharmacol. 63, 270-274
Onosaka, S., Tanaka, K., Doi, M., and Okahara, K. (1978) Eisei Kagaku 24, 128-133
Otsuka, F., Koizumi, S., Kimura, M., and Ohsawa, M. (1988) Anal. Biochem. 168, 184-192

Otvos, J.D., and Armitage, I.M. (1980) Proc. Natl. Acad. Sci. USA 77, 7094-7098
Patierno, S.R., Pellis, N.R., Evans, R.M., and Costa, M. (1983) Life Sci. 32, 1629-1636
Payan, J.P., Beydon, D., Ferrari, E., and de Ceaurriz, J. (1988) Toxicol. Lett. 44, 121-130
Piotrowski, J.R., Bolanowska, W., and Saporta, A. (1973) Acta Biochem. Pol. 20, 207-215
Richards, M.P. (1989) J. Chromatogr. 482, 87-97
Scheuhammer, A.M., and Cherian, M.G. (1986) Toxicol. Appl. Pharmacol. 82, 417-425
Scheuhammer, A.M., and Cherian, M.G. (1991) Methods Enzymol. 205, 78-83
Suzuki, K.T. (1980) Anal. Biochem. 102, 31-34
Suzuki, K.T. (1991) Methods Enzymol. 205, 198-205
Suzuki, K.T., Sunaga, H., Aoki, Y., and Yamamura, M. (1983) J. Chromatogr. 281, 159-166
Thomas, D.G., Linton, H.J., and Garvey, J.S. (1986) J. Immunol. Meth. 89, 239-247
Tohyama, C., and Shaikh, Z.A. (1978) Biochem. Biophys. Res. Commun. 84, 907-913
Vander Mallie, R.J., and Garvey, J.S. (1979) J. Biol. Chem. 254, 8616-8421
Vasak, M. (1991) Methods Enzymol. 205, 452-458
Vasak, M., Kägi, J.H.R., and Hill, H.A. (1981) Biochemistry 20, 2852-2856
Waalkes, M.P., Garvey, J.S., and Klaassen, C.D. (1985) Toxicol. Appl. Pharmacol. 79, 524-527
Webb, M. (1979) in: The Chemistry, Biochemistry and Biology of Cadmium (M. Webb, ed.) pp. 195-266, Elsevier, Amsterdam
Winge, D.R., and Miklossy, K.A. (1982) J. Biol. Chem. 257, 3471-3476
Zelazowski, A.J., and Piotrowski, J.R. (1977) Acta Biochem. Pol. 24, 97-103
Zelazowski, A.J., Gasyna, Z., and Stillman, M.J. (1989) J. Biol. Chem. 264, 17091-17099

Metallothionein III
K. T. Suzuki, N. Imura and M. Kimura (eds)

BIOLOGICAL FUNCTIONS OF METALLOTHIONEIN - A REVIEW.

M. George Cherian and Hing Man Chan

Department of Pathology The University of Western Ontario London, Ontario, Canada N6A 5C1

SUMMARY: This review describes the papers presented at the third international meeting of metallothionein on biological functions, induction and degradation of MT. The various physiological conditions and chemical agents which can induce MT are described. The presence of MT-like proteins and peptides in invertebrates and plants are discussed in relation to their biological function. The significance of the nuclear/cytoplasmic localization of MT in proliferating cells and tumor cells are discussed. Most of these studies support a metal storage function for MT during mammalian development and rapid growth and also a potential detoxification role in metal toxicity, oxidative stress and in inflammation.

The biological functions of metallothioneins (MTs) are closely related to their chemical structures which consist of two distinct adamantane-like metal binding clusters (Kagi and Kojima, 1987). During the last three decades of research work, following its isolation, significant progress has been made in understanding the structure of MT. For example, the three-dimensional structure of MT has been fully elucidated in crystals by X-ray diffraction (Furey *et al.*, 1986) and in solution by two dimensional NMR (Braun *et al.*, 1986). These studies reveal the dynamic nature of this molecule and its ability to exchange metals within clusters and also between MT molecules. However the biological functions of MTs are not yet understood clearly but are being investigated actively in several laboratories. A number of potential functions for MT were proposed from the time of its isolation (Kagi and Vallee, 1960) and these are described elsewhere in this volume (Vallee, 1994).

The nutritional and physiological significance of MT were reviewed in detail in several articles (Cousins, 1985; Hamer, 1986; Sato and Bremner, 1993) including one in a previous volume of this series (Bremner, 1987). The major focus of the Third Intertnational Meeting on MT was on the biological functions and medical implications of MT. This review is based on the papers presented at the conference on the role of MT in regulation of mineral metabolism, the induction of MT by organic compounds, oxidative stress and inflammatory conditions, the degradative pathways of mammalian MT, the significance of its nuclear/cytoplasmic cellular localization and the presence of MT like proteins and peptides in invertebrates and in plants. Thus the presentations and discussions on biological functions of MT mainly dealt with the induction and degradation of MT during various physiological conditions and expression of MT in proliferating cells during certain stage in cell cycle and also on its inducibility by a variety of agents under different experimental conditions. However, some of these observed effects are related to certain experimental conditions and cannot be considered as the normal physiological functions of MT.

REGULATION OF ZINC AND COPPER METABOLISM

Although the protective effects of MT in metal toxicity have been studied extensively both in animal experiments and in cell culture systems, detoxification is by no means universally accepted as the primary function of MT (Webb, 1987; Templeton and Cherian, 1991). The high endogenous hepatic levels of zinc and copper bound MT in fetal and early neonatal life in several mammals suggest that MT may serve as an intracellular storage protein for these essential metals during perinatal development. However the proportion of these essential metals associated with

MT may vary with the species as well as fetal and neonatal age. In rat, rabbit, mouse, syrian and chinese hamsters, the hepatic MT concentration is maximal at or soon after birth (Wong and Klaassen, 1979; Webb and Cain, 1982; Panemangalore *et al.*, 1983) while in human (Bakka and Webb, 1981; Nartey *et al.*, 1987), sheep (Bremner *et al.*, 1977) and guinea pig (Lui, 1987) the maximum amount of MT is found during mid or late gestation. The amount of zinc or copper associated with MT may also vary depending on the species. For example, in neonatal rat liver zinc is the major MT bound metal while in guinea pig it is mainly copper and in human fetal liver both zinc and copper are found in high proportion with MT (Ryden and Deutsch, 1978). Differences in the two major isoforms, MT-I and MT-II also have been reported in rat and mice liver (Suzuki *et al.*, 1983; Lehman-McKeeman, *et al.*, 1991). During development rat liver contained more MT-II than MT-I while mouse liver contained more MT-I than MT-II. Recent studies also report changes in distribution of these two isoforms in livers of various mammalian species and these differences may be related to changes in metabolism of different metals in these species (Chan and Cherian, 1993). Although MT is mainly localized in the cytoplasm in adult hepatocytes, its presence in nucleus and cytoplasm is observed in both fetal and neonatal livers (Panemangalore *et al.*, 1983; Nartey *et al.*, 1987). The significance in the nuclear-cytoplasmic localization of MT and the functions of specific isoforms of MT are not yet understood clearly. In general, studies suggest that MT may serve as a major zinc and copper storage protein, reducing the ionic concentrations of these metals in liver during fetal and neonatal period in several mammalian species. This " buffering" effect of MT is similar to the role of ferritin in binding of iron during development. Thus the marked changes in MT levels and cellular localization in mammalian livers during development and also its presence in certain undifferentiated tumours

(Cherian *et al.*, 1993) suggest a biological function for this protein in cellular differentiation and maturation.

Studies on analyses of MT gene expression in the mouse reproductive tract and embryo have established that from the time of implantation to late in gestation, the mouse embryo is surrounded by cells, interposed between the maternal and embryonic environments, that actively express MT genes (De *et al.*, 1989; Ouellette, 1982; McMaster *et al.*, 1994, this Volume). Developmentally-regulated high level expression of MT genes occurs within the reproductive tract (deciduum, placental spongiotrophoblasts) and in extraembryonic membranes (visceral endoderm) and also in male germ cells (De, et al, 1991). The precise roles and mechanisms of regulation of MT genes during mammalian development are not well defined. It has been suggested that both endogenous zinc and glucocorticoids may contribute to high expression of MT. Both MT and its mRNA levels can be decreased in rat fetal liver when the dams were fed on zinc deficient diet during late gestation (Gallant and Cherian, 1986; Andrews *et al.*, 1987). However, depletion of glutathione in the dams with BSO injection or with sulphur deficient diet did not affect the hepatic MT synthesis in rat fetus. But inhibition of cystathionase pathway by repeated injection of propargylglycine decreased the hepatic levels of MT in newborn rats (Gallant and Cherian, 1989). In human fetus, the cystathionase or transsulfuration pathway is absent and thus MT may play a role as a cysteine storage protein in early human development (Zlotkin and Cherian, 1988). Analysis of metal regulation of MT genes in midgestation mouse embryos <u>in vivo</u> and <u>in vitro</u> during the teratogenic period in mice (days 9 to 10 of gestation) suggested that following maternal injection, cadmium but not zinc was prevented from reaching the embryo. High affinity binding of cadmium by MT in deciduum and placenta may provide a barrier for cadmium from reaching the post-implantation embryo. Moreover, MT has been localized in various cell types in full-term human placenta (Waalkes, et al, 1984; Goyer *et al.*, 1991) and thus MT may play a role in preventing the teratogenic effects of cadmium.

It has been shown that incubation of the zinc-finger transcription factor Sp1 with apo-MT can remove zinc from the transcription factor to form zinc-MT (Zeng *et al.*, 1991). The displacement of zinc from zinc-finger transcription factors can affect its binding to DNA with decreased ability to induce transcription of functional genes. Thus it is possible that regulation of cellular levels of MT and apo-MT may be an important mechanism to alter cellular functions. Activation of certain apo-metalloenzymes also has been shown after addition of zinc-MT. However, intracellular apo-MT is known to have very short half life and therefore it is important to demonstrate changes in apo-MT levels in association with zinc-finger transcription factor dependent gene expression in in vivo systems.

The evidence for any specific roles for MT in the gastrointestinal absorption of zinc and copper and also in the hepatotoxicity of copper in certain genetic diseases is critically discussed by Dr. Bremner (Bremner, 1994,in this volume). At very high oral doses of zinc the intestinal MT content is increased and this in concert with another cysteine-rich intestinal protein (CRIP) can regulate zinc absorption (Hempe and Cousins, 1992). CRIP is a saturable intracellular zinc-binding protein that can facilitate the uptake and transport of zinc ions across the enterocyte and its transfer at the basolateral membrane. The low intestinal MT concentrations will allow high binding of zinc to CRIP and its absorption. But at high intestinal MT levels, both the binding of zinc to CRIP and its absorption are decreased because of the active sequestration of zinc by intestinal MT. However, the mechanism of this control depends on high mucosal MT concentrations with very high zinc intakes which occurs after the breakdown of normal homeostatic control. The induction of MT in intestinal mucosal cells can result in sequestration of dietary copper which is subsequently excreted as Cu-MT when the intestinal cells are desquamated. Therefore zinc supplementation has been used for an effective treatment for Wilson's disease patients (Brewer *et al.*, 1991). Although these effects in zinc and copper absorption are found at high levels of intestinal MT levels, the role of MT in the normal absorption of zinc and copper is not yet understood.

A direct evidence for a causative role of MT in metal resistance has come from studies of copper sensitivity. A series of copper-resistant hepatoma cell lines have been reported that demonstrate a level of resistance proportional to the cellular concentration of Cu-MT (Freedman and Peisach, 1989). Yeast require CUP1 gene, encoding a MT-like protein, to utilize Cu and avoid Cu-induced damage. Deletion of this gene renders yeast sensitive to Cu, and introduction of a plasmid construct containing a mammalian MT gene confers both copper and cadmium resistance in a copy dependent manner. Thus the resistant phenotype is consistently associated with MT, can be modified by removal of MT, and can be restored by reintroduction of the MT gene (Thiele, 1992). However, in mammals, increased accumulation of hepatic copper results in liver damage, although there is a concurrent increase in Cu-MT. These effects are found in genetic diseases such as Wilson's disease in human (Nartey *et al.*, 1987a), LEC rats (Sumi *et al.*, 1993) and toxic milk mutant mice (Mercer *et al.*, 1992). There is evidence that Cu-MT can increase lipid peroxidation initiated by an iron-chelate and cause oxidative cell damage in neonatal guinea pig liver (Suntre and Lui, 1991) and also in an in vitro liver microsomal system. These studies imply that excessive Cu-MT may act as a pro-oxidant under certain conditions but there is no evidence that Cu-MT itself can initiate lipid peroxidation (Stephenson *et al.*, 1993).

The recent studies on cloning of the Menkes gene (Vulpe *et al.*, 1993; Mercer *et al.*, 1993) show that this gene product has strong homology to P-type ATPase proteins, similar to those involved in inorganic ion transport in metal resistant bacteria (Silver and Walderhaug, 1992). The N-terminal domain of this protein contains 6 repeats of a CysXXCys motif which is similar to the bacterial metal detoxification proteins and shows a weak resemblance to the metal binding sites in MT.

INDUCTION BY ORGANIC COMPOUNDS AND OXIDATIVE STRESS

Although both metals and glucocorticoids can directly induce MT synthesis, divalent metals like cadmium and zinc are the best known inducers of MT synthesis. Metal responsive elements (MRE) and glucocorticoid responsive elements (GRE) have been identified in MT genes (Palmiter, 1987). In addition to these inducers, several other compounds (which do not even bind with MT), oxidative stress and inflammatory conditions have been shown to induce MT synthesis in various systems. Several mechanisms are proposed for such induction of MT synthesis and they may involve lipid peroxidation with free radical formation, release of cytokines during inflammation and changes in tissue distribution of zinc.

Presentations from the laboratories of Dr. K. Tanaka (Min *et al.*, 1994, this volume) and Dr.M.Sato (Sato *et al.*, 1994, this volume) have described the mechanisms involved in the induction of MT synthesis by various organic compounds, oxidative stress and inflammatory conditions along with the physiological role of MT as a free radical scavenger. Most of the organic compounds induced MT synthesis indirectly either by release of certain cytokines from macrophages, monocytes and other cell types or by increase of cellular zinc concentration. A role for glucocorticoid in induction of MT synthesis in inflammation also has been suggested. For example injection of turpentine to mice can cause both induction of hepatic MT synthesis and acute inflammatory responses, such as increase in plasma fibrinogen and ceruloplasmin; but all these effects were prevented by pretreatment with dexamethasone (Min *et al.*, 1994, this volume). It should be pointed out that the hepatic MT levels induced by glucocorticoids alone are usually lower than those induced by inflammation or non-metallic compounds. Therefore changes in glucocorticoids alone may not explain induction of MT in inflammation. Moreover, since induction of hepatic MT synthesis by these compounds has been demonstrated in adrenalectomized mice, glucocorticoids may not be the primary inducer of MT (Min, *et al.*, 1992). It was also shown in mice, injected with menadione or carbon tetrachloride and in rats, injected with paraquat that the induction of hepatic MT synthesis was independent of lipid peroxidation (Sato and Sasaki, 1991). Depletion of glutathione also did not affect the induced synthesis of MT, suggesting that cysteine pools for these two thiols compounds may be different. Both MT mRNA and protein were induced in livers of mice following whole body X-irradiation but no such changes could be demonstrated in cultured cells, suggesting an indirect effect such as in inflammatory response (Koropatnick et al, 1989). The cytokines released during oxidative stress and inflammation may affect various cellular functions.

The potential role of cytokines in induction of MT synthesis has been studied extensively and they demonstrate that interleukin (IL)-1, IL-6, tumor necrosis factor and interferon can induce MT synthesis in the liver (Friedman and Stark, 1985; Cousins and Leinart, 1988; De *et al.*, 1990). It has been established that IL-6 is a major mediator of acute phase protein synthesis in hepatocytes in response to infection and tissue injury. The induction of MT synthesis in hepatocytes by IL-6 is enhanced by both zinc and glucocorticoids (Schroeder and Cousins, 1990). In addition, differential regulation of the isoforms of MT by IL-6 has been reported in various tissues in rats, suggesting a direct role for IL-6 in the induction of MT in inflammation. Induction of several acute-phase protein genes, such as those for acid glycoprotein and C-reactive proteins by metals shows significant sequence homology (Yiangou *et al.*, 1991) to the metal responsive elements in MT gene. However, it is unclear whether there is any relation between this homology and induced synthesis of MT in inflammation.

Several studies suggest that induction of MT synthesis can play a role as a free radical scavenger in X-irradiation (Matsubara *et al.*, 1986) and after treatment with adriamycin (Nagamuma *et al.*, 1988), carbon tetrachloride and tumor necrosis factor (Sato *et al.*, 1994, this volume). Injection of zinc-MT to rats can suppress the formation of gastric ulcers induced by various stress conditions and exposure to chemicals (Mimura *et al.*, 1988). Although induction of zinc-MT in certain conditions can scavenge hydroxyl and other oxygen radicals and provide some protection against lipid peroxidation and DNA damage, the relative importance of MT in comparison with other antioxidant defense systems (such as GSH and SOD) in normal physiological conditions is not well understood. The co-expression of MT and Cu,Zn-superoxide dismutase (SOD) genes has been reported in yeast where MT may act both as a donor of metals to the enzymes and as an antioxidant (Carri *et al.*, 1991). In vitro studies have shown that in GSH depleted cells, induction of MT can protect against the cytotoxic effects of quinone (Chan *et al.*, 1992) and hydroperoxide (Ochi., 1988). It is important to consider the subcellular distribution of MT and the ability of cells to synthesize MT in evaluating its role as an antioxidant. However, it is unlikely that MT will provide a long term protection in continuous exposures to oxidant stress.

DIFFERENT PATHWAYS OF DEGRADATION

Although the mechanism of induced synthesis of MT including gene amplification has been elucidated (Hamer, 1986; Palmiter, 1987; Thiele, 1992), little information is available on the turnover rates and degradative pathways of MT. Earlier studies suggested that the turnover rate of MT depends on the species of metals bound to it. Thus the turnover rate of Cu,Zn-MT is faster than Cd-MT ($t_{1/2}$ 15-20hr Vs. 80hr) and in vitro the apo-MT is degraded faster than MT (Feldman and Cousins, 1976; Shaikh and Smith, 1976; Bremner *et al.*, 1978; Feldman *et al.*, 1978; Cain and Holt, 1979). In addition the half-life for Cu-MT is less in Zn-deficient rats than those fed with Zn-adequate diet (Bremner *et al.*, 1978) and the half life of MT isoform MT-1 is less than that of MT-2 (Anderson *et al.*, 1978; Suzuki and Maitani, 1983). Degradation of zinc-MT in monolayer cultures of rat hepatocytes is primarily regulated by cellular zinc content and the degradation occurs in both lysosomal and cytoplasmic compartments (Chen and Failla, 1989).

Recent studies on the effects of various protease inhibitors on the degradation of MT by lysosomal proteases suggest that cathepsin B, a cysteine protease may play an important role in the degradation of MT (Min *et al.*, 1992; McKim *et al.*, 1992; Klaassen *et al.*, 1994, this volume). The differences in degradation of isoforms of MT and their metal composition are described in detail elsewhere in this volume (Klaassen *et al.*, 1994). The structural changes in MT, the isoforms, the removal of metals and other cellular changes during induction of MT should be considered in extrapolation of the in vitro observations to in vivo degradation of MT. This may be particularly important in the degradation of MT in livers of fetal and neonates where most of the MT may be present in the cell nucleus, in addition to cytoplasm (Panemangalore *et al.*, 1983; Nartey *et al.*, 1987). The particulate forms of copper-MT has been detected in livers of copper-loaded animals of several species and they may be localized mainly in lysosomes in granular forms. The turnover rates of the particulate form of copper-MT is much less than that of the cytosolic form (Mehra and Bremner, 1985). Therefore the cellular distribution of MT may also play a major role in its degradation.

According to the suggested pathways of the degradation of MT, release of metals from MT with subsequent changes in its cluster structure to a random configuration may occur before its breakdown by proteases. This view is supported by the observation that selective removal of zinc from Cd,Zn-MT by EDTA treatment makes the protein susceptible to proteolytic cleavage (Winge and Miklossy, 1982). However, there is no good experimental evidence showing the presence of apo-MT within the cell. In a recent in vitro study of MT degradation in rat hepatoma cell line, the authors speculate the presence of apo-MT based on a compartmental analysis of the data (Steinebach and Wolterbeck, 1992). The observed two different rates of MT degradation can be explained by the different subcellular localization rather than the two forms of MT. It is difficult to assume that one of these compartment is apo-MT without its characterization.

Although the in vitro studies demonstrate that the major degradative pathway may involve lysosomal proteases, especially cysteine protease, the degradation of endogenous MT has not been fully elucidated. The involvement of non-lysosomal systems in the degradation of MT in vivo can not be ruled out because both biochemical and genetic evidences suggest that the covalent conjugation of ubiquitin to intracellular proteins in eukaryotic cells is essential for their selective degradation (Finley and Varshavsky, 1985) and also chloroquine, a lysosomotropic amine did not inhibit MT degradation completely (Chen and Failla, 1989). However, at least in yeast, the amino-terminal ubiquitination is not a signal for the proteolysis of MT (Butt *et al.*, 1988). The potential role of some signal proteins to regulate the proteolysis of endogenous MT in non-lysosomal systems needs further investigation.

MT-LIKE PROTEINS AND PEPTIDES IN INVERTEBRATES AND PLANTS

There are several reports on occurrence of MTs or MT-like proteins in most organisms ranging from prokaryotes, eukaryotic microorganisms, plants, invertebrates (echinoderms, annelids, molluscs, arthropods) to vertebrates (fish, amphibians, reptiles, birds, mammals) (Engel and Brouwer, 1989; Kagi and Kojima, 1987; Rauser, 1990; Roesijadi, 1992). The biology of MTs in non-mammalian species and plants with particular emphasis on their roles in metal regulation and toxicology is reviewed elsewhere in this volume (Roesijadi 1994; volume; Rauser, 1994). Unlike mammalian MTs which usually contain 61 amino acids, with 20 cysteinyl residues, MTs isolated from some invertebrate have different amino acid sequences. The proposed classification system by Fowler *et al.* (1987) subdivides MTs into three distinct classes. Type I include polypeptides related in primary structure to equine renal MT. Thus most of the MTs isolated from fish, lobster and oyster are classified as Type I and those from sea urchin and yeast which are only distantly related to that of equine renal MT are classified as Type II. Type III include the γ-glutamyl-cysteinyl oligopeptides isolated from plants (Grill *et al.*, 1985), microorganisms *Euglena gracilis* and *Schizosaccharomyces pombe*. Despite the inter-species diversity, the primary structures of MTs in species within a single phylum are usually very similar.

One of the most well characterized invertebrate MT is from crab Scylla serrata and is shown to contain two 3-metal clusters by NMR, rather than the 3-metal and 4-metal clusters of mammalian MTs (Otvos *et al.*, 1982). A previous study suggested the presence of two Cd-binding proteins, distinct from MT in rainbow trout exposed to environmental levels of cadmium (Thomas *et al.*, 1983). However, subsequent studies failed to show any such proteins and Cd and other metals were shown to bind with two isoforms of MT in gibel liver after exposure to environment levels of cadmium (Suzuki *et al.*, 1987).

Similar to mammalian MTs, multiple isoforms have been identified in MT from other species (Roesijadi, 1994, this volume). Moreover, polymeric isomers were also reported, particularly for molluscan species (Roesijadi, *et al.*, 1989). The significance of multiple MT isoforms is still unclear; but the specific forms may have unique functions. In lobster, one form of MT expressed constitutively may be involved in Cu exchange reactions, while the other two inducible MT isoforms may be involved in metal detoxification (Brouwer *et al.*, 1989, 1992; Brouwer and Brouwer-Hoexum, 1992).

The mechanism of induction of MT in non-mammalian species is not that well understood in comparison to mammalian MTs. However, it was found that cadmium can induce synthesis of MTs in a wide range of organisms including algae, sea urchin, polycheates, mussels, crabs, oysters, insects, fish, frogs, turtles, birds and lizards (Engel and Brouwer, 1987; Roesijadi, 1994, this volume). Hormonal modulation of MT synthesis has been suggested in fish (Olsson *et al.*, 1990). A recent study has shown four-fold induction of MT-mRNA in blue crab exposed to cadmium through diet (Schlenk and Brouwer, 1993).

Very little is known about the metal binding affinity of non-mammalian MTs. Petering and Fowler (1986) suggested that invertebrates bind cadmium less tightly than mammalian MTs. The limited information on the degradation of non-mammalian MT was reported by Brouwer *et al.* (1984) that in the gill and digestive gland of the blue crab, Cd remains bound to MT in a steady state of biosynthesis and biodegradation for a period of 12 days after Cd exposure. Unlike the many physiological functions proposed for mammalian MTs, non-mammalian MTs are shown to play a role only in metal metabolism and toxicity. Thus they may play a role in the homeostasis of essential metals such as Cu and Zn and also in the detoxification of non-essential metals such as Cd. The most direct evidence on the participation of MT in the physiological regulation of metals was reported in the lobster. During the moult cycle, Cu released from the degraded haemocyanin was sequestered by MT and stored in the hepatopancreas which can serve as a Cu (I) donor for the newly synthesized apohemocyanin via a glutathione-mediated mechanism (Brouwer *et al.*, 1989, 1992; Brouwer and Brouwer-Hoexum, 1992). Some evidences were also reported for the role of MT as Cu and/or Zn donor for certain metallo- enzymes (Roesijadi, 1994, this volume). Variations in MT levels have been reported in fish with season and also with a certain physiological status which is associated with increased metabolic requirement for metals. For example, the reproductive status in the rainbow trout (Olsson *et al.*, 1990) and the early development stages of the sea urchin (Wilkinson and Nemer, 1987).

Several studies have established a metal detoxification role for MT in non-mammalian species. Most of the Cd found in the gills and intestinal cells of aquatic animals were bound to MT (Roesijadi, 1992). However, induction of MT in gills or intestinal cells does not facilitate metal uptake. Metals ingested with food was found to be bound to MT in the digestive gland cells (Engel and Brouwer, 1984), and they can be transported intact to other organs. MT in the kidney of fish and molluscs can sequester metals such as Cd (Olsson and Hogstrand, 1987; George, 1983a, b; Hogstrand and Haux, 1991) which may eventually process into residual bodies for excretion through urine (Viariengo, 1989; Fowler and Gould, 1988). In continuous exposure to metals the rate of MT synthesis may be a limiting factor because as the exposure concentrations increase it may result in 'spillover' of metals to other intracellular compartments, and cause toxicity. In the brook trout, mortality is correlated to the concentration of non-MT bound Cd fractions in the cytosol of hepatocytes (Hamilton and Mehrle, 1987).

There are several similarities in the protective role of MT in metal toxicity in mammalian and non-mammalian species. A low dose of metal exposure confers resistance to subsequent high dose of metal exposure (Roesijadi, 1992). Moreover, induction of MT synthesis by one metal such as Zn or Cd can induce tolerance to another metal. The enhanced tolerance to metal toxicity is believed to be mediated by replacement of metals bound in the pre-induced MT and/or through amplification of MT gene expression. Amplification of MT gene was reported in natural populations of Drosophilia with increased tolerance to Cu toxicity (Maroni *et al.*, 1987). A cadmium-tolerant population of oligochaete worms were found in a Cd-rich, freshwater environment which was contaminated by past mining of cadmium and nickel (Klerks and Levinton, 1989). Larvae of certain species of mayfly which were Cd sensitive and lack the capability to synthesize MT were found to be absent in metal-contaminated environments (Aoki *et al.*, 1989). Similar to the

LEC rats and toxic milk mice, high concentrations of Cu-MT was found in white perch from Chesapeake Bay which are associated with an abnormal genetic defect in copper metabolism (Bunton *et al.*, 1987).However, the phenomenon of enhanced resistance is not universal. For example, after chronic exposure of the slipper limpet in the laboratory to low concentrations of Hg did not enhance tolerance to Hg, although enhanced binding of Hg to MT was reported (Harrison *et al.*, 1987).

The bioavailability of metals bound to MT in food is not understood clearly. Earlier studies using isolated Cd-MT suggested that it can be absorbed intact from the gastrointestinal tract in mice (Cherian *et al.*, 1978; Cherian, 1979). A study in New Zealand fishermen showed that in spite of a very high intake of cadmium containing oysters the concentration of cadmium in whole blood and urine was not elevated greatly in proportion to the intake (Sharma *et al.*, 1983). This suggested that the form of cadmium in oyster may not be bioavailable. However, more studies are needed to determine the bioavailablity of metals in seafood.

The role of γ-glutamylcysteinylglycine isopeptides, detected in yeast, algae and certain plants are classified as Class III MTs and they may play a role in metal tolerance (Rauser, 1994; this volume). Their contents may alter depending on the exposure to metals and the seasonal changes in plants. These peptides are related closely to glutathione and their synthesis is controlled at enzymatic level rather than directly at transcriptional level. However, induced synthesis of these group of peptides with different size has been demonstrated in several systems either by induction or activation of enzymes, involved in their synthesis (Grill *et al.*, 1987; 1989; Reese and Wagner., 1987; Gekeler *et al.*, 1988). The strength of Cd binding can be enhanced by increasing the length of the isopeptides (Hayashi *et al.*, 1991). Although these peptides can provide protection against cytotoxicity of metals in cultured plant cells and yeasts, the role of such complexes in metal tolerance of intact plants has not been demonstrated yet. The biosynthetic pathway of these isopeptides, their cellular localization and their role in metal homeostasis and tolerance are described in detail elsewhere in this volume (Rauser, 1994).

In addition to these isopeptides, MTs have been isolated from roots of copper tolerant plants (Rauser and Curvetto, 1980), pea and soybean (Kawashima *et al.*, 1991). Mature wheat embryos also contained a cysteine rich zinc binding protein with sequence homology to rat liver MT (Lane *et al.*, 1987) and it was found to be in the cell nucleus during germination.

One of the major pathway of development of metal resistance in prokaryotes is the mutation of enzyme ATPase in plasmid by which the organisms can eliminate metals (Silver and Misra, 1988). However, a metal binding protein has been isolated from cynobacteria and is classified as class II MT (Olafson, 1986). Recently, a prokaryotic MT locus, Smt, from the cynobacteria synechococcus PCC 7942 and PCC 6301 was isolated (Huckle *et al.*, 1993). This locus includes Smt A, which encodes a class II MT and a divergently transcribed gene Smt B which encodes a repressor of Smt A transcription (Morby *et al.*, 1993). Transcription from the Smt is stimulated by metal ions such as zinc, copper and cadmium. The possible role of these genes in the mechanisms of metal tolerance, and in the metabolism of essential trace metals such as zinc in cynobacteria requires further investigation.

SIGNIFICANCE OF NUCLEAR LOCALIZATION

The presence of MT in cell nucleus under certain physiological conditions and early mammalian development (Panemangalore *et al.*, 1983) was confirmed in several recent studies but its biological significance is not yet understood completely. Although MT has been considered generally as a ubiquitous metal binding protein, its specific expression in epithelial and myoepithelial cells of tumors (Cherian, 1994, this volume), its binding to rat spermatozoa nucleus (Nakajima, 1994, this volume) and its presence in the cell nucleus during cell proliferation (Tsujikawa *et al.*, 1994; Cherian, 1994) provide some important information on the biological functions of MT. The nuclear presence of MT may be related to transfer of zinc for several nuclear enzymes, modulation of transcription by zinc chelation from transcriptional factors by thionein (Zeng *et al.*, 1991) and protection of DNA damage by metals and oxygen free radicals (Sato and Bremner, 1993). The overexpresssion of MT in certain cells has been correlated to anti-cancer drug resistance and high concentration of MT has been detected in radioresistant adenoid cystic carcinoma of human salivary gland (Chauvin *et al.*, 1992) and cisplatinum resistant embryonal carcinomas (Kontozoglou, *et al.*, 1989; Chin *et al.*, 1993). Modulation of MT tumor levels in mice can alter the antitumor activity of cisplatinum (Satoh *et al.*, 1993). These studies provide indirect evidence that MT may be one of the factors involved in resistance to anti-cancer drugs (Lazo, 1994, this volume; Naganuma *et al.*, 1994, this volume).

Flow cytometric analysis of rat hepatocytes in culture demonstrated that MT was translocated into the nuclei from the cytoplasm during S to G_2M phases of the cell cycle following stimulation with growth factors such as epidermal growth factor and insulin (Tsujikawa *et al.*, 1994). These results are supported by the recent observation of the presence of MT in the nucleus of epithelial cells in S phase of cell cycle in certain tumors (Haile-Meskel *et al.*, 1992, 1993). It is unclear whether MT in the nucleus is bound to either DNA or any nuclear proteins. The mechanisms of transport of MT to the nucleus from cytoplasm are not understood and there is no evidence for any nuclear localization signal (Garcia-Bistos *et al.*, 1991) associated with MT which is synthesized as a single polypeptide on free polyribosomes (Cherian *et al.*, 1981). A zinc transport role for MT to the nucleus of sperm is suggested by <u>in vitro</u> studies (Nakajima, 1994).

During the last three decades of research work on MT has elucidated its dynamic structure and revealed some information on its potential functions as a metal storage protein. It is anticipated that major information on the biological roles of MT in cell differentiation and growth and its relation in tumor biology will be forthcoming during the next few decades.

REFERENCES;

Anderson, R.D., Winter, W.P., Maher, J.J., and Bernstein, I.A. (1978) Biochem. J. 174, 327-338.

Andrews, G.K., Gallant, K.R., and Cherian, M.G. (1987) Eur. J. Biochem. 166, 527-531.

Aoki, Y., Hatakeyama, S., Kobayashi, N., Sumi, Y., Suzuki, T., and Suzuki, K.T. (1989) Comp. Biochem, Physiol. 93c, 345-357.

Bakka, A. and Webb, M. (1981) Biochem, Pharmacol. 30, 721-725.

Braun, W., Wanger, G., Wöngötter, E., Vasak, M., Kägi, J.H.R., and Wüthrich, K. (1986) J. Mol. Biol. 187, 125-129.

Bremner, I., Williams, R.B., and Young, B.W. (1977) Brit. J. Nutr. 38, 87-92.

Bremner, I., Hoekstra, W.G., Davies, N.T., and Young, B.W. (1978) Biochem. J. 174, 883-892.

Bremner, I. (1987) In: Metallothionein II, J.H.R. Kägi and Y. Kojima, Eds, Birkhauser Verlag, Basal, pp. 81-107.

Brewer, G.J., Yuzbasiyan-Gurkan, V., and Johnson, V. (1991) J. Lab. Clin. Med. 118, 466-470.

Brouwer, M., Brouwer-Hoexum, T. and Engel, D.W. (1984) Mar. Environ. Res. 14, 71-88.

Brouwer, M., Winge, D.R., and Gray, W.R. (1989) J. Inorg. Biochem. 35, 289-303.

Brouwer, M., and Brouwer-Hoexum, T. (1992) Biochemistry 31, 4096-4102.

Brouwer, M., Schlenk, D., Ringwood, A.H., and Brouwer-Hoexum, T. (1992) Arch. Biochem, Biophys. 294, 461-468.

Bunton, T.E., Baksi, S.M., George, S.G., and Frazier, J.M. (1987) Vet. Pathol. 24, 515-524.

Butt, T.R., Khan, M.I., Marsh, J., Ecker, D.J., and Crooke, S.T. (1988) J. Biol. Chem. 263, 16364-16371.

Cain, K., and Holt, D.E. (1979) Chem. Biol. Interac. 28, 91-106.

Carri, M.T., Baliagzo, F., Ciriolo, M.R., and Rotilio, G. (191) FEBS Lett. 278, 263-266.

Chan, H.M., Tabarrok, R., Tamura, Y., and Cherian, M.G. (1992) Chem. Biol. Interac. 84, 113-124.

Chan, H.M., and Cherian, M.G. (1993) Toxicologist. 13, 165.

Chauvin, P.J., Cherian, M.G., Wysocki, G.P. and Pringle, G.A. (1992) Toxicologist 12, 379.

Chen, M.L., and Failla, M.L. (1989) Proc. Exp. Biol. Med. 191, 130-138.

Cherian, M.G. (1994) this volume.

Cherian, M.G. (1979) Environ. Hlth. Perspect. 28, 127-130.

Cherian, M.G., Goyer, R.A., and Valberg, L.S. (1978) J.Toxicol.Environ.Hlth. 4,861-868.

Cherian, M.G., Yu,S. and Redman,C.M. (1981) Cand.J.Biochem. 59, 301-306.

Cherian, M.G., Huang, P.C., Klaassen, C.D., Liu, Y.P., Longfellow, D.G., and Waalkes, M.P. (1993) Cancer Res. 53, 922- 925.

Chin, J.L., Banerjee, D., Kadhim, S.A., Kontozoglou, T.E., Chauvin, P.J. and Cherian, M.G.(1993) Cancer (In Press).
Cousins, R.J. (1985) Physiol. Rev. 65, 238-309.
Cousins, R.J., and Leinart, A.S. (1988) FASEB J. 2, 2884-2890.
De, S.K., McMaster, M.T., Dey, S.K., and Andrews, G.K. (1989) Development 107, 611-621.
De, S.K., McMaster, M.T., and Andrews, G.K. (1990) J. Biol. Chem. 265, 15267-15274.
De, S.K., Enders, G.C. and Andrews, G.K. (1991) Mol. Endocrin. 5, 628-636.
Engel, D.W., and Brouwer, M. (1984) Mar. Environ. Res. 14, 139-151.
Engel, D.W., and Brouwer, M. (1987) Biol. Bull. 173, 239-251.
Engel, D.W., and Brouwer, M. (1989) Adv. Comp. Environ. Physiol. 5, 54-74.
Feldman, S.L., and Cousins, R.J. (1976) Biochem. J. 160, 583-588.
Feldman, S.L., Failla, M.L., and Cousins, R.J. (1978) Biochim. Biophys. Acta 544, 638-646.
Finley, D., and Varshavsky, A. (1985) Trends Biochem. Sci. 10, 343-347.
Fowler, B.A., Hildebrand, C.E., Kojima, Y., and Webb, M. (1987) In: "Metallothionein II", J.H.R. Kagi and Y. Kojima Eds. Birkhauser Verlag, Basal pp. 19-22.
Fowler, B.A., and Gould, D. (1988) Mar. Biol. 97, 207-216.
Freedman, J.H., and Peisach, J. (1989) Biochim. Biophys, Acta 992, 145-
Friedman, R.L., and Stark, G.R. (1985) Nature 314, 637-639.
Furey, W.F., Robbins, A.H., Clancy, L.L., Winge, D.R., Wang, B.C., and Stout, C.D. (1986) Science 231, 704-710.
Gallant, K.R., and Cherian, M.G. (1986) Biochem. Cell Biol. 64, 8-12.
Gallant, K.R., and Cherian, M.G. (1989) J. Pharmacol. Exp. Therap. 249, 631-637.
Garcia-Bustos, J. Heitman, J., and Hall, M.N. (1991) Biochim. Biophys. Acta 1071, 83-101.
Gekeler, W., Grill,E., Winnacker, E.L. and Zenk, M.H. (1988) Arch. Microbiol. 150, 197-202.
George, S.G. (1983a) Comp. Biochem. Physiol. 76c, 53-57.
George, S.G. (1983b) Comp. Biochem. Physiol. 76c, 59-65.
Goyer, R.A., Haust, M.D., and Cherian, M.G. (1992) Placenta 13, 349-355.
Grill, E., Winnacker, E.L., and Zenk, M.H. (1985) Science 230, 674-676.
Grill, E., Winnacker, E.L., and Zenk, M.H. (1987) Proc.Natl.Acad.Sci. USA 84, 439-443.
Grill, E., Loffler, S., Winnacker, E.L. and Zenk, M.H. (1989) Proc.Natl.Acad.Sci. USA 86, 6838-6842.
Haile-Meskel, H., Cherian, M.G., Martinez, V.J., Veinot,L.A. and Frei,J.V. (1992) Proc.Cand.Assoc.Pathol, Toronto,Ontario.
Haile-Meskel, H., Cherian, M.G., Martinez, V.J., Veinot, L.A. and Frei, J.V. (1993) Modern.Path. (In Press).

Hamer, D.H. (1986) Annu. Rev. Biochem. 55, 913-951.
Hamilton, S.J., and Mehrle, P.M. (1987) Trans. Am. Fish Soc. 116, 551-560.
Harrison, F.L., Watnes, K., Nelson, D.A., Miller, J.E., and Calabrese, A. (1987) Estuaries 10, 78-83.
Hayashi, Y., Nakagawa, C.W., Mutoh, N., Isobe,M. and Goto, T. (1991) Biochem. Cell. Biol. 69, 115-121.
Hempe, J.M., and Cousins, R.J. (1992). J. Nutr. 122, 89-95.
Hogstrand, C., and Haux, C. (1991) Comp. Biochem. Physiol. 100c, 137-141.
Huckle, J.W., Morby, A.P., Turner, J.S. and Robinson, N.J.(1993).Mole.Microbiol.7, 177-187.
Kägi, J.H.R. and Vallee, B.L. (1960) J. Biol. Chem. 235, 3460-3465.
Kägi, J.H.R., and Kojima, Y. (1987) In: Metallothionein II, J.H.R. Kägi and Y. Kojima, Eds, Birkhauser Verlag, Basal, pp. 25-61.
Kawashima, I., Inokuchi, Y., Chino,M., Kimura, M. and Shimizu,N. (1991) Plant.Cell.Physiol. 32, 913-916.
Klaassen, C.D., Choudhuri, S., McKim, J.M., Lehman-McKeeman, L.D., and Kershaw, W.C. (1994). This volume
Klerks, P.L., and Levinton, J.S. (1989) In: Ecotoxicology: Problems and Approaches, M.A.H.S.A. Levin, J.R. Kelly and K.D. Kimball Eds. Springer-Verlag, N.Y. pp. 41-67.
Kontozoglou, T.E., Banerjee, D., Cherian, M.G. (1989) Virchows. Archiv. A. Pathol. Anat. 415, 545-549.
Koropatnick, J., Leibbrandt, M., and Cherian, M.G. (1989) Radiat. Res. 199, 356-365.
Lane, B.L., Kajioka, R. and Kennedy, T. (1987) Biochem. Cell. Biol. 65,1001-1005.
Lazo, J.S. (1994), this volume.
Lehman-McKeeman, L.D., Kershaw, W.C., and Klaassen, C.D. (1991) In: Metallothionein in Biology and Medicine, C.D. Klaassen and K.T. Suzuki, eds. CRC Press, Boca Raton, pp. 121-132.
Lui, E.M.K. (1987) Comp. Biochem. Physiol. 86, 173-183.
Maroni, G., Wise, J., Young, J.E. and Otto, E. (1987) Genetic 117, 739-744.
Matsubara, J., Shida, T., Ishioka, K., Egawa, S., Inada, T., and Machida, K. (1986) Environ. Res. 41, 558-567.
McKim, J.M., Choudhuri, S. and Klaassen, C.D. (1992) Toxicol. Appl. Pharmacol. 116, 117-124.
McMaster, M.T., De, S.K., Paria, B.C., Huet-Hudson, Y., Dey, S.K., Andrews, G.K. (1994). This volume pp.
Mehra, R.K., and Bremner, I. (1985) Biochem. J. 227, 903-908.
Mercer, J.F.B., Grimes, A., and Rauch, H. (1992) J. Nutr. 122, 1254-1259.
Mercer, J.F.B., Livingston, J., Hall, B., Paynter, J.A., Begy, C., Chandrasekharappa, S., Lockhart, P., Grimes, A., Bhave, M., Siemieniak, D., and Glover, T.W. (1993) Nature Genet. 3, 20-25.

Mimura, T., Tsujikawa, K., Yasuda, N., Nakajima, H., Haruyama, M., Ohmura, T., and Okabe, M. (1988) Biochem. Biophys. Res. Comm. 151, 725-729.
Min, K.S, Terano, Y., Onosaka, S., and Tanaka, K. (1992) Toxicol. Appl. Pharmacol. 113, 74-79.
Min, K.S., Itoh, N., Okamoto, H. and Tanaka, K. (1994), this volume.
Morby, A.P., Turner, J.S., Huckle, J.W. and Robinson, N.J.(1993).Nucl.Ac.Res.21, 921-925.
Naganuma, A., Satoh,M., and Imura,N. (1994), this volume.
Naganuma, A., Satoh, M., and Imura, N. (1988) Jpn. J. Cancer Res. 79, 406-411.
Nakajima,K. (1994), this volume.
Nartey, N.O., Banerjee, D., and Cherian, M.G. (1987) Pathology 19, 233-238.
Nartey, N.O., Frei, J.V., and Cherian, M.G. (1987a) Lab. Invest. 57, 397-401.
Ochi, T. (1988) Toxicology 50, 257-268.
Olafson, R.W. (1986) Environ. H9lth. Perspect. 65, 71-75.
Olsson, P.E., and Hogstrand, C. (1987) Environ. Toxicol. Chem. 6, 867-874.
Olsson, P.E., Zafarullah, M., Foster, R., Hamor, T., and Gedamu, L. (1990) Eur. J. Biochem. 193, 229-235.
Otvos, J.D., Olafson, R.w., and Armitage, I.H. (1982) J. Biol. Chem. 257, 2427-2431.
Ouellette, A.J. (1982) Dev. Biol. 92, 240-246.
Palmiter, R.D. (1987) In: Metallothionein II, J.H.R. Kagi and Y. Kojima Eds. Birkhauser Verlag, Basal pp. 63-80.
Panemangalore, M., Banerjee, D., Onosaka, S., and Cherian, M.G. (1983) Dev. Biol. 97, 95-102.
Petering, D.H., and Fowler, B.A. (1986) Environ. Hlth. Perspect. 65, 217-226.
Rauser, W.E. and Curvetto, N.R. (1980) Nature. 287, 563-564.
Rauser, W.E. (1990) Annu. Rev. Biochem. 59, 61-86.
Rauser, W.E. (1994) This volume
Reese, R.N. and Wagner, G.J. (1987) Plant.Physiol. 84, 574-577.
Roesijadi, G., Kielland, S.L., and Klerks, P. (1989) Arch. Biochem. Biophys. 273, 403-413.
Roesijadi, G. (1992) Aquat. Toxicol. 22, 81-114.
Roesijadi, G. (1994) This volume
Ryden, L., and Deutsch, H.F. (1978) J. Biol. Chem. 253, 519-524.
Sato, M., and Sasaki, M. (1991) Chem. Biol. Interac. 78, 143-154.
Sato, M., and Bremner, I. (1993) Free. Rad. Biol. Med. 14, 325-337.
Sato, M., Sasaki, M. and Honjo, H. (1994), this volume.
Satoh, M., Kloth, D.M., Kadhim, S.A., Chin, J.L., Naganuma,A., Imura,N. and Cherian, M.G. (1993) Cancer Res. 53, 1829-1832.
Schlenk, D., and Brouwer, M. (1993) Comp. Biochem. Physiol.104c, 317-321.
Schroeder, J.J., and Cousins, R.J. (1990) Proc. Natl. Acad. Sci. USA 87, 3137-3141.

Shaikh, Z.A., and Smith, J.C. (1976) Chem. Biol. Interac. 15, 327-336.
Sharma,R.P., Kjellstrom,T., and McKenzie, J.M. (1983) Toxicology.29, 163-171.
Silver, S., Misra, T.K. (1988) Ann. Rev. Microbiol. 42, 717-743.
Silver, S., and Walderhaug, M. (1992) Microbiol. Rev. 56, 195-228.
Steinebach, O.M., and Wolterbeck, B.T. (1992) Biochim. Biophys. Acta 1116, 155-165.
Stephenson, G.F., Chan, H.M., and Cherian, M.G. (1993) Toxicol. Appl. Pharmacol.(In Press).
Sumi, Y., Kawahara, S., Kikuchi, Y., Sawada, J., Suzuki, T. and Suzuki, K.T. (1993) Acta Histochem. Cytochem. 26, 5-9.
Suntre, Z.E., and Lui, E.M.K. (1991) J. Pharmacol. Exp. Therap. 258, 797-806.
Suzuki, K.T., Ebihara, Y., Akitomi, H., Nishikawa, M., and Kawamura, R. (1983) Comp. Biochem. Physiol. 76, 33-38.
Suzuki, K.T., and Maitani, T. (1983) Chem. Pharm. Bull. 31, 4469-4475.
Suzuki, K.T., Sunaga, H., Kobayashi, E., and Hatakayama, S. (1987) Comp. Biochem. Physiol. 87, 87-93.
Templeton, D.M., and Cherian, M.G. (1991) Meth. Enzymol. 205, 11-24.
Thiele, D.J. (1992) Nucl. Acid. Res. 20, 1183-1191.
Thomas, D.D., Solbe, J.F., De, L.G., Kay, J., and Cryer, A. (1983). Biochem. Biophys. Res. Comm. 110, 584-592.
Tsujikawa, K., Kakutani,M., Sagawa,K., Suzuki,N., Mimura,T., Otaki,N. and Kimura,M. (1994), this volume.
Vallee, B.L. (1994) This volume, pp.
Viarengo, A.M. (1989) Rev. Aquat. Sci. 1, 295-317.
Vulpe, C., Levinson, B., Whitney, S., Packman, S., and Gitschier, J. (1993) Nature Genet. 3, 7-13.
Waalkes, M.P., Poisner, A.M., Wood, G.W., and Klaassen, C.D. (1984) Toxicol. Appl. Pharmacol. 74, 179-184.
Webb, M., and Cain, K. (1982) Biochem. Pharmacol. 31, 137-142.
Webb, M. (1987) In: Metallothionein II, J.H.R. Kagi and Y. Kojima eds. Birkhauser Verlag, Basal, pp. 109-134.
Wilkinson, D.G. and Nemer, M. (1987) Mol. Cell. Biol. 7, 48-58.
Winge, D.R., and Miklossy, K.A. (1982) J. Biol. Chem. 257, 3471-3476.
Wong, K.L., and Klaassen, C.D. (1979) J. Bio. Chem. 254, 12399-12403.
Yiangou, M., Ge, X., Carter, K.C., and Papaconstantinou, J. (1991) Biochemistry 30, 3798-3806.
Zeng, J., Vallee, B.L., and Kagi, J.H.R. (1991) Proc. Natl. Acad. Sci. USA 88, 9984-9988.
Zlotkin, S.H., and Cherian, M.G. (1988) Ped. Res. 24, 326-329.

Metallothionein III
K. T. Suzuki, N. Imura and M. Kimura (eds)

INVOLVEMENT OF METALLOTHIONEIN IN THE REGULATION OF MINERAL METABOLISM

I. Bremner

Rowett Research Institute, Greenburn Road, Bucksburn, Aberdeen AB2 9SB, UK

SUMMARY: There has been much speculation about the biological roles of metallothionein, particularly in the control of zinc and copper metabolism. Many aspects of its occurrence and chemical properties imply a role in fundamental biological processes but there have also been suggestions that it is specifically involved in the control of metal absorption, transport, storage and excretion. However, many of these claims are based on experiments where non-physiological conditions have been imposed. It is now known that intestinal metallothionein concentrations do change in response to fluctuations in dietary zinc supply and are inversely related to changes in the efficiency of zinc absorption. However, it is not yet certain that this is a causal relationship. The role of metallothionein in the hepatic metabolism and toxicity of copper is discussed, with particular reference to the effects of species, age and genetic factors.

INTRODUCTION

Ever since the discovery that metallothionein binds the essential trace elements copper and zinc as well as cadmium, there has been intense speculation that it plays an important regulatory role in homeostasis (see Bremner, 1987). This has been stimulated by observations that tissue concentrations of the protein frequently reflect those of zinc and copper and also the dietary intakes of the metals. Moreover both metals can induce synthesis of the protein by a common mechanism involving increased gene transcription. The protein is highly conserved in evolutionary terms and is present in most tissues, apparently always with a full complement of metal atoms, bound in a complex stereo-specific manner.

By analogy with its putative role in the detoxification

of cadmium and other heavy metals, it has therefore been proposed that metallothionein is involved also in the cellular detoxification of copper and zinc, or at least provides a mechanism whereby ionic concentrations of the metals can be reduced, while still maintaining them in a form where they can be readily utilised. This concept of metallothionein in the "buffering" of zinc and possibly copper ions has much to commend it, since a ready supply of zinc is required for a multitude of biological processes and zinc-finger motifs are a characteristic feature of many transcription factors (Chesters, 1989). It is perhaps significant that zinc can transfer from some transcription factors to thionein, at least *in vitro* (Zeng, *et al.*, 1991*a,b*). Such fundamental actions of metallothionein would not therefore be specific for a particular cell type or tissue, and are consistent with the ubiquitous distribution of the protein.

Nevertheless, there have been frequent suggestions that metallothionein does have certain tissue-specific functions and is involved in the regulation of particular processes involving the absorption, transport, storage and excretion of zinc and copper. The validity of these claims is difficult to ascertain as it is only recently that serious attempts have been made to investigate these hypotheses under conditions of physiological relevance. In much of the earlier work, pharmacological doses of metals were given, often by parenteral routes, with little regard to the fact that homeostatic control mechanisms are often of limited capacity and can breakdown under conditions of excessive metal loading.

Another complication in trying to establish the physiological roles of metallothionein is that we still know very little of the fate of the protein or of the metals bound to it. Much is now known of the mechanisms whereby synthesis of MT is induced, with identification of key regulatory elements on the metallothionein gene which are responsive to metals, glucocorticoids and other inducers, and characterisation of the nuclear metal-binding proteins which interact with these promoter elements (Palmiter, 1987). However, our knowledge of the mechanisms controlling degradation is still extremely limited. It may depend on the nature of the bound metal and there may be differences between isoforms of metallothionein, but it is still not clear if degradation occurs in lysosomes or the cytosol. In cases of copper overload, the protein may even aggregate rather than degrade, with eventual formation of polymeric and insoluble products. Secretion of metallothionein into the extracellular space can occur and could contribute to the transfer of the protein and bound metals between tissues. Direct transfer of metals to apoenzymes or other molecules is another possibility. Until such questions are resolved, the uncertainties as to the functions of the protein will remain.

In this review, existing evidence for specific roles for metallothionein in the control of zinc and copper metabolism is considered.

INTESTINAL ABSORPTION OF ZINC AND COPPER

Cousins and his colleagues have done most to establish whether metallothionein plays a critical role in the regulation of the intestinal absorption of zinc. They reported an inverse relationship between the efficiency of zinc absorption and the binding of zinc to metallothionein in the intestinal mucosa of rats. This suggested that the reduced efficiency of zinc absorption in animals or humans given diets with high zinc content might reflect increased binding of the zinc to metallothionein and concomitant reduction in its transfer across the basolateral membrane into the plasma (Richards and Cousins, 1976). However, other workers suggested that metallothionein facilitated rather than inhibited zinc absorption (Starcher *et al.*, 1980) or failed to find any relationship between metallothionein concentrations and zinc absorption in mice (Flanagan *et al.*, 1983). It was also reported that significant increases in intestinal metallothionein content only occurred at dietary zinc intakes well outside a normal physiological range (Hall *et al.*, 1979).

Because many of the early experiments were carried out in animals that were severely zinc-deficient or were given massive parenteral or oral doses of zinc, Cousins has systematically re-assessed the impact of moderate changes in dietary zinc content on the concentrations of metallothionein and of metallothionein mRNA in the intestines and other tissues of rats. No significant changes were initially detected in either analyte when the dietary zinc content was increased from 5 to 30 mg/kg, although increases were detected at 180 mg/kg provided the dietary copper intake was very low (Blalock *et al.*, 1988); this increase in metallothionein concentration was not associated with reduced efficiency of zinc absorption. However, in more recent studies using improved methodologies, differences in metallothionein protein and mRNA concentrations were detected in rats given <1, 5, 30 or 180 mg/kg diet, and at the extremes of intake were inversely related to the efficiency of zinc absorption (Cousins and Lee-Ambrose, 1992; Hempe and Cousins, 1992). It has now been proposed that intestinal metallothionein acts in concert with a cysteine-rich intestinal protein (CRIP) to regulate zinc absorption. CRIP is a saturable intracellular zinc-binding protein that facilitates the uptake and transport of zinc across the enterocyte and its subsequent export at the basolateral membrane. Binding of zinc to CRIP and subsequent zinc absorption are reduced in the presence of high metallothionein concentrations, which occur when the dietary zinc concentration is elevated. Conversely, in zinc-deficient animals, the low metallothionein concentrations result in high binding of zinc to CRIP and increased efficiency of zinc absorption.

Attractive though this hypothesis is, it is still true that it is difficult to detect changes in intestinal metallothionein protein or mRNA concentrations at dietary zinc intakes within a normal nutritional range. The concentrations used (<1 to 180 mg/kg) range from <10% to 15 times the dietary requirements for zinc in the rat. Such extreme variations would not normally be encountered in humans or domestic animals. Dose-dependent changes in metallothionein protein and/or mRNA are sometimes more readily detectable in other tissues, such as kidneys, liver and pancreas, over a narrow range of zinc intakes (Cousins and Lee-Ambrose, 1992; Bremner *et al.*, 1987). Presumably the main determinant of whether metallothionein concentrations change in response to changes in dietary zinc supply is the intracellular zinc concentration. It is likely that the dramatic increases in mucosal metallothionein concentrations at very high zinc intakes reflect the breakdown of normal control mechanisms that regulate cellular zinc accumulation (Hall *et al.*, 1979).

Another aspect that needs to be borne in mind is the possibility that metallothionein is involved in the secretion of zinc into the lumen of the gut, as has been indicated by studies on the immunolocalisation of the protein (Clarkson *et al.*, 1985). Moreover, little is known of the fate of zinc after its incorporation into intestinal metallothionein and its rate of flux through this pool. This is more important than the absolute concentration of the protein. The implicit assumption is that the protein remains within the cell until it is lost by desquamation but this has never been proved. The half-life of intestinal metallothionein-I in zinc-injected rats was reported to be about 15 hours, which is less than the half-life of mucosal cells, suggesting that the protein is degraded or secreted from the cell before desquamation occurs (Mehra and Bremner, 1987). It is possible also that the released zinc could be available for reabsorption.

The evidence that metallothionein plays a role in the control of copper absorption is even less convincing. No changes in mucosal metallothionein concentrations occur in response to substantial changes in copper supply (Hall *et al.*, 1979; Oestreicher and Cousins, 1985; Blalock *et al.*, 1988). Although the decreased efficiency of copper absorption in brindled mice is associated with excessive accumulation of copper bound to intestinal metallothionein, this is almost certainly a consequence of a defect in the efflux of copper from the cell (Crane and Hunt, 1983). The decrease in copper absorption in zinc-supplemented animals or human subjects may in some circumstances reflect zinc-induced synthesis of intestinal metallothionein, since this can result in increased and preferential binding of copper to the protein, with consequent reduction in transfer of copper across the basolateral membrane into the plasma (Hall *et al.*, 1979). However, other mechanisms may also operate, as it is often necessary to use excessive amounts of dietary zinc before these changes become evident (Oestreicher and Cousins, 1985). The uncertainty as to the ultimate fate of the copper bound to metallothionein also remains. From a teleological point of view, it appears strange that a system for homeostatic control of zinc absorption would result automatically in the inhibition of copper absorption.

TISSUE-SPECIFIC EXPRESSION OF METALLOTHIONEIN

Although the evidence for induction of intestinal metallothionein synthesis in response to dietary metal administration is equivocal, changes in metallothionein protein and mRNA levels do occur consistently in other tissues in response to changes in zinc supply. Thus, metallothionein-I concentrations in liver, kidneys, pancreas and blood cells decreased in a dose-dependent manner as dietary zinc concentrations were reduced from 12 to 6 or 3 mg/kg (Bremner *et al.*, 1987). Similar dose-dependent changes were observed in metallothionein mRNA levels in kidneys and on occasion in the liver, spleen and heart (Blalock *et al.*, 1988; Cousins and Lee-Ambrose, 1992). The effects of zinc supply on induction of metallothionein synthesis are also evident in the livers of neonatal animals, since moderate maternal zinc deficiency during pregnancy and lactation results in reduced expression of metallothionein in their pups (Morrison and Bremner, 1987; Gallant and Cherian, 1986, 1987). Increases in tissue zinc concentrations are almost invariably associated with an increase in metallothionein concentrations and in the binding of most of the additional zinc to the protein. Exceptions include copper-loaded liver of some species and the kidneys of sulphur-deficient sheep, where little of the zinc may be bound to metallothionein. Conversely, the decreases in liver zinc content in protein-deficient rats may be associated with increased metallothionein concentrations (Morrison and Bremner, 1990).

Such relationships are less evident in response to changes in dietary copper supply. Thus hepatic metallothionein-I concentrations were actually increased rather than decreased in copper-deficient rats (Bremner *et al.*, 1987), possibly because of their anorexic state, and metallothionein protein and mRNA concentrations were unchanged in pups reared on copper-deficient dams (Gallant and Cherian, 1986, 1987). Renal copper-metallothionein concentrations in rats were found to decrease in one study as dietary copper availability was reduced (Mehra and Bremner, 1983) but Blalock *et al.* (1988) found no increase in either metallothionein protein or mRNA concentrations as Cu intake was increased from 1 to 36 mg/kg. However, copper-metallothionein does accumulate in the kidneys of rats and sheep subjected to copper overload. The renal accumulation of the protein appears to result from the hepatic unloading that occurs in sheep before the onset of the haemolytic crisis (Bremner, 1987) and in rats as they adapt to copper toxicosis (Evering *et al.*, 1991*a,b*). Renal copper and metallothionein concentrations are also elevated in brindled mice but, as with intestinal metallothionein in these animals, this is probably a consequence of a defect in the efflux of copper from the cell (Prins and van der Hamer, 1980).

Although injection of copper results in induction of metallothionein synthesis in the liver of rats (Wake and Mercer, 1985), dietary supplementation with copper does not always elicit a similar response. Thus it was only at liver copper concentrations in excess of about 600 µg/g wet weight that increases in liver metallothionein-I concentrations were observed in Hooded Lister rats and even then only a small proportion of the total copper was bound in this form (Bremner *et al.*, 1986). Similar results were obtained in copper-loaded Wistar rats except that the threshold liver Cu concentration above which metallothionein appeared was about 140 µg/g wet weight (Evering *et al.*, 1991a). The amount of metallothionein present was, however, only sufficient to bind about 2% of the total hepatic copper at maximum copper concentration of 1000 µg/g wet weight. In contrast, practically all the cytosolic copper was bound to metallothionein in the livers of LEC rats, which have an inherited defect that leads to extensive hepatic accumulation of copper, with about 70% of the total copper occurring in the cytosol (Sugawara *et al.*, 1991). The reasons for these differences are not clear but copper may need to be in a relatively labile form before it is able to bind to the nuclear factors that then react with the metal regulatory elements on the DNA. Alternatively part of the increased synthesis which occurs after intraperitoneal injection of copper could be an inflammatory response, as has been reported for iron-induced metallothionein synthesis (McCormick and Dietert, 1991). However, it is only the second part of the bi-phasic response to copper injection, where hepatic uptake of zinc is also stimulated, that can be attributed to induction by released cytokines.

In considering the importance of metallothionein in the binding of hepatic copper, it has to be recognised that the rat is not typical of all species and that there are major differences between species in the hepatic distribution of copper. For example in both the pig and dog most of the hepatic copper appears to be bound to metallothionein in the cytosol and also in particulate fractions, regardless of how the copper has accumulated, although this relationship does not apply in zinc-deficient animals (Bremner, 1992). In copper-loaded human liver, as is found in neonates and in subjects with Wilson's Disease and primary biliary cirrhosis, much of the liver copper is again bound to metallothionein although there may be differences in its distribution between the cytosol and lysosomal and nuclear fractions (Nartey *et al.*, 1987*a,b*). However, it does not follow that copper is always responsible for induction of synthesis of the protein since it may sometimes displace zinc from pre-existing protein. It has also been reported that only 36% of the cytosolic copper in human liver with molecular weight of about 10,000 is present as metallothionein (Hunziker and Sternlieb, 1991).

HEPATOTOXICITY OF COPPER

The absence of any simple relationship between the hepatic concentration of copper and its cytotoxic effects has implied that the intracellular distribution and speciation of the metal influences its toxicity. Observations that species which are relatively copper tolerant tend to have high liver copper-metallothionein concentrations, whereas copper-intolerant species like the sheep have low copper-metallothionein concentrations, has prompted the view that metallothionein plays an important role in the cellular detoxification of copper. This is supported by the finding that a copper-resistant hepatoma cell line is characterised by increased metallothionein gene transcription and gene amplification, possibly with some post-transcriptional gene regulation (Czaja *et al.*, 1991). Moreover, prolonged exposure of rats to high-copper diets leads to an acquired tolerance after about 4 weeks, with a gradual reduction in liver copper and increase in kidney copper concentrations. This is associated with changes in the intracellular distribution of copper and increases in the amount of metal bound to metallothionein and its increased excretion in bile and plasma, although the proportion of hepatic copper in this form is still relatively low (Evering *et al.*, 1991*a*,*b*). This is difficult to reconcile with the situation in LEC rats which develop fulminant liver damage and hepatitis once liver copper levels reach 300 μg/g wet weight, despite the fact that most of the cytosolic copper is bound to metallothionein (Sugawara *et al.*, 1991). Similarly, toxic milk mice which also have an inherited defect that leads to excessive liver copper accumulation and the development of gross morphologic, histologic and ultrastructural changes (Biempica *et al.*, 1988), exhibit greatly increased hepatic metallothionein mRNA levels (Mercer *et al.*, 1992).

No definitive conclusions can be drawn from these studies as to the hepatotoxic effects of copper-metallothionein, as the lesions could have been induced by other copper-containing compounds. Nevertheless there is evidence that copper-metallothionein exacerbates the oxidative cell damage and lipid peroxidation induced by iron-nitrilotriacetate in neonatal guinea pig liver (Suntres and Liu, 1991). This is consistent with the enhancement by copper-metallothionein of the lipid peroxidation of liver microsomes incubated with xanthine and xanthine oxidase *in vitro* (Arthur *et al.*, 1987). The implication of these studies is that copper-metallothionein may act as a pro-oxidant and that its effects in biological systems may well depend on the antioxidant status of the cell.

Another important consideration is the intracellular distribution of copper and copper-metallothionein. As liver copper concentrations increase, the metal often accumulates in granular form in lysosomes, the nucleus and other organelles.

The lysosomal copper is often relatively inert and its localisation appears to constitute part of the mechanism for detoxifying copper. The changes in hepatic copper distribution during the progression of Wilson's Disease are consistent with this view, as the apparent transfer of copper from the cytoplasmic to the lysosomal pool is associated with a reduction in its hepatotoxic effects (Goldfischer and Sternlieb, 1968). Similarly the liver regeneration that occurs in copper-loaded rats as they become tolerant of copper is associated with its accumulation in stable lysosomes (Fuentealba *et al.*, 1989). In contrast, increase in the nuclear concentrations of copper is associated with irreversible nuclear damage, possibly because of binding of copper to DNA and the resultant changes in DNA structure (Sagripanti *et al.*, 1991). However, no such hepatocellular injury occurs in fetal liver despite extremely high copper concentrations. In fetal deer liver, liver copper concentrations can be 1000 μg/g dry matter, with the metal concentrated mainly in the nucleus and definitely not in the lysosomes, without any adverse effects (Leighton *et al.*, 1990).

The speciation of the copper in these particulate fractions appears to vary considerably. Thus the copper in lysosomes in Bedlington Terriers (Johnson *et al.*, 1981), in the particulate fractions of copper-loaded pigs and bovine and human fetal liver (see Bremner, 1992), and probably in the nuclei of fetal deer (Leighton, 1987) is bound to metallothionein. Immunocytochemical studies have frequently shown the presence of metallothionein in the nucleus in copper-loaded liver (e.g. Nartey *et al.*, 1987*a,b;* Williams *et al.*, 1989) but do not provide quantitative information. However, none of the copper in lysosomes of copper-loaded rats and sheep and at most only a small proportion of that in the nuclei of copper-loaded rats is bound to metallothionein. Although electron probe microanalysis frequently shows the presence of both copper and sulphur in granules within these organelles, this is not proof of the existence of metallothionein; there are only three major ligands for copper, *viz.* sulphur, oxygen and nitrogen, so it is not surprising that sulphur is often present in copper loaded granules.

Further studies on the binding of copper within isolated lysosomes and nuclei are clearly warranted. The increased density of copper-containing lysosomes can lead to their co-sedimentation with nuclei and it is sometimes difficult to ascertain from earlier investigations where differential centrifugation techniques were used which organelle contained the copper. Information is also required on other ligands present within the organelles, particularly the nucleus. The oxidative damage to DNA induced by copper is greatly exacerbated in the presence of reducing agents (Aruoma *et al.*, 1991) and may involve hydroxyl radical generation on the DNA. However, thiols are also known to stabilise cuprous copper, and glutathione can inhibit copper-dependent formation of free radicals (Hanna and Mason, 1992). Moreover glutathione concentrations are relatively high in mammalian cell nuclei (Bellomo *et al.*, 1992) and there is evidence that this peptide plays an important role in the cellular uptake of copper and possibly also in its release from metallothionein (Freedman *et al.*, 1989). There is therefore considerable need for detailed investigation of the speciation of copper within the nucleus and on the impact of this on free radical generation and on subsequent DNA damage.

The financial support of the Scottish Office Agriculture and Fisheries Department is gratefully acknowledged.

REFERENCES

Arthur, J.R., Bremner, I., Morrice, P.C. and Mills, C.F. (1987) Free Rad. Res. Commun. 4, 15-20.

Aruoma, D.I., Halliwell, B., Gajewski, E. and Dizdaroglu, M. (1991) Biochem. J. 273, 601-604.

Bellomo, G., Vairetti, M., Stivala, L., Mirabelli, F., Richelmi, P. and Orrenius, S. (1992) Proc. Natl. Acad. Sci. USA 89, 4412-4416.
Biempica, L., Rauch, H., Quintana, N. and Sternlieb, I. (1988) Lab. Invest. 59, 500-508.
Blalock, T.L., Dunn, M.A. and Cousins, R.J. (1988) J. Nutr. 118, 222-228.
Bremner, I. (1987) In Metallothionein-II. Edited by J.H.R. Kägi and Y. Kojima. Birkhauser Verlag, Boston, pp. 81-107.
Bremner, I. (1992) Methods in Enzymology (Metallobiochemistry Part B) 205, 25-35; 584-591.
Bremner, I., Mehra, R.K., Morrison, J.N. and Wood, A.M. (1986) Biochem. J. 235, 735-739.
Bremner, I., Morrison, J.N., Wood, A.M. and Arthur, J.R. (1987) J. Nutr. 117, 1595-1602.
Chesters, J.K. (1989) Proc. Nutr. Soc. 50, 123-129.
Clarkson, J.P., Elmes, M.E., Jasani, B. and Webb, M. (1985) Histochem. J. 17, 343-352.
Cousins, R.J. and Lee-Ambrose, L.M. (1992) J. Nutr. 122, 56-64.
Crane, I.J. and Hunt, D.M. (1983) Chem.-Biol. Interactions 45, 113-124.
Czaja, M.J., Weiner, F.R. and Freedman, J.H. (1991) J. Cell. Physiol. 147, 434-438.
Evering, W.E.N.D., Haywood, S., Bremner, I. and Trafford, J. (1991*a*) Chem.-Biol. Interactions 78, 283-295.
Evering, W.E.N.D., Haywood, S., Bremner, I., Wood, A.M. and Trafford, J. (1991*b*) Chem.-Biol. Interactions 78, 297-305.
Flanagan, P.R., Haist, J. and Valberg, L.S. (1983) J. Nutr. 113, 962-972.
Freedman, J.H., Ciriolo, M.R. and Peisach, J. (1989) J. Biol. Chem. 264, 5598-5605.
Fuentealba, I., Haywood, S. and Foster, J. (1989) Exp. Molecular Pathol. 50, 26-37.
Gallant, K.R. and Cherian, M.G. (1986) Biochem. Cell Biol. 64, 8-12.
Gallant, K.R. and Cherian, M.G. (1987) J. Nutr. 117, 709-716.
Goldfischer, S. and Sternlieb, I. (1968) Am. J. Pathol. 53, 883-899.
Hall, A.C., Young, B.W. and Bremner, I. (1979) J. Inorg. Biochem. 11, 57-66.
Hanna, P.M. and Mason, R.P. (1992) Arch. Biochem. Biophys. 295, 205-213.
Hempe, J.M. and Cousins, R.J. (1992) J. Nutr. 122, 89-95.
Hunziker, P.E. and Sternlieb, I. (1991) Eur. J. Clin. Invest. 21, 466-471.
Johnson, G.F., Morell, A.G., Stockert, R.J. and Sternlieb, I. (1981) Hepatology 1, 243-248.
Leighton, M. (1987) PhD thesis, Univ. London, UK.
Leighton, M.J., Peters, T.J., Hill, R., Smith, G.D., Holt, W. and Jones, D.M. (1990) Res. Vet. Sci. 49, 298-305.
McCormick, C.C. and Dietert, R.D. (1991) In Metallothionein in Biology and Medicine. Edited by C.D. Klaassen and K.T. Suzuki. CRC Press, Boca Raton, pp. 209-220.

Mehra, R.K. and Bremner, I. (1983) Biochem. J. 213, 459-465.
Mehra, R.K. and Bremner, I. (1987) In Metallothionein-II. Edited by J.H.R. Kägi and Y. Kojima. Birkhauser Verlag, Boston, pp. 565-572.
Mercer, J.F.B., Grimes, A. and Rauch, H. (1992) J. Nutr. 122, 1254-1259.
Morrison, J.N. and Bremner, I. (1987) J. Nutr. 117, 1588-1594.
Morrison, J.N. and Bremner, I. (1990) Proc. Nutr. Soc. 49, 68A.
Nartey, N.O., Banerjee, D. and Cherian, M.G. (1987*a*) Pathology 19, 233-238.
Nartey, N.O., Frei, J.V. and Cherian, M.G. (1987*b*) Lab. Invest. 57, 397-401.
Oestreicher, P. and Cousins, R.J. (1985) J. Nutr. 115, 159-166.
Palmiter, R.D. (1987) In Metallothionein-II. Edited by J.H.R. Kägi and Y. Kojima. Birkhauser Verlag, Boston, pp. 63-80.
Prins, H.W. and van der Hamer, C.J.A. (1980) J. Nutr. 110, 151-157.
Richards, M.P. and Cousins, R.J. (1976) J. Nutr. 106, 1591-1599.
Sagripanti, J.-L., Goering, P.L. and Lamanna, A. (1991) Toxicol. Appl. Pharmacol. 110, 477-485.
Starcher, B.C., Glauber, J.G. and Madaras, J.G. (1980) J. Nutr. 110, 1391-1397.
Sugawara, N., Sugawara, C., Sato, M., Katakura, M., Takahashi, H. and Mori, M. (1991) Res. Commun. Chem. Pathol. Pharmacol. 72, 353-362.
Suntres, Z.E. and Lui, E.M.K. (1991) J. Pharmacol. Exp. Ther. 258, 797-806.
Wake, S.A. and Mercer, J.F.B. (1985) Biochem. J. 228, 425-432.
Williams, L.M., Cunningham, H., Ghaffar, A., Riddoch, G.I. and Bremner, I. (1989) Toxicology 55, 307-316.
Zeng, J., Vallee, B.L. and Kägi, J.H.R. (1991*a*) Proc. Natl. Acad. Sci. USA 88, 9984-9988.
Zeng, J., Heuchel, R., Schaffner, W. and Kägi, J.H.R. (1991*b*) FEBS Lett. 279, 310-312.

Metallothionein III
K.T. Suzuki, N. Imura and M. Kimura (eds)

INDUCTION OF METALLOTHIONEIN SYNTHESIS BY OXIDATIVE STRESS AND POSSIBLE ROLE IN ACUTE PHASE RESPONSE

Masao Sato[1], Misao Sasaki[1], Hiroshi Hojo[2]

[1]Division of Environmental Pollution Research, Research Institute, Fukushima Medical College, 1 Hikarigaoka, Fukushima 960-12, Japan; [2]Department of Hygienic Chemistry, Pharmaceutical Institute, Tohoku University, Aobayama, Aoba-ku, Sendai 980, Japan

SUMMARY: The induction of metallothionein (MT) synthesis by oxidative stress and the biological roles of the protein have been studied. MT was induced by a variety of compounds which produce oxygen free radicals, such as paraquat, carbon tetrachloride and glutathione depletors. Attention was focussed on possible mediators of MT synthesis. Lipid peroxidation was not necessary for MT synthesis in the liver induced by oxidative stress. Since pretreatment of rats with dexamethasone prevented paraquat-induced MT synthesis, MT synthesis by oxidative stress may be, at least partly, mediated by cytokines. Release of cytokines by oxidative stress and MT synthesis induced by cytokines were investigated. Tissue specific induction of MT synthesis by interleukin-6 (IL-6) and tumor necrosis factor- α (TNF) was observed. Induction of MT synthesis by IL-6 seemed to be restricted to the liver, whereas TNF induced MT synthesis in the liver, lung and heart. It is probable that locally released IL-6 and TNF are responsible for MT synthesis in the liver and TNF is also responsible in some other tissues than liver. Pretreatment with TNF prevented the liver injury and lipid peroxidation caused by carbon tetrachloride. Since TNF administration also increased activities of manganese-superoxide dismutases, the enzyme and MT may cooperatively exert antioxidative roles in tissues. However TNF pretreatment did not affect the increased levels of liver MT and plasma fibrinogen. Thus, MT induced by cytokines can play a role as a radical scavenger in the liver but does not affect acute phase protein synthesis. The role of MT at the local sites of inflammation is also discussed.

INTRODUCTION

Metallothionein (MT) synthesis can be stimulated in many organs not only by various metals such as Cd, Zn, and Cu, but also by non-metallic compounds, including alkylating agents, alcohols, chelators and glucocorticoid hormones. Some of these inducers produce reactive oxygen radicals, such as superoxide- and hydroxy-radicals, and hydrogen peroxide. Exposure to such reactive oxygen radicals results in tissue or cell injury and inflammaion, which elicit acute phase responses. Acute phase responses are the local and systemic reactions of the organism to disturbances of its homeostasis due to tissue injury, infection, immunological disorder, or stress exposure (Heinrich et al., 1990). A variety of cytokines secreted by macrophages and other cell types as the initial local reaction at the injured site mediate the acute phase responses. Increases in the synthesis of acute phase proteins including fibrinogen, α_2-macroglobulin, ceruloplasmin and α_1-acid glycoprotein, occur in the liver, as one of the systemic responses following the initial local reaction.

Recently some cytokines, namely α-interferon (IFN) (Friedman and Stark, 1985), interleukin-1 (IL-1) (Cousins and Leinart, 1988), interleukin-6 (IL-6) (Schroeder and Cousins, 1990) and tumor necrosis factor-α (TNF) (De et al, 1990, Sato et al., 1992) have been shown to induce MT synthesis. IL-1, IL-6 and TNF are the three major mediators responsible for acute phase protein synthesis. It has been suggested that the induction of MT synthesis by endotoxin (De et al., 1990, Liu et al., 1991) and inflammation by turpentine (Min et al., 1992a) is mediated by these cytokines. The increase in synthesis of MT by

IL-1 was tissue specific, and peaked in liver at 36 hr after the injection (Cousins and Leinart, 1988).

Although MT synthesis is induced by a wide range of radical generating agents or conditions such as X-irradiation (Shiraishi et al., 1989), carbon tetrachloride (Oh et al., 1978) and paraquat (Sato, 1991), its specific mechanism and biological roles are still unknown (Sato and Bremner, 1993). In addition, there is a little information on MT synthesis _in vivo_ by TNF and IL-6, and on possible interaction between the two cytokines.

This paper describes (1) induction of MT synthesis by oxidative stress, (2) induction of MT synthesis by cytokines, which are possible mediators for the induction, (3) possible biological roles of MT during the acute phase response.

MATERIALS AND METHODS

Bolton-Hunter reagent was obtained from ICN Biochemicals Inc. (Irvine, CA, USA). Anti-sheep/goat IgG donkey serum was obtained from Scottish Antibody Production Unit (Carluke, Lanarkshire, U. K.). Anti-rat liver metallothionein-I (MT-I) sheep serum was kindly provided from Dr. Ian Bremner, Rowett Research Institute (Aberdeen, U. K.). Recombinant human TNF-α or IL-6 was given by Asahi Kasei Corp., Ltd., Tokyo and Ajinomoto Corp., Ltd., Yokohama, Japan, respectively.

Male Wistar rats (5-6 weeks of age) were used. Rats were injected subcutaneously with TNF (10^5 units/rat) or IL-6 (5 x10^4 units/rat) respectively. Blood was collected by cardiac puncture under pentobarbital anesthesia, then rats were sacrificed. The liver, kidney, lung, heart, thymus and spleen were rapidly excised, and homogenized in 9 vol of cold 50 mM Tris-HCl buffer, pH 8.0. In other experiments, 10 ml of air was injected subcutaneously on the back to make an air pouch. After 24 hr, 4 ml of 2% kappa-carrageenan was injected into the air pouch (Adachi and Yamashita, 1986). After 1, 2 or 5 days the inflammatory exudate in the air pouch and the air pouch wall were removed. The degree of lipid peroxidation in the liver homogenate was estimated by formation of thiobarbituric acid reactive substances (TBA-RS) (Uchiyama and Mihara, 1978). The supernatants of the homogenates (1500 x g, 10 min) were used for estimation of total MT by the Cd-heme method (Onosaka and Cherian, 1981) or of MT-I by radioimmunoassay (Mehra and Bremner, 1983). The plasma concentration of fibrinogen was determined using a Wako Fibrinogen B

Test Kit (Swaim and Feders, 1967). Ceruloplasmin was measured from its oxidase activity in plasma using p-phenylenediamine as substrate.

RESULTS AND DISCUSSION

Metallothionein as an acute phase protein

Acute phase responses are characterized by a decrease in serum levels of Zn and Fe, increases in secretion of ACTH and glucocorticoids, activation and accumulation of granulocytes and mononuclear cells, and by dramatic changes in serum concentrations of acute phase proteins. Decreases in Zn levels were typically observed in rats treated with carbon tetrachloride and endotoxin (Sato et al., 1984). Increases in plasma concentrations of Cu, ceruloplasmin and fibrinogen were also observed in carbon tetrachloride- and paraquat-treated rats (not shown). Table 1 shows the changes in plasma concentrations of acute phase proteins in rats (Heinrich et al., 1990).

The significance of metallothionein in plasma has not been greatly studied so far because of the difficulties in its measurement. However in recent years, increases in plasma levels of MT have been reported in rats exposed to oxidative stress, including injection of carbon tetrachloride, endotoxin (Sato et al., 1984), paraquat (Sato, 1991) and stilbene oxide (not shown), and subjected to the stresses of restraint (Hidalgo et al., 1989) and water-immersion (Tsujikawa et al., 1991). The increment in plasma MT concentrations was preceded by a corresponding alteration in hepatic MT concentrations in carbon tetrachloride-treated rats (Sato et al., 1984) and paraquat-treated rats (Fig. 1). TNF, an important inflammatory cytokine, also significantly increased plasma concentrations of MT-I (Sato et al., 1992). These data suggest that MT is an acute phase protein whose concentrations change substantially in response to a challenge (Table I).

Metallothionein synthesis by oxidative stress

Induction of MT synthesis occurs in response to a wide range of chemical stimuli, including a group of compounds which produce oxygen free radicals, and of physical stresses which accelerate lipid peroxidation (Table II). The induction of MT synthesis by chemical stress is tissue-specific, with the liver being to the most responsive of tissues (Bauman et al., 1991).

The question of whether induction of MT synthesis by oxidative stress is related to lipid peroxidation has been investigated. Pretreatment with vitamin E prevented paraquat-induced lipid peroxidation but not MT synthesis in the liver of rats, suggesting that enhanced lipid peroxidation is not necessary for induction of MT synthesis by paraquat in the liver (Sato and Sasaki, 1991). Min et al. (1992b) confirmed that MT synthesis upon oxidative stress was independent of lipid peroxidation in the liver of menadione- and carbon tetrachloride-treated mice. Whole body X irradiation induced MT mRNA transcription and protein expression and its accumulation in mouse liver, but did not induce MT in cultured cells (Koropatnick et al., 1988), suggesting that radiation-induced MT expression is indirect.

Oxidative stress triggers the release of a variety of mediators such as glucocorticoids, glucagon and cytokines. However, since exposure to oxidative stress such as X-irradiation

Table I. Acute phase plasma proteins in rats

Increase			
10-100-fold	2-10-fold	<2-fold	Decrease
α_2-Macroglobulin α_1-Acid glycoprotein	Fibrinogen Haptoglobin Cysteine proteinase inhibitor Metallothionein	α_1-Proteinase inhibitor Ceruloplasmin Prekallikrein Hemopexin C-Reactive protein	α_1-Inhibitor Transferrin Prealbumin Albumin

No change: α_1-Macroglobulin, Antithrombin III, Serum amyloid P, Prothrombin

(Adapted from Heinrich et al., 1990)

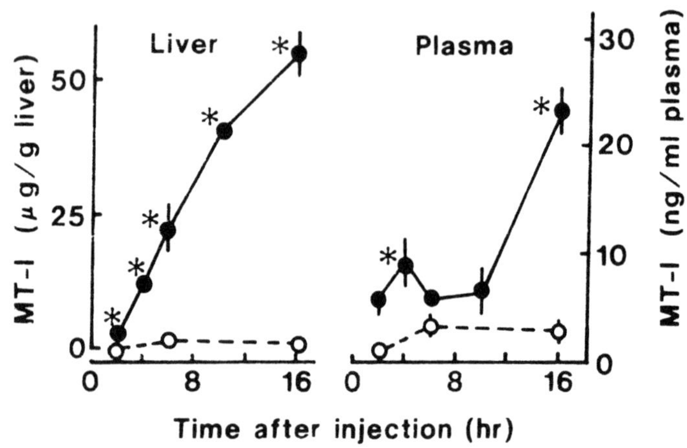

Fig. 1. Concentrations of metallothionein-I (MT-I) in the liver and plasma at various times following paraquat injection. Rats were injected s.c. with paraquat (40 mg/kg). MT-I concentration was determined by radioimmunoassay. Values represent mean ± SE of 3-4 rats and asterisks indicate values significantly different ($p<0.05$) from control. O, control; ●, paraquat.

(Shiraishi et al., 1986) and menadione (Mins et al., 1991) increased hepatic MT concentrations in adrenalectomized rats, glucocorticoid does not contribute to the induction of MT. Pretreatment of rats with dexamethasone, a strong inhibitor of cytokine production, prevented the increase of hepatic MT induced by turpentine, menadione (Min et al., 1992a) and paraquat (Fig. 2). This suggests that cytokines could be an important factor involved in the induction of MT by oxidative stress.

<u>Release of cytokine during acute phase response</u>

At the site of injury, local reactions of acute phase response occur with release of acute phase cytokines. Iijima et al. (1989) showed that a factor released by lipopolysaccharide (LPS) from macrophages was different from other MT inducers such as glucocorticoid hormone, IFN and IL-1, and identical to transforming growth factor. The combined administration of endotoxin and D-galactosamine induced hepatitis and increased serum levels of TNF but no TNF was detectable in the sera of mice pretreated with superoxide dismutase, allopurinol and catalase (Table III) (Neihoster et al., 1992). These findings indicate that the formation of extracellular reactive oxygen species is linked with the release of TNF into the serum. Remick et al. (1989) also showed that TNF production was induced by LPS, and it was suppressed by dexamethasone. Endotoxin injection in mice caused a coordinate, transient increase in the mRNA levels of IL-1α, IL-1β and IL-6, and TNF which peaked by 1 hr (De et al., 1990). It is well known that these cytokines have interrelated effects and form a complicated network.

<u>Induction of metallothionein synthesis by cytokines</u>

IL-1, IL-6 and TNF are major inducers for acute phase proteins in the liver. Induction of MT synthesis by these cytokines has been shown both <u>in vivo</u> and <u>in vitro</u> (Table IV).

The first observation of MT induction by cytokines was the

Table II. Induction of metallothionein by oxidative stress

Carbon tetrachloride	Acetaminophen	Diquat
Chloroform	Adriamycin	Paraquat
Ethanol	Bleomycin	
	Cephaloridine	Diamide
tert-Butyl hydroperoxide	Cisplatin	Diethyl maleate
	Menadione	Phorone
Dimethyl sulfoxide	Metronidazole	Stilbene oxide
L-Ascorbic acid	Fasting	
Carrageenan	Food restriction	
Endotoxin	High oxygen tension	
Methyl indole	Restraint	
Streptozotocin	UV-irradiation	
	X-irradiation	

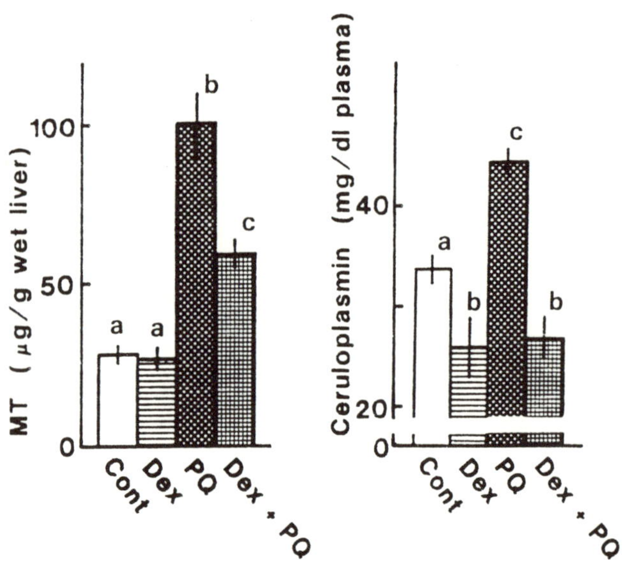

Fig. 2. Effect of pretreatment with dexamethasone on concentrations of liver metallothionein (left) and plasma ceruloplasmin (right) in paraquat-treated rats. Rats were injected with dexamethasone (Dex, 100 mg/kg) 6 hr before injection of paraquat (PQ, 35 mg/kg, s.c.) or saline. Rats were killed 18 hr after the last injection. Values represent mean ± SE of 4 rats. Groups without a common superscript letter are significantly different at $p < 0.05$.

report that administration of IL-1 increased hepatic concentrations of Zn-MT in rats (DiSilvestro and Cousins, 1984). It was then found that the increase in MT synthesis induced by IL-1, was associated with tissue specific redistribution of Zn, with a transient depression of Zn in the plasma and concomitant uptake of Zn by the liver, bone marrow and thymus (Cousins and Leinart, 1988). Liu et al. (1991) suggested that endotoxin-induced MT synthesis is mediated by IL-1, and IL-6, and TNF. In fact, IL-1α, IL-1β and IL-6, IFN-γ and TNF were all effective inducers of hepatic MT-I and II genes (De et al., 1990).

IL-6 has been shown to be a major inducer of acute phase proteins (Castell et al., 1988). In hepatocytes IL-6 rather than IL-1 appears to be the mediator of MT production and of changes in Zn metabolism (Schroeder and Cousins, 1990). However, the interactions between IL-6 and TNF are still unclear. Recently, we demonstrated that IL-6 induced MT synthesis in rat liver and had minimal effects on MT induction in organs other than liver, whereas TNF maintained high levels of MT in the heart (Fig. 3) and lung. During the acute phase response, IL-6 and TNF may be responsible for MT induction in the liver whereas in some other organs TNF is mainly responsible.

<u>Role of metallothionein during acute phase response</u>

The physiological roles of most acute phase proteins are not yet fully known. Ceruloplasmin, which is induced by adjuvant treatment, has functional copper atoms which may allow the protein to scavenge free radicals (Cousins and Swerdel, 1985).
MT is similarly induced by oxidative stress and can also scavenge

Table III. Serum TNF levels in mice protected against galactosamine (GalN)/endotoxin-induced hepatitis by pretreatment with allopurinol, catalase or superoxide dismutase (SOD)

Pretreatment	ALT(U/L)	TNF (pg/mL)	M/N
None	4320±640	4940±620	3/6
Allopurinol(100 mg/kg)	180±40*	<50	0/6
Catalase(10^6 U/kg)	95±45*	<50	0/6
SOD(3.3 x 10^4 U/kg)	75±20*	<50	0/6

*$P < 0.05$ vs hepatic injury control (no pretreatment prior to administration of 700 mg/kg GalN plus 5 g/kg endotoxin). Allopurinol was administered intraperitoneally. Catalase or SOD was given intravenously 1 hr prior to induction of liver injury. Plasma alanine aminotransferase (ALT) was determined 8 hr after challege, serum TNF concentrations 90 min after GalN/endotoxin administration. Data: means ± SE; N = number of animals, M = number of animals which died within 8 hr. (Neihorster et al., 1992).

Table IV. Metallothionein synthesis induced by cytokine

Cytokine	Organ, cells	Reference
Interleukin-1α	Rat liver	DiSilvestro and Cousins, 1984
	Hep G2	Karin et al., 1985
	Rat liver, bone marrow, thymus	Cousins and Leinart, 1988
	Mouse liver	De et al., 1990
	Mouse liver	Liu et al., 1991
Interleukin-1β	Mouse liver	De et al., 1990
Interleukin-6	Rat hepatocyte	Schroeder and Cousins, 1990
	Mouse liver	De et al., 1990
	Mouse liver	Liu et al., 1991
	Rat liver, lung	Sato et al., unpublished
Tumor necrosis factor-α	Rat liver	Grimble and Bremner, 1989
	Mouse liver	De et al., 1990
	Rat liver, lung heart, plasma	Sato et al., 1992
Interferon-α	Neuroblastoma cells (human)	Friedman and Stark, 1985
Interferon-γ	HeLa cells	Kusari et al., 1987
	Mouse liver	De et al., 1990

free hydroxyl- and superoxide radicals produced by xanthine-xanthine oxidase reaction, at least *in vitro* (Thornalley and Vasak, 1985). Recently, turpentine-induced inflammation accompanied with MT synthesis was shown to inhibit the hepatotoxic effects of carbon tetrachloride (DiSilvestro and Carlson, 1992). It was speculated that the preventive effect was mediated by cytokines. In accord with this view, pretreatment of rats with TNF increased the hepatic concentration of MT and prevented the liver damage and lipid peroxidation caused by carbon tetrachloride (Fig. 4). This *in vivo* study suggests that hepatic MT induced by cytokines plays a role in scavenging free radicals.

TNF induces not only MT synthesis but also the expression of mitochondrial manganese superoxide dismutase, an important anti-oxidative enzyme (Wong and Goeddel, 1988). We also observed that TNF administration increased activities of manganese superoxide dismutase in rat liver (not shown). As to subcellular localization of MT, i.e. cytoplasmic, is different from that of

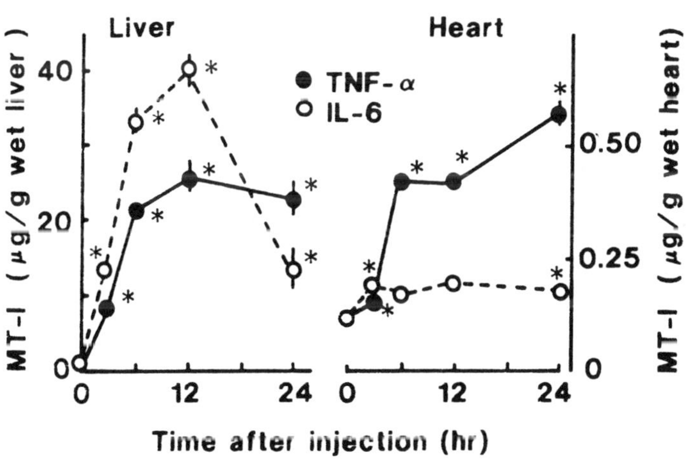

Fig. 3. Changes in concentrations of metallothionein-I (MT-I) in the liver (left) and heart (right) at various times following injection of tumor necrosis factor-α (TNF, ●) or interleukin-6 (IL-6, ○). Male Wistar rats were injected either s.c. with 10^5 units of TNF or s.c. with 5×10^4 units of IL-6. MT-I concentration was determined by radioimmunoassay. Each point represents mean ± SE of 4 rats and asterisks indicate values significantly different ($p < 0.05$) from control.

manganese superoxide dismutase, i.e. mitochondria, these cell constituents may cooperatively exert anti-oxidative roles in different compartments of cells, and hence boost the anti-inflammatory activity of the tissue.

If hepatic MT induction by oxidative stress results from the free radical production, animals with much preexisting MT reserves would not be expected to synthesize more MT in the liver when exposed to further oxidative stress. However, pretreatment with TNF did not affect paraquat-induced MT synthesis or the increment in plasma fibrinogen levels (not shown), although it did prevent lipid peroxidation in the liver (Fig. 4). This suggests that MT in the liver is not directly induced by free radicals and does not affect the synthesis of acute phase proteins mediated by cytokines, which are released from macrophages at the local sites of inflammation or tissue injury.

In addition to its putative role as a radical scavenger in liver, MT may play some roles at local sites of inflammation or tissue injury. MT concentrations increased in gastric mucosal cells of rats stressed by water-immersion (Tsujikawa et al., 1991) and in fibroblast cells in granulation tissue local to the base of chronic human mucosal ulcers (Douglas-Jones et al., 1992). We also observed high contents of MT both in the liver and at the inflammatory site in carrageenan-treated rats; the MT contents in granulation tissue increased with days post-injection (not shown). The data suggest a role of MT in the growth of granulation tissue, although the origin of MT in the local sites is unknown. The locally accumulated MT may also play a role as a cytoprotective radical scavenger. Recently _in vitro_ studies demonstrated that MT inhibited superoxide generation in human granulocytes (Miesel et al., 1990; Nakajima et al., 1991). Zn-MT reacted with three principal neutrophil oxidants such as hydrochlorous acid, superoxide and hydrogen peroxide *in vitro*, cleaving thiolate bonds of MT and mobilizing Zn ions (Fliss and Merland, 1992). This process may prevent other Zn-metalloproteins from damage. Furthermore, injected Zn-MT suppressed the formation of gastric ulcer induced by stress or HCl-ethanol (Tsujikawa et al., 1991), and injected Cu-MT showed anti-inflammatory reactivity *in vivo* (Miesel et al., 1990). The contribution of circulating MT to these effects of MT at local sites of injury or inflammation have not yet been established.

Finally, Fig. 5 shows a schematic representation of the regulation and possible roles of MT during the acute phase re-

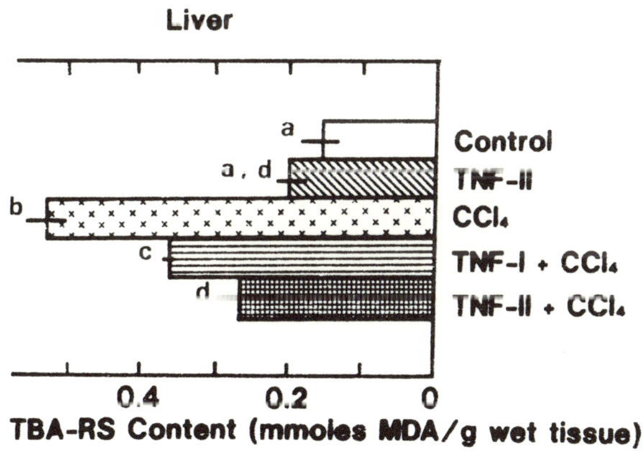

Fig. 4. Effect of TNF pretreatment on lipid peroxidation in the liver of CCl_4-treated rats. Rats were injected with TNF (I, $4x10^4$; II, $1x10^5$ units/rat) or saline (5 ml/rat) before injection of CCl_4 (5 ml/kg, s.c.). Rats were killed 12 hr after the last injection. Values represent mean $\pm$ SE of rats. Groups without a common superscript letter are significantly different at $p < 0.05$.

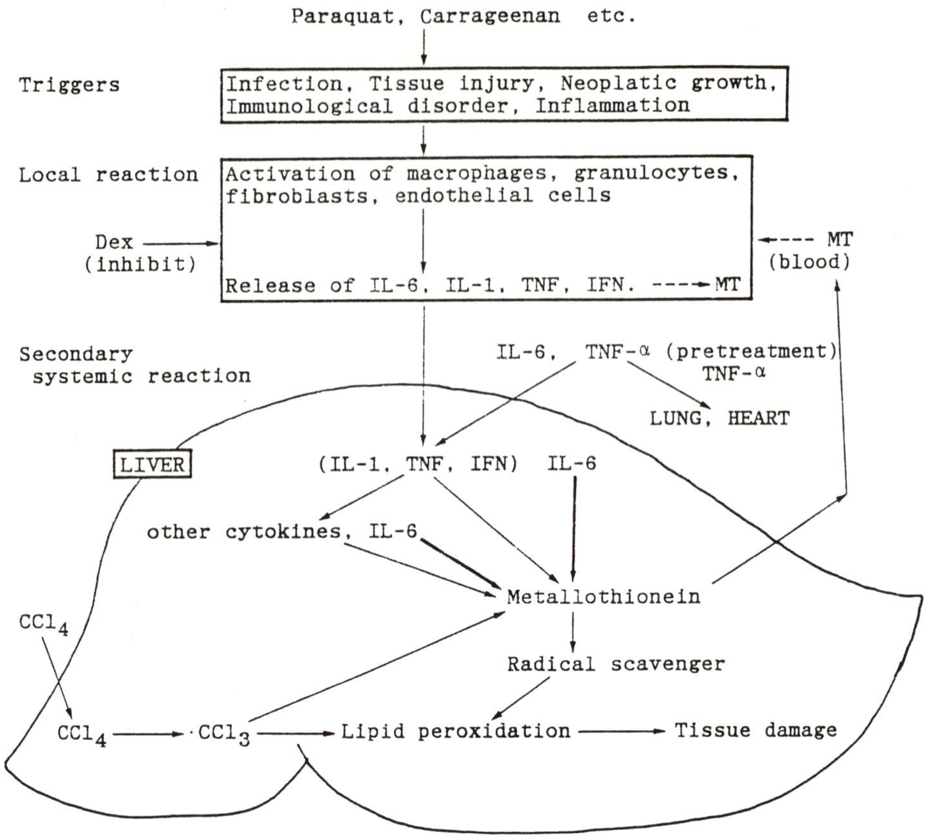

Fig. 5. Schematic representation of metallothionein production and its possible roles: Effects of TNF pretreatment on carbon tetrachloride-induced lipid peroxidation and on paraquat-induced metallothionein synthesis. The dotted lines represent lack of supporting experimental data.

sponse. TNF pretreatment induces MT synthesis in the liver. This may be mediated through IL-6, TNF itself and/or other cytokines. When paraquat or other radical generating agent are given, tissue injury or inflammation occurs. This elicits release of cytokines such as IL-1, IL-6, TNF and IFN, which also induce MT synthesis. Therefore, the total amount of MT in the liver is the sum of that induced by TNF pretreatment and by paraquat. Increased MT prevents the lipid peroxidation caused by carbon tetrachloride in the liver.

Acknowledgements

The authors are grateful to Ms. Emiko Yoshioka for secretarial work, and to the staff of the Experimental Animal Laboratory of the Fukushima Medical College for the care of the animals. The authors express appreciation to Asahi Kasei Corp. Ltd. and Ajinomoto Corp. Ltd. for supplying the rhTNF-α and rhIL-6 respectively.

REFERENCES

Adachi, K., and Yamashita,T. (1986) Oyo-Yakuri-Pharmacometrics. 32, 711-713.

Bauman, J. W., Liu, J., Liu, Y. P., and Klaassen, C. D. (1991) Toxicol. Appl. Pharmacol. 110, 347-354.

Castell, J. V., Andus, T., Geiger, T., Gomez-Lechon, M., David, M., Klapproth, J., Hirano, T., Kishimoto, T., and Heinrich, P.C. (1988) Adv. Immunopharmacol. 4 191-201.

Cousins, R. J., and Leinart, A. S. (1988) FASEB J. 2, 2844-2890.

Cousins, R. J., and Swerdel M. R. (1985) Proc. Soc. Exp. Biol. Med. 176, 168-172.

De, S. K., McMaster, M. T., and Andrews, G. K.(1990) J. Biol. Chem. 265, 15267-15274.

DiSilvestro, R. A., and Carlson, G. P. (1992) Toxicol. Lett. 60 175-181.

DiSilvestro, R. A., and Cousins, R.J. (1984) Life Sci. 35, 2113-2118.

Douglas-Jones, A. G., Thomas, N. D., Elmes, M. E., and Jasani, B. (1992) Histochem. J. 24, 40-50.

Fliss, H., and Merland, H. (1992) Arch. Biochem. Biophys. 293, 195-199.

Friedman, R. L., and Stark, G. R. (1985) Nature. 314, 637-639.

Grimble, R. F., and Bremner, I. (1989) Proc. Nutr. Soc. 48, 64A.

Heinrich, P. C., Castell, J. V., and Audus, T. (1990) Biochem. J. 265, 621-636.

Hidalgo, J., Campany, L., Borras, M., Garvey, J. S., and Armario, A. (1988) Amer. J. Physiol. 255, E518-E524.

Iijima, Y., Fukushima, T., and Kosaka, F.(1989) Biochem. Biophys. Res. Commun. 164, 114-118.

Karin, M., Imbra, R.J., Heguy, A., and Wong, C. (1985) Mol. Cell. Biol. 5, 2896-2890.

Koropatnick, J., Leibbrandt, M., and Cherian, M. G.(1989) Radiat. Res. 199, 356-365.

Kusari, J., Tiwari, R.J., Kumar, R., and Sen, G.C. (1987) J. Virol. 61, 1524-1531.

Liu, J., Liu, Y. P., Sendelbach, L. E., and Klaassen, C. D.(1991) Toxicol. Appl. Pharmacol. 109, 235-240.

Mehra, R. K., and Bremner, I. (1983) Biochem. J. 213, 459-465.

Miesel, R., Hartman, H. J., and Weser, U. (1990) Inflamm. 14, 471-483.

Min K-S., Mukai, S. Ohta, M., Onosaka, S., and Tanaka, K. (1992a) Toxicol. Appl. Pharmacol. 113, 293-298.
Min, K.-S., Terano, Y., Onosaka, S., and Tanaka, K. (1991) Toxicol. Appl. Pharmacol. 111, 152-162.
Min, K-S., Terano, y., Onosaka, S., and Tanaka, K. (1992b) Toxicol. Appl. Pharmacol. 113, 74-79.
Nakajima, K., Adachi, M., Otaki, N., Kimura, M., Tani, T., Igawa, T., Kawaharada, U., and Suzuki, K. (1991) pp. 359-373. in Metallothionein in Biology and Medicine, eds by C. D. Klaassen and K. T. Suzuki, CRC Press, Boca Raton, FL.
Neihoster, M., Inoue, M., and Wendel, A., (1992) Biochem. Pharmacol. 43, 1151-1154.
Oh, S. H., Deagen, J. T., Whanger, P. D., and Weswig, P. H. (1978) Am. J. Physiol. 234, E282-E285.
Onosaka, S., and Cherian, M. G. (1981) Toxicology. 22, 91-101.
Remick, D. G., Steieter, R. M., Lynch III, J. P, Nguyen D., Eskandan, M., and Kunkel, S L. (1989) Lab. Invest. 60 766-771.
Sato, M. (1991) Toxicol. Appl. Pharmacol. 107, 98-105.
Sato, M., and Bremner, I. (1993) Free Rad. Biol. Med., in press
Sato, M., Mehra, R. K., and Bremner, I. (1984) J. Nutr. 114, 1683-1689.
Sato, M., and Sasaki, M. (1991) Chem.-Biol. Interact. 78, 143-154.
Sato, M., Sasaki, M.,and Hojo, H. (1992) Res. Commun. Chem. Path. Pharmacol. 72, 353-362.
Schroeder, J. J., and Cousins, R. J. (1990) Proc. Natl. Acad. Sci. USA. 87, 3137-3141.
Shiraishi, N., Hayashi, H., Hiraki, Y., Aono, K., Itano, Y., Kosaka, F., Noji, S., and Taniguchi, S. (1989) Toxicol. Appl. Pharmacol. 98, 501-506.
Shiraishi, N., Yamamoto, h., Takeda, Y., Kondoh, S., Hayashi, H., Hashimoto, K., and Aono, K. (1986) Toxicol. Appl. Pharmacol. 85, 128-134.
Swaina, W, R., and Feders, M. B. (1967) Clin. Chem. 13, 1026-1028.
Thornalley, P. J., and Vasak, M. (1985) Biochim. Biophys. Acta 827, 36-44.
Tsujikawa, K., Imai, T., Mimura, T., Otaki, N., and Kimura, M. (1991) pp. 339-346. in Metallothionein in Biology and Medicine ed by C.D. Klaassen and K.T. Suzuki, CRC Press, Boca Raton. FL.
Uchiyama, M., and Mihara, M. (1978) Anal Biochem. 86, 271-278.
Wong, G. H. W.; Goeddel, V. (1988) Science. 242, 941-944.

Metallothionein III
K. T. Suzuki, N. Imura and M. Kimura (eds)

RESPONSE OF INVERTEBRATE METALLOTHIONEINS AND MT GENES TO METALS AND IMPLICATIONS FOR ENVIRONMENTAL TOXICOLOGY

G. Roesijadi

University of Maryland, Chesapeake Biological Laboratory, Box 38, Solomons, MD 20688, USA

SUMMARY:

Metallothioneins (MTs) have been identified in at least 45 invertebrate species from six phyla. Those of echinoderms, arthropods, molluscs, and nematodes have been assigned to either the class I or class II MTs. Invertebrates have served as models for the developmental regulation, evolution, and cellular function of MTs. However, the primary emphasis in most studies on invertebrate MTs has been the toxicological implications of MT induction and metal-binding. As in higher animals, MTs in invertebrates are induced by cadmium, copper, mercury, and zinc. Induction by metals is related to detoxification and protection against metal toxicity. Increased levels of MTs occur in organisms in metal-contaminated environments. Recent work implicating MT in key reactions associated with regulating the expression of genes suggests that examination of MT in rapidly differentiating and growing systems such as early developmental stages will be instructive in understanding both the normal and toxicological function of MT.

INTRODUCTION

Invertebrates occupy key positions in the ecological structure of natural environments and serve as food resources for human consumers. They have been of interest as ecologically-sensitive targets for metal toxicity and vectors for metal contamination of humans through the food chain. Invertebrates have played a major role in efforts to understand the environmental toxicology of metals. The potential use of metallothioneins (MTs) in asessing environmental metal exposure, a possibility that has been of interest for some time, has highly influenced research on invertebrate MTs (Brown et al., 1977; Engel and Roesijadi, 1987).

Table I. Invertebrate groups for which metallothioneins have been reported. Modified from Roesijadi and Fowler (1991) and Roesijadi (1992).

Phylum	Common Name	No. of Species	Environment[1]
Echinoderms	sea urchin, starfish	3	M
Pogonophora	hydrothermal vent worm	1	M
Arthropods	insect, crustacean	13	M,FW,T
Molluscs	bivalve; gastropod; chiton	22	M,FW,T
Annelids	oligochaete; polychaete	7	M,FW,T
Nematodes	soil nematode	1	T

[1]M, marine; FW, freshwater; T, terrestrial

Invertebrates also serve as models for probing fundamental biological processes. The study of invertebrates has been useful in topics related to molecular evolution, developmental expression, and metallobiochemical interactions of MTs.

OCCURRENCE OF INVERTEBRATE MTs

The occurrence of MT in an invertebrate was first reported for the oyster *Crassostrea virginica* (Casterline and Yip, 1975) and has since been extended to at least 45 species that represent molluscs, echinoderms, arthropods, annelids, pogonophora, and nematodes (Table I). These species occur in diverse marine, freshwater, and terrestrial environments. Arthropods and molluscs have received the greatest attention.

CHARACTERIZATION OF INVERTEBRATE MTs

The structure of the model equine renal MT is highly conserved in the vertebrates. The positions of the cysteines are invariant, and minor changes at other positions distinguish

Table II. Amino acid sequences[1,2] of invertebrate MTs. Those of horse kidney included for comparison. Gaps are introduced to optimize fit for MTs within a single phylum.

	10 20 30 40 50 60 70
CLASS I	| | | | | | |
horse MT-1A	MDPNCSCPTGGSCTCAGSCKCKECRCTSCKKSCCSCCPGGCARCAQGCVCKGASDKCSCCA
<u>Molluscs</u>	
oyster[3]	SDPCNCIETGTCACSDSCPATGCKCGPGCKC-GDDCKCAGCKVKCSCTSEGGCKCGEKCTGPATCKCGSGCSCKK
mussel MT10-1	PAPCNCIETNUCICDTGCSGEGCRCGDACKCSGADCKCSGCKVVCKCS--GRCECGKGCTGPSTCKCAPGCSCK
mussel MT20-1	PGPCNCIETNUCICGTGCSGKCCRCGDACKC-ASGCGCSGCKVVCKCSGTCKCGCD--CTGPTNCKCESGCSCK
snail MT	xXXXXXXXKCTSACRSEPCQCGSKCQC-GEGCTCAACK-TCNCTSDG-CKCGKGCTGPDSCKCGSSCSCK
<u>Arthropods</u>	
crab MT1	PGPCCNDKCVCKEGGCKEGCQCTSCRCSPCEKCSSGCKCANKEECSKTCSKACSCCPT
crab MT2	PDPCCNDKCDCKEGECKTGCKCTSCRCPPCEQCSSGCKCANKEDCRKTCSKPCSCC
lobster MT	PGPCCKDKCECAEGGCKTGCKCTSCRCAPCEKCTSGCKCPSKDECAKTCSKPCXCCXX
Drosophila Mtn[1]	MPCPCGSGCKCASQATKGSCNCGSDCKCGGDKKSACGCSE
Drosophila Mto	MVCKGCGTNCQCSAQKCGDNCACNKDCQCVCKNGPKDQCCSNK
CLASS II	
<u>Echinoderm</u>	
sea urchin MTB	MPDVKCVCCTEGKECACFGQDCCVTGECCKDGTCCGICTNAACKCANGCKCGSGCSCTEGNCAC
sea urchin MTA	MPDVKCVCCKEGKECACFGQDCCKTGECCKDGTCCGICTNAACKCANGCKCGSGCSCTEGNCAC
<u>Nematode</u>	
nematode CeMT1	ACKCDCKNKQCKC--GDK-CECSGDKCCEKYCCEEASEKKCCPAGCKGDCKCANCHCAEQKQCGDKTHQHQGTAAAH
nematode CeMT2	VCKCDCKNQNCSCNTGTKDCDCSDAKCCEQYCCPTASEKKCCKSGCAGGCKCANCECAQ---------------AAH

[1]Citations: horse (Kagi and Kojima, 1987); oyster (Roesijadi et al., 1989; Unger et al., 1991); mussel (Mackay et al., 1990); slug (Dallinger, in press); crab (Lerch et al., 1982); lobster (Brouwer et al., 1989); Drosophila (Lastowski-Perry et al., 1985; Mokdad et al., 1987); sea urchin (Nemer et al., 1985; Wilkinson and Nemer, 1987); (Imagawa et al., 1990; Slice et al., 1990)

[2]"X", unidentified amino acid; "x", unidentified N-terminal block; "-", gap.

[3]N-acetylated and nonacetylated forms of the protein exist in the oyster. Methionine is the coded N-terminal amino acid.

one MT from another. The invertebrate MTs exhibit much greater diversity (Table II).

Like their vertebrate counterparts, the invertebrate MTs are cysteine-rich, lack aromatic amino acid residues, and are induced by and bind metals. Invertebrate MTs are assigned to class I and class II MTs (nomenclature of Fowler et al., 1987) on the basis of alignment of cysteinyl residues and the presence of characteristic CXC and CXXC and CC motifs (Table II). They can exhibit considerable differences from the equine model and from each other. MTs in species within a single phylum are usually very similar in primary structure (Table II).

In the crab *Scylla serrata*, the MTs are organized into metal clusters similar to that in the mammalian MTs (Otvos et al.,

1982). However, the crab MTs have two 3-metal clusters, rather than the one 3-metal and one 4-metal clusters of mammalian MTs (Otvos et al., 1982). This difference is suggestive of differences in cellular function between the crab and mammalian MTs since the 3-metal cluster possesses a lower affinity for metals than the 4-metal cluster. In the crab *Cancer pagurus*, the affinity of cadmium for MT was shown to be considerably lower than that for rabbit liver MT (Higham et al., 1986).

The molluscan and equine model MTs exhibit a relatively high degree of similarity in the N-terminal half of the proteins, particularly with regard to placement of the cysteines (Roesijadi et al., 1989; Unger et al., 1991). This half of the molluscan MT may bind metals in a 3-metal-cluster configuration (Roesijadi et al., 1989). The resemblance in the C-terminal half is not so strong, however. In the sea urchin and nematode MTs, the primary structures are sufficiently different from the equine form that they have been classified as class II MTs (Kägi and Kojima, 1987; Imagawa et al., 1990).

Multiple forms of MT frequently appear in invertebrate species. For example, crustacean MTs exist as two (Lerch et al., 1982), three (Brouwer and Brouwer-Hoexum, 1991) or four (Brouwer et al., 1992) isoforms, *D. melanogaster* possesses two (Lastowski-Perry et al., 1985; Mokdad et al., 1987), mussels possess seven (Mackay et al., 1990), the earthworm *Eisenia foedita* possesses up to four (Suzuki et al., 1980), a nematode possesses two (Imagawa et al., 1990), and a sea urchin possesses at least three (Nemer et al., 1991).

There is evidence for the differential function of the invertebrate MT isoforms. The induction of crustacean MTs by copper and their involvement in detoxification or copper donation to hemocyanin appears to be isoform-specific (Brouwer and Brouwer-Hoexum, 1991). In the echinoderm, expression of the different isoforms is tissue-specific during embryogenesis and larval development and may be related to changing functional requirements for metals (Wilkinson and Nemer, 1987; Nemer et al., 1991). In the oyster *C. virginica*, what appears to be a single transcript (Unger et al., 1991) is responsible for the synthesis of two 10 kDa forms that differ only in N-acetylation status (Roesijadi et al., 1991; Unger et al., 1991). The acetylated and nonacetylated forms exhibit different time-dependent patterns of induction by cadmium (Roesijadi et al., 1991), with the acetylated form being induced sooner.

Two molecular-mass variants of 10 and 20 kDa are known to occur in molluscs (George and Pirie, 1979; Fowler et al., 1986; Roesijadi, 1986; Langston and Zhou, 1987). They have been reported for the mussel *Mytilus edulis*, snail *Littorina littorea*, and oyster *C. virginica*. They have also been reported for the earthworm *Lumbricus variegatus* (Bauer-Hilty et al., 1989). In the oyster the distribution of the variants may be population-dependent with those from more northern latitudes of the eastern United States expressing only the 10 kDa MT and those from the southern United States expressing both 10 and 20 kDa forms (Frazier, 1979; Engel and Brouwer, 1982). A possible structural basis for the natural existence of the two forms has been identified in the mussel (Mackay et al., 1990). Both can be purified to 10 kDa monomeric units. However, the subunits of the 20 kDa proteins possess an additional two cysteine residues that are proposed to serve as bridging ligands for dimer formation. Cadmium-binding in the snail *L. littorea* is initially to the 20 kDa form (Langston and Zhou, 1987). The 10 kDa MT is induced after the binding of cadmium to the 20 kDa form is maximized. Conversely, disappearance of cadmium from the 10 kDa protein following cessation of exposure is related to apparent redistribution of cadmium from the 10 to the 20 kDa form. These observations suggest a coordinated expression of the two proteins in response to cadmium.

INVERTEBRATES AS MODELS FOR MT FUNCTION

MTs are considered to function in the cellular regulation of metals. They are involved in activating metalloenzymes such as carbonic anhydrase (Udom and Brady, 1980), hemocyanin (Brouwer et al., 1989; Brouwer and Brouwer-Hoexum, 1991; Brouwer and Brouwer-Hoexum, 1992), and pyridoxal kinase (Churchich et al., 1989) *in vitro*. The metal-free apothionein can modulate the activity of zinc-finger transcriptional factors by competing for zinc, thereby altering their ability to bind cis-acting, regulatory elements in genes (Zeng et al., 1991a,b). To date, however, the direct evidence for the involvement of metallothionein in normal biological processes apart from response to metal exposure has been limited. The study of invertebrates has been instructive in deducing the function and behavior of MTs and MT genes.

Crustaceans and MT-Metalloprotein Interactions

Engel and Brouwer and coworkers (Engel, 1987; Engel and Brouwer, 1987; Brouwer et al., 1989; Brouwer and Brouwer-Hoexum, 1991; Brouwer et al., 1992) examined the behavior of MT and hemocyanin, the copper-binding oxygen carrier, as part of efforts to delineate the role of MT in regulating copper and zinc during the crustacean molt cycle. Molting in crustaceans is characterized by a cycle of degradation and resynthesis of hemocyanin, which appears to be synchronized with a cycle of increased then decreased binding of copper to MT (Engel and Brouwer, 1987). A direct linkage of the two events was hypothesized (Engel and Brouwer, 1989) whereby copper released during hemocyanin degradation is partly sequestered by MT, and copper required for activation of newly-synthesized hemocyanin is donated by MT. For activation of hemocyanin, Cu(I) must be bound at the oxygen-binding site. *In vitro*, the copper bound to specific MT isoforms can be transferred to apohemocyanin via a glutathione-mediated mechanism (Brouwer and Brouwer-Hoexum, 1991; 1992), resulting in activation to hemocyanin. This mechanism of glutathione-mediated transfer of Cu(I) from MT to apohemocyanin may be fundamental to copper transfer and useful as a model for the study of copper-activated systems (Brouwer and Brouwer-Hoexum, 1992). In accordance with the hypothesis noted above, MT expression and synthesis should be regulated as part of the overall regulation of the molting process, which is under the control of steroid molting hormones. Glucocorticoid responsive elements in the vertebrate MT gene (Hamer, 1986) exist as a precedent for the discovery of analogous steroid-sensitive elements in other species.

Echinoderms and the Developmental Regulation of MTs

Zinc is required by DNA and RNA polymerases (Mildvan and Loeb, 1979) and numerous transcription factors (Miller et al., 1985; Berg, 1990) and is essential for growth and development (Rubin, 1972). Reports that apothionein can remove zinc from zinc-finger proteins implicate MT in regulating transcription (Zeng et al., 1991a,b). Such a function would be expected to be particularly significant in rapidly differentiating and growing systems. Thus, the involvement of MT in normal biological function may be heightened in early development. The sea urchin has been useful in deducing general principles related to the developmental expression of MT.

ZnMT and MT mRNA of maternal origin are present in the unfertilized sea urchin egg (Ohtake et al., 1983; Nemer et al., 1984). The MT may be involved in regulating zinc ion availability. Zinc in the incubation medium can prolong the life of unfertilized eggs in seawater (Nakamura et al., 1991) and inhibit formation of the fertilization membrane (Nakamura et al., 1989).

The maternal MT mRNA in the egg is poly(A)-deficient and follows a program for polyadenylation immediately following fertilization (Nemer et al., 1984). This mRNA is presumed to be responsible for the basal-level synthesis that occurs in early embryogenesis. Between the 7th and 8th cleavage, MT mRNA accumulates as a result of increased transcription and can be induced by zinc.

Basal levels of MT mRNA are greater in the ectoderm than in endo/mesoderm (Nemer et al., 1984; Wilkinson and Nemer, 1987; Nemer et al., 1991) Thus, the blastula, which is comprised of ectodermal cells, has a higher level of basal MT expression than the gastrula. The latter is characterized by proliferation of cells of endomesodermal origin. Induction by zinc is greater in these cells.

The patterns of synthesis and induction by zinc exhibit temporal- and tissue-specificity for individual MT isoforms. MT mRNA in the egg and basal expression in the embryo and larva are those of MTA (Wilkinson and Nemer, 1987). Exposure to zinc results in induction of both MTA and members of an MTB gene family (Wilkinson and Nemer, 1987; Nemer et al., 1991). The former is more sensitive to metals and is induced at much lower concentrations of metal than MTB. Precise localization of MT mRNA in the pluteus larva has identified MTA in cells of the aboral ectoderm (Nemer et al., 1991). Little is found in the oral ectoderm or gut. The MTB family of mRNAs are uniformly expressed in the pluteus, with one isoform MTB1 restricted to the oral ectoderm and gut (Nemer et al., 1991).

It appears from these specific patterns of localization and expression that MTs are regulated in response to cellular requirements for metals during development. Concentrations of zinc in seawater are generally very low (in the low nanomolar range); thus, it would seem unlikely that endogenous patterns of MT mRNA synthesis are responding to protect against toxicity exerted by external zinc concentrations. The differential expression of MT during development may reflect changing requirements for the metal-free apothionein as a mechanism for modulating the activity of zinc transcription factors and, thus, gene expression (Zeng et al., 1991b).

The pattern of expression of MT during sea urchin development is paralleled in the mouse (Andrews et al., 1991) and supports use of the sea urchin as a model for the developmental expression of MTs. The presence of maternal MT mRNAs in the ovum, lack of metal-responsiveness of MT mRNA transcription during the early cleavages; and induction by metals later in embryonic development in the mouse are also seen in the sea urchin. Both MT I and II isoforms are present in the ovum and embryo of the mouse and appear to be coordinately expressed following fertilization. As in the sea urchin, MT gene expression in the mouse is proposed to serve a vital function during embryogenesis (Andrews et al., 1991).

Drosophila and the Evolution of MTs

Two MT genes that appear to be distantly related, Mtn (Lastowski-Perry et al., 1985) and Mto (Mokdad et al., 1987), exist in *D. melanogaster*. Although they exhibit only 26% identity in amino acid sequence, the two are considered to be homologous (Mokdad et al., 1987; Laurent et al., 1991).

Duplications of Mtn are seen in natural populations of *D. melanogaster* and are associated with enhanced MT mRNA transcription and copper resistance (Maroni et al., 1987). These duplications are considered to be recent and the result of increased industrialization and release of metals to the environment. Mto duplications have not been observed in intact individuals (Lange et al., 1990; Laurent et al., 1991), although they may occur in cell culture (Mokdad et al., 1987).

Mtn duplications have been reported in individuals from Asia (Japan), Australia, Europe (France), North America (U.S.), and North Africa (Egypt) (Maroni et al., 1987). They were not observed in individuals from tropical Africa (Botswana, Congo, Zambia) (Maroni et al., 1987; Laurent et al., 1991). Recent work has now attributed the Mtn duplications to a single allelic variant designated Mtn^{1} (Laurent et al., 1991). This allele is absent in tropial African flies. A second allele Mtn$^{\cdot 3}$, which appears not to undergo duplication, was observed in the African flies. Mtn^{1} differs from Mtn$^{\cdot 3}$ by four single-base substitutions, one in the coding region that changes the C-terminal glu in Mtn^{1} to lys in Mtn$^{\cdot 3}$ and three in the promoter or 3'-untranslated regions. In seven other species of *Drosophila* from various geographically-disperse locales world-wide (Laurent et al., 1991), five typed positive for Mtn$^{\cdot 3}$ but not Mtn^{1} (two species being uninformative). In one species examined in greater detail (*D. simulans*), 32 tested lines exhibited restriction patterns and an insert in the 3'-noncoding region characteristic for Mtn·3. Mtn·3 appears to be the more primitive of the two, being in common with tropical African species of *D. melanogaster* and other species of *Drosophila*.

A scenario proposed for the current status of the Mtn locus in *Drosophila* describes an African origin for the Mtn1 ancestral gene (Laurent et al., 1991), presumably one similar to Mtn·3. Following radiation of *D. melanogaster* out of Africa, duplication of Mtn·3 is believed to have resulted in Mtn1. This was considered to have occurred relatively recently and related to a specialized function of Mtn1 in metal detoxification in response to increased agricultural and industrial contamination. This hypothesis has broad relevance with respect to organisms in the natural environment because of the widespread contamination of both terrestrial and aquatic environments by metals. Increased involvement of MTs in binding metals has been reported in metal-resistant natural populations of other invertebrate species (Johansson et al., 1986; Klerks and Levinton, 1989). The presence of resistant populations is considered a potentially-useful indicator of contamination severe enough to cause ecologically-significant changes (Luoma, 1977). MT gene duplications in natural populations may have practical significance with respect to the long-term implications of the environmental toxicology of metals and novel approaches for assessing the environmental consequences of metal contamination.

TOXICOLOGY OF INVERTEBRATE MTs

Metal contamination of the environment and the role of invertebrates as potential target species have served as the primary context for many studies of invertebrate MTs (Roesijadi, 1992). Invertebrates are capable of accumulating metals to high concentrations (Zaroogian and Cheer, 1972), and, for those that are consumed by humans (e.g., shellfish such as bivalve molluscs and crustaceans), there is interest in the chemical speciation and bioavailability of the accumulated metals.

Induction of Invertebrate MTs by Experimental Metal Exposure

The model for regulation of MT transcription through interaction of metal-binding transcriptional factors (MRF) and metal responsive elements (MRE) in the promoter region of the MT gene (Thiele, 1992) applies to invertebrates. In this model, a metal-induced conformational change in the MRF enables it to bind to the appropriate element and initiate transcription. In the sea urchin and *Drosophila*, nucleotide sequences similar to MREs in mammals have been described (Otto et al., 1987; Harlow et al., 1989). A *Drosophila* Mtn promoter and *Herpes simplex* thymidine kinase reporter fusion gene transfected into baby hamster kidney cells can be induced by zinc (Otto et al., 1987). This demonstrates responsiveness of the Mtn gene to mammalian MT regulatory factors. Based on these observations, the general mechanism of MT induction for most invertebrates will likely be similar to that proposed in other systems (Thiele, 1992).

In the invertebrates, MT induction in response to metal exposure has most often been deduced by measuring the amount of bound metal, protein, and, less frequently due to the current lack of appropriate hybridization probes, the responsible mRNA. Cadmium (George et al., 1979; Olafson et al., 1979), copper (Viarengo et al., 1980), mercury (Roesijadi, 1982), and zinc (Nemer et al., 1984) can induce invertebrate MTs. Interestingly, although zinc is bound by MTs of numerous species, it may not be as strong an inducer of MTs in many adult invertebrates in comparison with higher animals (Maroni and Watson, 1985; Roesijadi and Fellingham, 1987; Dallinger, in press).

Whereas, MT induction in animals such as rodents can be detected within a few hours of administration of metal, in invertebrates a lag of four to 12 days is not uncommon (Roesijadi, 1982; Young and Roesijadi, 1983). The basis for this difference may reside in factors related to temperature-dependence of either MT induction or, indirectly, rates of metal uptake and transport in poikilothermic animals. Body temperatures of 10 °C or lower are not uncommon for aquatic animals, and it is possible that the lag is due to the rate-limiting effects of low temperature. Work on fish hepatocytes (Hyllner et al., 1989) indicates an inverse correlation between time to induction and temperature. The amount of cadmium-binding to induced MT in a shrimp is positively correlated with temperature (Howard and Hacker, 1990). The lag in MT induction has toxicological implications because acute exposures may not enable individuals to respond in a timely fashion. Toxic effects may manifest themselves prior to induction of MT. This has practical significance due to the common use of the 96 hour LC-50 in determining metal toxicity. Routine toxicity tests with invertebrates can be confounded by varying times for the onset of MT induction.

Induction of MT is considered to spare other cellular structures from binding metals, providing protection against their toxicity. In the invertebrates, induction does not prevent such binding, however, and metal concentrations in nonthionein pools increase together with increases on MT (Sanders and Jenkins, 1984; Roesijadi and Klerks, 1989). Instead, MT induction appears to confer kinetic protection in that increased rates of metal binding to induced MT are related to decreased rates of binding to other structures (Roesijadi and Klerks, 1989). MT induction can reduce the rate of accumulation by nonthionein ligands. Metal ions initially bound to nonthionein ligands are redistributed to MT following induction (Roesijadi and Klerks, 1989; Roesijadi et al., 1991).

As in higher animals, prior induction of MTs by low level metal exposure in invertebrates confers protection against challenge by toxic concentrations (Roesijadi et al., 1982; Roesijadi and Fellingham, 1987; Aoki et al., 1989). In the case of a polychaete worm, regeneration of tissues previously damaged by exposure to copper was associated with apparent induction of MT (Young and Roesijadi, 1983). In the preexposure/challenge experiments, the basis for the increased resistance probably resides in formation of an enhanced pool of preexisting MT, which is available for metal-substitution reactions, or increased rates of MT transcription and protein synthesis. The tissue regeneration that occurred in the copper-exposed polychaete was probably related to a protective function of MT induction and its relationship to the onset of repair mechanisms.

In crustaceans (Brouwer and Brouwer-Hoexum, 1991; Schlenk and Brouwer, 1991; Brouwer et al., 1992c), fruit fly (Laurent et al., 1991), and the sea urchin (Nemer et al., 1991), specific forms of MT are preferentially-induced by metals and may play a greater role in the response to elevated environmental concentrations than others. In the oyster *C. virginica* in which MT occurs in N-acetylated and unacetylated forms, the acetylated MT responds more rapidly to metal exposure (Roesijadi et al., 1991) and is also dominant in oysters that inhabit a metal-contaminated environment (Roesijadi, G., unpublished data). A greater understanding of MT polymorphism will undoubtedly clarify the role of MTs in the toxicological response to elevated cellular concentrations of certain metals.

<u>MTs in Natural Populations</u>

Virtually all known metals are natural components of the environment and occur at concentrations that vary as a function of natural geochemical processes and anthropogenic activities. MTs have been detected in invertebrates from diverse natural habitats such as the deepsea hydrothermal vents, where metals concentrations are extremely high ((Roesijadi et al., 1985; Cosson-Mannevy et al., 1986), and environments considered to be relatively pristine by contemporary standards (Roesijadi et al., 1984), in addition to environments contaminated by human activities (Noël-Lambot et al., 1980; Engel and Brouwer, 1984; Roesijadi et al., 1984; Langston and Zhou, 1986; Janssen and Dallinger, 1991; Klerks and Bartholomew, 1991). The levels of MT or the MT-bound metals usually reflect the degree of contamination of the environment (Roesijadi, 1992). In one case, an apparently MT-deficient species of the mayfly is restricted in distribution and not found in metal-contaminated environments (Aoki et al., 1989), whereas a sibling species with the ability to express MTs is metal-resistant and found in such environments.

The bioavailability and toxicity of the various chemical forms of metals accumulated by invertebrates are problematic. The available information indicates that the toxicity to mammalian consumers may be of concern. For example, the chemical form of cadmium that occurs in the tissues of molluscs such as oysters and clams is reported to be more bioavailable and toxic than cadmium from other sources (Siewicki et al., 1983). Additionally, the binding of metals such as cadmium to MT tends to extend the biological half-life and increase the likelihood of food chain transfer. Additional research is needed on this topic.

MTs and Environmental Biomonitoring

An early incentive for the study of MTs in invertebrate species was to achieve a better understanding of organisms that inhabit metal-contaminated habitats for the purpose of assessing the magnitude of environmental degradation (Brown et al., 1977). This perspective has resulted in a large body of work that relates to the environmental toxicology of invertebrate MTs (Roesijadi, 1992). As a result of increased understanding of the behavior and regulation of MTs and improved analytical techniques, there has been renewed interest in the use of MT as an environmental biomarker for metal pollution (Stegeman et al., 1992). Continued research on MTs in invertebrate species is warranted due to the need for a better understanding of factors associated with metal-induced cell injury and organismal impairment in ecologically important species. In the environmental context, interest in the toxicological function of MTs extends beyond an understanding of specific cellular phenomena to significance at higher levels of biological organization, particularly with respect to organismal and population health. There is a need to demonstrate connectivity between events at cellular and more integrated levels of organization. Our current understanding of MT function and the relationship between MT induction and metal toxicity precludes facile interpretations regarding the functional significance of MT induction in individuals *in situ*. Thus, focused research that addresses these matters is needed (Engel and Roesijadi, 1987; Roesijadi, 1992; Stegeman et al., 1992). Because recent advances are implicating differentiating and growing systems as those in which MT function may have particular relevance, focus on early life history stages may prove to be particularly useful for understanding the toxicological relevance of MT function.

Acknowledgements

This study was supported in part by grants from the U.S. Geological Survey, Department of Interior (14-08-0001-G1900); State of Maryland, The Department of Natural Resources (CB91-001-004); and the U.S. Environmental Protection Agency (R819401-01).

REFERENCES

Andrews, G. K., Huet-Hudson, Y. M., Paria, B. C., McMaster, M. T., De, S. K. and Dey, S. K. (1991). Dev. Biol. 145, 13-27.

Aoki, Y., Hatakeyama, S., Kobayashi, N., Sumi, Y., Suzuki, T. and Suzuki, K. T. (1989). Comp. Biochem. Physiol. 93C, 345-357.

Bauer-Hilty, A., Dallinger, R. and Berger, B. (1989). Comp. Biochem. Physiol. 94C, 373-379.

Berg, J. M. (1990). Ann. Rev. Biophys. Biophys. Chem. 19, 405-421.

Brouwer, M. and Brouwer Hoexum, T. (1991). Arch. Biochem. Biophys. 290, 207-213.

Brouwer, M. and Brouwer-Hoexum, T. (1992). Biochemistry 31 , 4096-4102.

Brouwer, M., Schlenk, D., Ringwood, A. H. and Brouwer-Hoexum, T. (1992). *Arch. Biochem. Biophys.* 294, 461-468.

Brouwer, M., Winge, D. R. and Gray, W. R. (1989). J. Inorg. Biochem. 35, 289-303.

Brown, D. A., Bawden, C. A., Chatel, K. W. and Parsons, T. R. (1977). Environ. Conserv. 4, 213-216.

Casterline, J. L. and Yip, G. (1975). Arch Environ. Contam. Toxicol. 3, 319-329.

Churchich, J. E., Scholz, G. and Kwok, F. (1989). Bioch. Biophys. Acta 996, 181-186.
Cosson-Mannevy, M.-A., Cosson, R. and Gaill, F. (1986). C. R. Acad. Sc. Paris 302, Ser III, 347-352.
Dallinger, R. (in press 1992). In: "Ecotoxicology of Metals in Invertebrates", R. Dallinger and P. S. Rainbow eds., Lewis Publishers, Boca Raton, Florida.
Engel, D. W. (1987). Biol. Bull. 172, 69-82.
Engel, D. W. and Brouwer, M. (1982). In: "Physiological Mechanisms of Marine Pollutant Toxicity", W. B. Vernberg Calabrese, A., Thurberg, F.P., Vernberg, F.J. eds., pp. 89-107, Academic Press, New York.
Engel, D. W. and Brouwer, M. (1984). Mar. Environ. Res. 14, 139-151.
Engel, D. W. and Brouwer, M. (1987). Biol. Bull. 173, 239-251.
Engel, D. W. and Brouwer, M. (1989). Adv. Comp. Environ. Physiol. 5, 54-75.
Engel, D. W. and Roesijadi, G. (1987). In: "Pollution Physiology of Estuarine Organisms", A. C. W.B. Vernberg F.P. Thurberg, & F.J. Vernberg eds., pp. 421-438, University of South Carolina Press, Columbia.
Fowler, B. A., Engel, D. W. and Brouwer, M. (1986). Environ. Hlth Perspect. 65, 63-69.
Fowler, B.A., Hildebrand, C.E., Kojima, Y. and Webb, M. (1987 In: "Metallothionein II", J.H.R. Kägi and Y. Kojima eds., pp. 19-22, Birkhauser-Verlag, Basel.
Frazier, J. M. (1979). Environ. Hlth Perspect. 28, 75-79.
George, S. G., Carpene, E., Coombs, T. L., Overnell, J. and Youngson, A. (1979). Biochim. Biophys. Acta 580, 225-233.
George, S. G. and Pirie, B. J. S. (1979). Biochim. Biophys. Acta 580, 234-244.
Hamer, D. H. (1986). Ann. Rev. Biochem. 55, 913-951.
Harlow, P., Watkins, E., Thornton, R. D. and Nemer, M. (1989). Molec. Cell Biol. 9, 5445-5455.
Higham, D. P., Nicholson, J. K., Overnell, J. and Sadler, P. J. (1986). Environ. Hlth Perspect. 65, 157-165.
Howard, C. L. and Hacker, C. S. (1990). Arch. Environ. Contam. Toxicol. 19, 341-347.
Hyllner, S. J., Andersson, T., Haux, C. and Olsson, P.-E. (1989). J. Cell. Physiol. 139, 24-38.
Imagawa, M., Onozawa, T., Okumura, K., Osada, S., Nishimura, T. and Kondo, M. (1990). Biochem. J. 268, 237-240.
Johansson, C., Cain, D. J. and Luoma, S. N. (1986). Mar. Ecol. Prog. Ser 28, 87-97.
Janssen, H.H. and Dallinger, R. (1991). Arch. Environ. Contam. Toxicol. 20, 132-137.
Kägi, J.H.R. and Kojima, Y. (1987). In: "Metallothionein II", J.H.R. Kägi and Y. Kojima eds., pp. 25-61, Birkhauser-Verlag, Basel.
Klerks, P. L. and Bartholomew, P. R. (1991). Aquat. Toxicol. 19, 97-112.

Klerks, P. L. and Levinton, J. S. (1989). In: "Ecotoxicology: Problems and Approaches", M. A. H. S.A. Levin J.R. Kelly, and K.D. Kimball eds., pp. 41-67, Springer-Verlag, New York.
Lange, B. W., Langley, C. H. and Stephan, W. (1990). Genetics 126, 921-932.
Langston, W. J. and Zhou, M. (1986). Mar. Biol. 92, 505-515.
Langston, W. J. and Zhou, M. (1987). J. Mar. Biol. Ass. UK 67, 585-601.
Lastowski-Perry, D., Otto, E. and Maroni, G. (1985). J. Biol. Chem. 260, 1527-1530.
Laurent, T., Ho, A.-S. and Maroni, G. (1991). Genet. Res., Camb. 58, 203-210.
Lerch, K., Ammer, D. and Olafson, R. W. (1982). J. Biol. Chem. 257, 2420-2426.
Luoma, S. N. (1977). J. Fish. Res. Board Can. 34, 436-439.
Mackay, E. A., Dunbar, B., Davidson, I., Fothergill, J. E., Overnell, J., Hunziker, P. E. and Kägi, J. H. R. (1990). Experientia 46, A36 (Abstract No. 310).
Maroni, G., J.Wise, Young, J. E. and Otto, E. (1987). Genetics 117, 739-744.
Maroni, G. and Watson, D. (1985). Insect Biochem. 15, 55-63.
Mildvan, A. S. and Loeb, L. A. (1979). CRC Critical Rev. Biochem. 6, 219-244.
Miller, J., McLachlan, A. D. and Klug, A. (1985). EMBO J. 4, 1609-1614.
Mokdad, R., Debec, A. and Wegnez, M. (1987). Proc. Natl. Acad. Sci. USA 84, 2658-2662.
Nakamura, S., Oda, S., Nakamura, K., Kagotani, R. and Kojima, M. K. (1991). Experientia 47, 734-736.
Nakamura, S., Ohmi, C. and Kojima, M. K. (1989). Zool. Sci. 6, 329-333.
Nemer, M., Thornton, R. D., Stuebing, E. W. and Harlow, P. (1991). J. Biol. Chem. 266, 6586-6593.
Nemer, M., Travaglini, E. C., Rondinelli, E. and D'Alonzo, J. (1984). Dev. Biol. 102, 471-482.
Nemer, M., Wilkinson, D. G., Travaglini, E. C., Sternberg, E. J. and Butt, T. R. (1985). Proc. Natl. Acad. Sci. USA 82, 4992-4994.
Noël-Lambot, F., Bouquegneau, J. M., Frankenne, F. and Disteche, A. (1980). Mar. Ecol. Prog. Ser. 2, 81-89.
Ohtake, H., Suyemitsu, T. and Koga, M. (1983). Biochem. J. 211, 109-118.
Olafson, R. W., Sim, R. G. and Boto, K. G. (1979). Comp. Biochem. Physiol. 62B, 407-416.
Otto, E., Allen, J. M., Young, J. E., Palmiter, R. D. and Maroni, G. (1987). Mol. Cell. Biol. 7, 1710-1715.
Otvos, J. D., Olafson, R. W. and Armitage, I. M. (1982). J. Biol. Chem. 257, 2427-2431.
Roesijadi, G. (1982). Mar. Biol. 66, 151-157.
Roesijadi, G. (1986). Environ. Hlth Perspect. 65, 45-48.
Roesijadi, G. (1992). Aquat. Toxicol. 22, 81-114.

Roesijadi, G. and Fellingham, G. W. (1987). Can. J. Fish. Aquat. Sci. 44, 680-684.

Roesijadi, G. and Fowler, B. (1991). In: "Methods in Enzymology vol. 205, Metallobiochemistry, Part B: Metallothionein and Related Molecules", J. F. Riordan and. B. L. Vallee eds., pp. 263-273, Academic Press., San Diego.

Roesijadi, G. and Klerks, P. (1989). J. Exp. Zool. 251, 1-12.

Roesijadi, G., Drum, A. S., Thomas, J. M. and Fellingham, G. W. (1982). Mar. Pollut. Bull. 13, 250-253.

Roesijadi, G., Kielland, S. L. and Klerks, P. (1989). Arch. Bioch. Biophys. 273, 403-413.

Roesijadi, G., Vestling, M. M., Murphy, C. M., Klerks, P. L. and Fenselau, C. (1991). Biochim. Biophys. Acta 1074, 230-236.

Roesijadi, G., Young, J. S., Crecelius, E. A. and Thomas, L. E. (1985). Biol. Soc. Wash. Bull. 6, 311-324.

Roesijadi, G., Young, J. S., Drum, A. S. and Gurtisen, J. M. (1984). Mar. Ecol. Prog. Ser. 18, 155-170.

Rubin, H. (1972). Proc. Natl. Acad. Sci. U.S.A. 69, 712-716.

Sanders, B. M. and Jenkins, K. D. (1984). Biol. Bull. 167, 704-712.

Schlenk, D. and Brouwer, M. (1991). Aquat. Toxicol. 20, 25-34.

Siewicki, T. C., Balthrop, J. E. and Sydlowski, J. S. (1983). J. Nutr. 113, 1140-1149.

Slice, L. W., Freedman, J. H. and Rubin, C. S. (1990). J. Biol. Chem. 265, 256-263.

Stegeman, J. J., Brouwer, M., Di Giulio, R. T., Förlin, L., Fowler, B. A., Sanders, B. M. and Van Veld, P. A. (1992). In: "Biomarkers: Biochemical, Physiological, and Histological Markers of Anthropogenic Stress", R. J. Huggett, R. A. Kimerle, P. M. Mehrle and H. L. Bergman eds., pp. 235-335, Lewis Publishers, Chelsea, MI.

Suzuki, K.T., Yamamura, M. and Tadahiro, M. (1980). Arch. Envir. Contam. Toxicol. 9, 415-424.

Thiele, D. J. (1992). Nucleic Acids Res. 20, 1183-1191.

Udom, U. O. and Brady, F. O. (1980). Biochem. J. 187, 329-335.

Unger, M. E., Chen, T. T., Fenselau, C. C., Murphy, C. M., Vestling, M. M. and Roesijadi, G. (1991). Biochim. Biophys. Acta 1074, 371-377.

Viarengo, A., Pertica, M., Mancinelli, M., Zanicchi, G. and Orunesu, M. (1980). Comp. Biochem. Physiol. 67C , 215 218.

Wilkinson, D. G. and Nemer, M. (1987). Molec. Cell. Biol. 7, 48-58.

Young, J. S. and Roesijadi, G. (1983). Mar. Pollut. Bull 14, 30-32.

Zaroogian, G. E. and Cheer, S. (1972). Nature 261, 408-410.

Zeng, J., Heuchel, R., Schaffner, W. and Kägi, J. H. R. (1991a). FEBS Lett. 279, 310-312.

Zeng, J., Vallee, B. L. and Kägi, J. H. R. (1991b). Proc. Natl. Acad. Sci. USA 88, 9984-9988.

Metallothionein III
K. T. Suzuki, N. Imura and M. Kimura (eds)

INDIRECT INDUCTION OF METALLOTHIONEIN BY ORGANIC COMPOUNDS

Kyong-son Min[1], Norio Itoh[2], Hiroshi Okamoto[2], and Keiichi Tanaka

[1]Faculty of Nutrition and [2]Faculty of Pharmaceutical Sciences, Kobe-Gakuin University, Kobe 651-21 and Faculty of Pharmaceutical Sciences, Osaka University, 1-6, Senrioka, Suita, Osaka 565, Japan

SUMMARY: Various organic compounds indirectly induces the synthesis of hepatic metallothionein (MT) which contains zinc (Zn) only. We investigated whether *in vivo* hepatic MT induction by these compounds is mediated by either glucocorticoid, a Zn-associated process, radical production, glutathione metabolism or through an inflammatory response. It was shown that induction of MT synthesis by at least certain organic compounds is mediated by a cytokine, probably IL-6, as an acute-phase protein in inflammation.

INTRODUCTION

Metallothionein (MT) is known to be induced directly by heavy metals such as Zn and Cd (Searle, 1987), and glucocorticoid (Mayo and Palmiter, 1981). But intraperitoneal injection of other metals like iron (Fe) (McCormick, 1984; Petro and Hill, 1987) caused induction of hepatic MT synthesis. Fleet et al. (1990a,b) reported that Fe does not affect the induction of MT directly and the induction of hepatic MT by the parenteral administration of several metals is associated with an inflammatory response. In addition to metals, endotoxin lipopolysaccharide (LPS) and turpentine oil (TUR), which initiate inflammatory responses, also cause induction of hepatic MT synthesis (Sobocinski and Canterbury, 1982). Besides these inflammatory reagents, a number of nonmetallic compounds i.e., radical inducing compounds (Cagen and Klaassen, 1979; Wormser and Calp, 1988; Sato et al., 1989), organic solvents (Onosaka et al., 1988), alkylating agents (Kotsonis and Klaassen, 1979), chelating agents (Goering et al., 1985), alcohols (Waalkes et al., 1984; Swerdel and Cousins, 1984) and carbamate compounds (Brzeznicka et al., 1987), also cause the accumulation of MT in the liver, but not in the kidney. These compounds appear to stimulate MT induction indirectly. However, an important question remains: why do a number of xenobiotics induce MT when they are neither metals that bind to MT nor glucocorticoids ?

Injection of LPS and TUR causes the release of some cytokines from macrophages, monocytes, and other cell types. Cytokines mediate the acute-phase response in inflammation. Recent works have demonstrated that the cytokines, interleukin(IL)-1 (Cousins and Leinart, 1988; De et al., 1990), IL-6 (Schroeder and Cousins, 1990; De et al., 1990), tumor necrosis factor (TNF) (De et al., 1990) and interferon (Friedman and Stark, 1985; De et al., 1990), induce MT synthesis in the liver. Attention has been focused on the cytokine IL-6 as a major mediator of acute-phase protein (APP) synthesis in hepatocytes in response to infection and tissue injury (Audus et al., 1988). However no reports have shown which type of cytokines mediate MT synthesis by nonmetallic compounds *in vivo*.

There would be other possibilities in a indirect mechanism of MT induction by nonmetallic compounds. Certain stress situations cause an increase in hepatic Zn concentrations with a reduction of plasma Zn. An increase in uptake of Zn by the liver might bring about a decrease in plasma Zn level (Disilvestro and Cousins, 1984; Fleet et al., 1990a,b). Some compounds which induce hepatic MT, such as Fe, markedly, cause a reduction in serum zinc accompanied by an accumulation of this metal in the liver (Sobocinski and Canterbury, 1982; Swerdel and Cousins, 1984). It is possible that hepatic MT is induced through a Zn-associated process since Zn is a direct inducer of MT synthesis.

Glucocorticoids are also one of the classical inducing agents for hepatic MT synthesis. Since the concentration of glucocorticoid in the plasma is increased in inflammation and some kinds of stresses, early reports suggested induction of hepatic MT accumulation by an injection of nonmetallic compounds and initiators for inflammation might be mediated by glucocorticoids (Karin et al., 1980; Brackman and Klaassen, 1987). Although hepatic MT levels induced by glucocorticoid are usually lower than those induced by inflammation or nonmetallic compounds (Min et al., 1992b), it is necessary to consider an induction mechanism possibly involving the adrenal gland.

On the other hand, it has been suggested that MT plays an antioxidative role, since MT has multiple sulfhydryl groups, in a fashion similar to glutathione (GSH). Studies *in vitro* have demonstrated that MT can scavenge free hydroxyl and superoxide radicals produced by the xanthine-xanthine oxidase reaction (Thornalley and Vasak, 1985) and inhibits lipid peroxidation (Thomas, et al., 1986). Oxidative stresses, such as radiation and high oxygen tention, cause induction of MT synthesis (Shiraishi et al., 1983, 1986; Hidalgo et al., 1988). These stresses also increase lipid peroxidation levels through free radical production (Hildalgo et al., 1988). An injection of a prooxidant, such as carbon tetrachloride (CCl_4), paraquat, and acetaminophen, induces MT synthesis (Oh et al., 1978; Sato et al., 1989; Wormser and

Calp, 1988). However, the mechanism of MT induction under these conditions and the role of MT as a free radical scavenger and/or inhibitor of lipid peroxidation have not yet been demonstrated *in vivo.*

Furthermore, glutathione (GSH) protects against free radicals or lipid peroxides in cultured cells (Dethmers and Meister, 1981; Arrick et al., 1982). Some compounds like menadione cause a marked decrease in intracellular thiols such as GSH and protein sulfhydryl groups because of radical production and its conjugation with GSH (Thor et al., 1982; Takahashi et al., 1987). If radicals are inducers of MT synthesis in oxidative stress, MT should be increased for the protection of cells against the radicals when the concentration of GSH is decreased.

Therefore, in the present study, we investigated whether hepatic MT induction by nonmetallic compounds *in vivo* is associated with such Zn-mediated process, glucocorticoid release, lipid peroxidation and radical production, GSH metabolism or an inflammatory response.

RESULTS AND DISCUSSION

Induction of hepatic MT accumulation by nonmetallic compounds:

The concentration of hepatic MT in mice 24 hr after an injection of several nonmetallic compounds is shown in Fig. 1. Each nonmetallic compound caused marked increase in hepatic MT accumulation. Likewise several inflammatory initiators increased hepatic MT as shown in Fig. 2. These MTs contained only Zn. The hepatic MT concentration in the menadione, n-hexane (HX), and n-heptane treated groups was increased by about 13 times above the control group. However, renal MT accumulation was not affected in any treated groups.

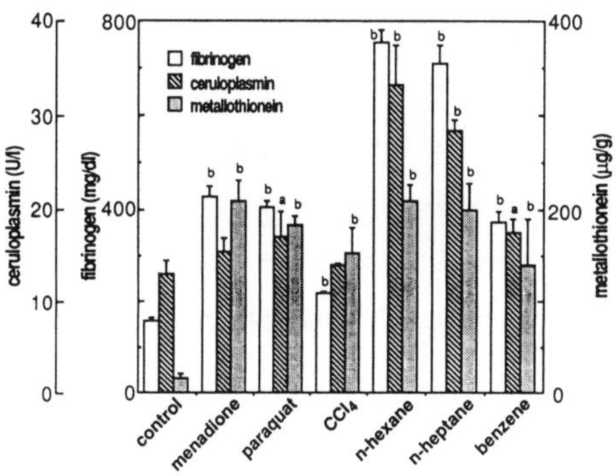

Fig. 1. Concentration of hepatic MT and acute-phase proteins in plasma 24 hr after injection of nonmetallic compounds.
ap<0.05, bp<0.01

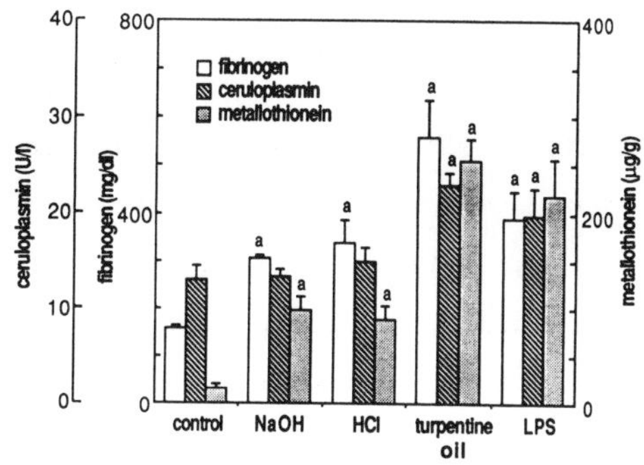

Fig. 2. Concentration of hepatic MT and acute-phase proteins in plasma 24 hr after injection of inflammatory initiators.
ap<0.05, bp<0.01

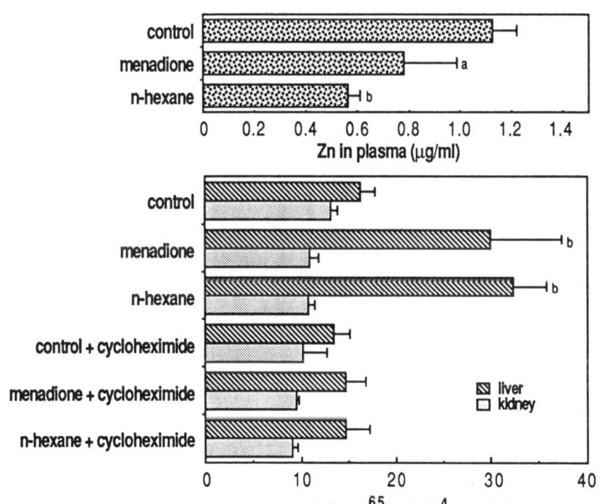

Fig. 3. Concentraion of Zinc in plasma, and uptake of 65Zn by liver and kidney 24 hr after injection of menadione (50 mg/kg) and n-hexane (10 ml/kg). ap<0.05, bp<0.01

Effect of nonmetallic compounds on Zn metabolism:

The concentration of Zn in plasma was decreased significantly 24 hr after an injection of menadione or HX, which caused the greatest increase in hepatic MT (Fig. 3). In contrast, hepatic Zn concentration was increased by these treatment (data not shown). Most of Zn increased following the injection was bound to MT in the liver. Uptake of ^{65}Zn and MT concentration were increased only in the liver not in the kidney. This significant increase by menadione and HX was not observed when cycloheximide was administered after these injections (Fig. 3). Disilvestro and Cousins (1984) reported cycloheximide fully blocked endotoxin-induced alterations in plasma Zn and hepatic Zn bound to MT. These results are consistent with the possibility that hepatic MT synthesis promotes the uptake of Zn by the liver, but not with the possibility that the uptake of Zn triggers induction of MT synthesis in this case.

Effect of adrenalectomy on the induction of hepatic MT by nonmetallic compounds:

The hepatic MT concentrations 24 hr after an injection of menadione, paraquat, HX or TUR to control and adrenalectomized mice are examined. In adrenalectomized mice, these nonmetallic compounds induced hepatic MT accumulation similar to that in the control mice (Min et al., 1992a). This result suggests that glucocorticoid is not a "primary inducer" of hepatic MT following an injection of nonmetallic compounds.

Induction of MT synthesis and lipid peroxidation in the liver by some organic compounds:

It is known that menadione (Powis and Appel, 1980; Thor et al., 1982) and CCl_4 (Poyer et al., 1978) are metabolized by a variety of flavoenzymes to radicals in liver. These radicals produced by the metabolism of CCl_4 or menadione initiate lipid peroxidation in the liver. Moreover, both menadione and CCl_4 induce synthesis of hepatic MT, which contains only Zn. In the present study, it was investigated whether induction of MT synthesis is related to lipid peroxidation or to free radical production in liver following the administration of menadione or CCl_4 in mice.

As shown in Table 1, both compounds caused a significant increase of the thiobarbituric acid reacting substances (TBARS) as well as MT concentrations in the liver. If MT plays an antioxidative role in vivo, the induction of MT synthesis must be related to subsequent radical production or lipid peroxidation.

The time courses of changes in the Zn concentration of the MT fraction and the TBARS in the liver following the administration of menadione at 50 mg/kg or CCl_4 at 10 ml/kg were examined (Min et al., 1992a). In the menadione group, Zn in the MT fraction had increased significantly at 4 hr and then reached a maximum 24 hr after administration. However, the TBARS were increased

Table I Effect of vitamin E on MT induction by menadione or CCl_4

Pretreatment	Challenge	TBARS(mmol/g)	Zn-MT(μg Zn/g)
None	None	0.15 ± 0.00	2.53 ± 1.03
Vitamin E	None	0.12 ± 0.00[a]	2.55 ± 1.03
None	Menadione	0.47 ± 0.14[b]	17.84 ± 2.03[b]
Vitamin E	Menadione	0.17 ± 0.01[c]	16.08 ± 2.75
None	CCl_4	0.73 ± 0.20[b]	11.50 ± 0.93[b]
Vitamin E	CCl_4	0.31 ± 0.04[c]	11.08 ± 1.83

Mice received vitamin E sc at a dose of 100 mg/kg/day for 2 days and then menadione or CCl_4 was administered sc one hr after the second injection at a dose of 75 mg/kg or 10 ml/kg. [a]Significantly different from control group (no pretreatment, no challenge), $p < 0.05$. [b]Significantly different from control group, $p < 0.01$. [c]Significantly different from nonpretreated group in each challenge group, $p < 0.01$.

significantly only at 24 hr. A significant increase of both the Zn concentration of the MT fraction and TBARS was observed 8 hr after the administration of CCl_4. The ratio of the Zn concentration in the MT fraction to the TBA value in the menadione group was higher than that in CCl_4 group. The TBA value did not significantly correlate with the hepatic MT in either the menadione or CCl_4 groups (Min ei al., 1992a).

Both menadione and CCl_4 cause lipid peroxidation, as indicated by an increased TBARS shown in Table 1. When vitamin E, a natural antioxidant, was given prior to menadione or CCl_4, TBARS were decreased significantly compared with the nonpretreated group 24 hr after administration. The hepatic MT concentration was increased by both menadione and CCl_4. However, pretreatment with vitamin E did not decrease the concentration of MT in the liver caused by either menadione or CCl_4. Moreover, sodium phenobarbital, which promotes radical production by menadione and CCl_4 through the induction of NADPH-cytochrome P-450 reductase (Yasukochi and Masters, 1976), did not increase the concentration of MT in the liver (Min et al., 1992a). These data indicate that radical production does not trigger the induction of MT synthesis by these prooxidative agents.

Effect of GSH depletors on induction of MT synthesis by menadione:

The effect of pretreatment with buthionine sulfoximine (BSO) and 2-cyclohexen-1-one (CH) on the concentration of hepatic MT was examined. Both GSH depletors did not affect the basal level of MT. Moreover, neither BSO nor CH affected the level of increase caused by menadione (Min et al., 1992a). These results suggest that induction of MT synthesis is independent of GSH metabolism.

Although MT contains many sulfhydryl groups, menadione caused a significant increase of MT concentration and not GSH. Our results together with previous reports indicate that MT might be involved not only in antioxidant defense mechanism but also in the metabolism of cysteine (Gallant and Cherian, 1989), when GSH is depleted.

Therefore, induction of MT synthesis by menadione and CCl_4 is independent of lipid peroxidation, radical production, and GSH metabolism in the liver. We suggest that *in vivo* MT does not play a protective role against oxidative stress directly. Recent studies *in vivo* have demonstrated that induction of MT by an intraperitoneal injections of Fe, which produces radicals, is associated with an inflammatory response (Fleet et al., 1990a,b). The concentration of zinc in the plasma was decreased significantly after administration of menadione, like Fe injection or inflammation (Fig. 3). We surmised that pro-oxidative compounds produce radicals to initiate inflammation at the injection site and then hepatic MT synthesis is induced by cytokine(s) released from inflamed sites (De et al., 1990; Min et al. 1990).

Concentration of acute-phase proteins in plasma and hepatic MT following an injection of HX and TUR:

All nonmetallic compounds which produced the increase of hepatic MT, including inflammatory reagents caused a significant increase in the concentration of fibrinogen in plasma (Fig. 1 and 2). The concentration of ceruloplasmin in plasma was increased significantly after an injection of several nonmetallic compounds. An sc injection of each nonmetallic compound caused induction of hepatic MT and acute-phase response in inflammation.

Time course of changes in the plasma concentration of fibrinogen and ceruloplasmin, and of changes in the concentration of hepatic MT following an injection of HX or TUR were investigated (Min et al., 1992b). The concentration of fibrinogen and ceruloplasmin in plasma was increased significantly 8 hr after an injection of TUR. A significant increase in the concentration of hepatic MT was observed at earlier time (after 4 hr) than the increase of these plasma proteins. This time lag was also observed after an injection of HX. In both groups, the concentration of hepatic MT and these plasma proteins were increased with time.

In TUR and HX treated mice, the concentration of hepatic MT showed a significant, positive correlation with the concentration of fibrinogen in plasma (Fig. 4).

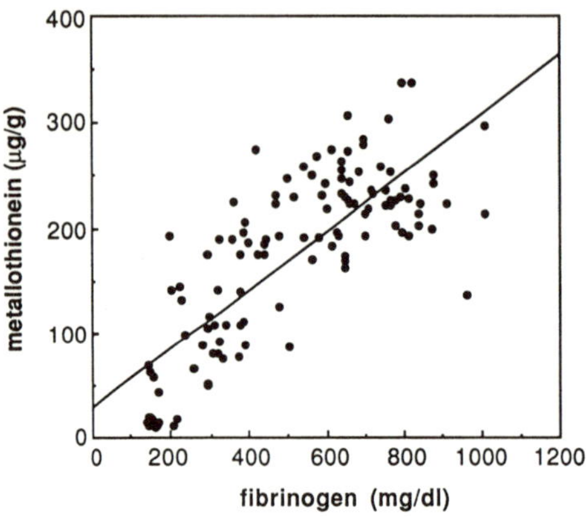

Fig. 4. Correlation of fibrinogen concentration in plasma with hepatic MT concentration. Data were obtained from mice injected various doses of and at various time after injection of nonmetallic compounds.

<u>Effects of glucocorticoid on the increase of plasma fibrinogen and hepatic MT synthesis by organic compounds:</u>

Fig. 5 shows effects of DEX on the increase in plasma fibrinogen and hepatic MT synthesis during inflammation initiated by TUR. Pretreatment with DEX inhibited the TUR-induced increases in hepatic MT and plasma fibrinogen at doses of 12.5 to 100 mg /kg. The concentration of MT in DEX-pretreated (25 mg/kg) mice was 20 % of nonpretreated group. However, both the concentrations of hepatic MT and plasma fibrinogen were increased gradually by pretreated dose with DEX at doses > 50 mg/kg. These inhibitive effects on hepatic MT and plasma fibrinogen were observed also after pretreatment of other glucocorticoid, prednisolone as well as DEX, but not after that of nonsteroid anti-inflammatory agent, salicylate. These results suggest that glucocorticoids have apparently dose-related, opposite effects on the induction of MT synthesis during inflammation. Recent reports show that glucocorticoids can suppress the production of cytokines (Lew et al., 1988; Nishida et al., 1989). In contrast to this effect, it has also been suggested that glucocorticoid enhances the acute-phase protein synthesis induced by IL-6 (Castell et al., 1988). These observations of opposite effects of glucocorticoid on acute-phase protein synthesis agree with the present results showing both enhancement and suppression of induction of MT synthesis during inflammation.

Hepatic MT concentrations after the administration of Cd(0.5 mg/kg) or Zn(5 mg/kg) in DEX-pretreated and non-pretreated mice were also studied. Pretreatment with DEX did not reduce the induction of hepatic MT synthesis by heavy metals.

Induction of hepatic MT by the inflammatory exudate:

When the inflammatory exudate and its supernatant, which were obtained from the air pouch of mice injected with TUR, were injected into other mice, the concentration of fibrinogen in plasma was increased significantly (Min et al., 1991). This result indicates inflammatory exudate containing cytokines initiates inflammatory response. In this case, the concentration of hepatic MT was increased significantly as well as the concentration of fibrinogen in plasma. MT concentration after an injection of the exudate obtained from the pouch of a mouse treated with DEX and TUR was lower than after an injection of the exudate obtained from the pouch of a mouse treated with TUR alone (Min et al., 1992b). Therefore, it appears that the suppression of cytokine production by glucocorticoids accounts for the inhibitive effects

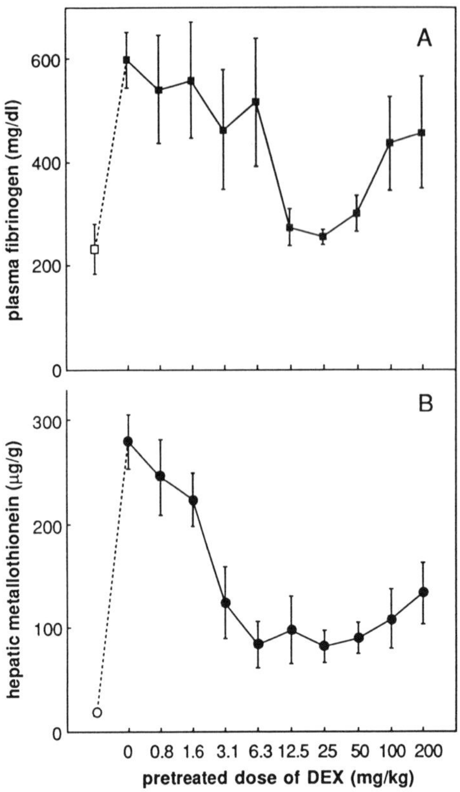

Fig. 5. Effect of DEX pretreatment on the concentration of plasma fibrinogen(A) and hepatic MT(B) 24 hr after TUR(5 ml/kg) administration. Open square and circle represent the value in intact mice.

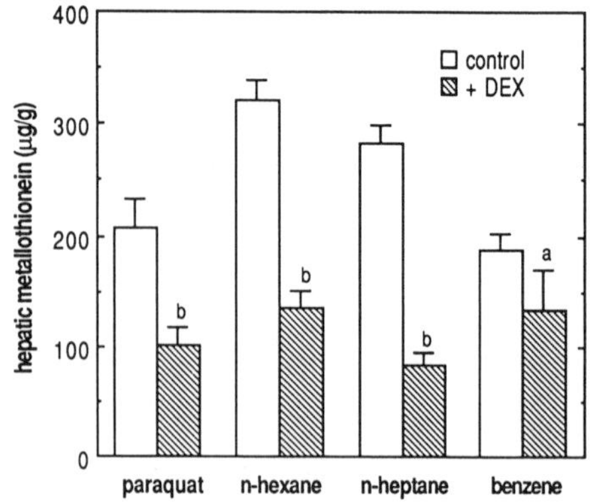

Fig. 6. Effect of pretreatment with DEX on the concentration of hepatic MT by the injection of four nonmetallic compounds. ap<0.05, bp<0.01

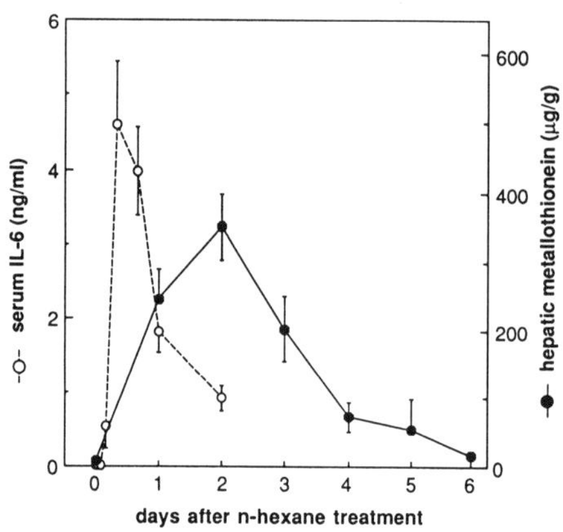

Fig. 7. Time course of changes in serum IL-6 and hepatic MT after sc injection of n-hexane (5 ml/kg).

on inflammation-induced MT synthesis, and MT may be a kind of acute phase protein (APP), at least in this case.

Mechanism of MT synthesis in the liver by sc administration of HX:

Effect of DEX on increased MT in the liver by paraquat and three organic solvents are shown in Fig. 6. The increase of MT by these compounds was also inhibited by DEX pretreatment as shown in the case of TUR (Fig 5). These results indicate the possibility of the participation of cytokine(s), especially IL-6, since, among cytokines, IL-6 has been shown to directly regulate the synthesis and secretion of variety of APP in the liver (Koj, 1986; Audus et al., 1988).

Fig. 7 shows the concentration of serum IL-6 and hepatic MT after sc administration of a potent inducer of hepatic MT, HX. Serum IL-6 was increased soon after Hx, reaching a maximal at 8 - 16 hr, then decreased. Even 48 hr after Hx administration, serum levels of IL-6 were still much higher than normal level. It is obvious that the elevation of serum IL-6 preceded that of MT synthesis in the liver. When MT synthesis induced by HX was inhibited by DEX pretreatment, the concentration of IL-6 in the serum was suppressed to a very low level (Table II).

Table II Effect of pretreatment of dexamethasone on serum IL-6 and hepatic MT after sc administration of HX

Group	IL-6 (ng/ml)	Metallothionein (μg/g)	
	16 hr	16 hr	48 hr
HX	6.58 ± 0.89	138.7 ± 23.3	486.3 ± 48.0
DEX + HX	0.07 ± 0.02[a]	63.4 ± 6.8[a]	174.8 ± 33.6[a]

[a]Significantly different from HX alone, $p < 0.01$.

Furthermore, the effect of anti-mouse IL-6 monoclonal antibody was examined. Five hr after HX administration, the antibody was injected iv, and the concentration of hepatic MT and plasma fibrinogen were determined. Increase of both proteins by HX was significantly inhibited by anti IL-6. These results suggest that MT synthesis resulting from HX is induced indirectly through cytokine(s) production, especially IL-6. Concentration of plasma IL-6 and hepatic MT was higher in the case of sc administration of Hx than iv. Even if the total amount of injected HX was the same. IL-6 and hepatic MT levels were higher when the concentration of HX in the olive oil vehicle was higher. These phenomena show the cytokine(s) is produced probably by macrophage and fibroblast through injury and inflammation at the site of administration.

Table III Effect of anti-mouse IL-6 on the concentration of hepatic MT and plasma fibrinogen 24 hr after sc administration of HX

Group	Metallothionein (μg/g)	Fibrinogen (mg/dl)
None treated control	18 ± 7	197 ± 10
HX	224 ± 33	712 ± 125
HX + Anti-IL-6	54 ± 26[a]	188 ± 3[a]

[a]Significantly different from HX alone, $p < 0.01$.

We surmise that nonmetallic compounds or radicals produced by them initiate inflammation at the injection site and then hepatic MT synthesis is induced by cytokines released from inflamed site. It was suggested that IL-6 induces MT synthesis directly, unlike other cytokines (De et al., 1990). These results suggest that MT may be a kind of APP, although the role of MT in the inflammatory response has not been identified.

REFERENCES

Arrick, B. A., Nathan, C. F., Griffith, O. W., and Cohn, Z. A. (1982) J. Biol. Chem. 257, 1231-1237.

Audus, T., Geiger, T., Hirano, T., Kishimoto, T., Tran-Thi, T.A., Decker, K., and Heinrich, P. C. (1988) Eur. J. Biochem. 173, 287-293.

Brakman, W. M. and Klaassen, C. D. (1987)Toxicol. Appl. Pharmacol. 87, 257-263.

Brzeznicka, E. A., Lehman, L. D., and Klaassen, C.D.(1987) Toxicol. Appl. Pharmacol. 87, 457-463.

Cagen, S. Z., and Klaassen, C. D. (1979)Toxicol. Appl. Pharmacol. 51, 107-116.

Castell, J. V., Gomez-Lechon, M. J., David, M., Hirano, T., Kishimoto, T., and Heinrich, P. C. (1988) FEBS Lett. 232, 347-350.

Cousins, R. J., AND Leinart, A. S. (1988) FASEB J. 2, 2884-2890.

DE, S. K., McMASTER, M. T.,and ANDREWS, G. K. (1990) J. Biol. Chem. 265, 15267-15274.

Disilvestro, R. A. and Cousins, R. J. (1984) Life Sci. 35, 2113-2118.

Friedman, R. L., and Stark, G. R. (1985) Nature 314, 637-639.

Fleet, J. C., Golemboski, K. A., Dietert, R. R., Anrews, G. K., and McCormick, C. C. (1990a) Am. J. Physiol. 258, G926-933.

Fleet, J. C., Andorews, G. K., and McCormick, C. C. (1990b) J. Nutr. 120, 1214-1222.

Gallant, K. R. and Cherian, M. G. (1989) J. Pharmacol. Exp. Ther. 249, 631-637.

Goering, P. L., Tandon, S. K., and Klaassen, C. D. (1985) Toxicol. Appl. Pharmacol. 80, 467-472.

Hidalgo, J., Camppany, L., Borras, M., Garvey, J. S., and Armario, A. (1988) Am. J. Physiol. 255, E518-524.

Karin, M., Anderson, R. D., Slater, E., Smith, K., and Herschman, H. R. (1980) Nature 286, 295-297.

Koj, A. (1986) *In* Modulation of Liver Cell Expression (Reutter, W. ed.) pp.331-342. MTP Press, Lancaster.

Kotsonis, F. N., and Klaassen, C. D. (1979) Toxicol. Appl. Pharmacol. 51, 19-27.

Lew, W., Oppenheim, J. J., and Matsushima, K. (1988) J. Immunol. 140, 1895-1902.

Mayo, K. E. and Palmiter, R. D. (1981) J. Biol. Chem. 256, 2621-2624.

McCormick, C. C. (1984) Proc. Soc. Exp. Biol. Med. 176, 392-402.

Min, K., Terano, Y., Onosaka, S. and Tanaka, K.(1991) Toxicol. Appl. Pharmacol. 111, 152-162.

Min, K., Terano, Y., Onosaka, S. and Tanaka, K.(1992a) Toxicol. Appl. Pharmacol. 113, 74-79.

Min, K., Mukai, S., Ohta, M., Onosaka, S. and Tanaka, K.(1992b) Toxicol. Appl. Pharmacol. 113, 293-298.

Nishida, T., Kawakami, T., Aihara, K., Nishino, N., Nakai, S. and Hirai, Y. (1989) Jap. J. Inflam. 9, 397-402.

Onosaka, S., Ochi, Y., Min, K. Fujita, Y. and Tanaka, K. (1988) Eisei Kagaku 34, 440-445.

Oh, S.H., Deagan, J. T., Whanger, P. D. and Weswig, P. H. (1978) Am. J. Physiol. 234, E282-285.

Petro, A., and Hill, C. H. (1987) Trace Elem. Res. 14, 255-263.

Powis, G. and Appel, P. L. (1990) Biochem. Pharmacol. 29, 2567-2572.

Poyer, L. J., Floyd, A. R., McCay, B. P., Janzen, G. E. and Davis, R. E.(1978) Biochim. Biophys. Acta 539, 402-409.

Sato, M., Ohtake, K., Mizunuma, H., and Nagai, Y. (1989) Toxicol. Lett. 45, 41-47.

Satoh, M., Naganuma, A., and Imura, N. (1988) Toxicology 53, 231-237.

Schroeder, J. J., and Cousins, R. J. (1990) Proc. Natl. Acad. Sci. USA 87, 3137-3141.

Searle, P. F. (1987) Biochem. Soc. Trans. 15, 584-586.

Shiraishi, N., Aono, K., and Utsumi, K. (1983) Radiat. Res. 95, 298-302.

Shiraishi, N., Yamamoto, H., Takeda, Y., Kondoh, S., Hayashi, K., Hashimoto, K., and Aono, K. (1986) Toxicol. Appl. Pharmacol. 85, 128-134.

Sobocinski, P. Z. and Canterbury, W. J. (1982) Ann. N. Y. Acad. Sci. 210, 354-367.

Schroeder, J. J., and Cousins, R. J. (1990) Proc. Natl. Acad. Sci. USA 87, 3137-3141.

Swerdel, M. R., and Cousins, R. J. (1984) Proc. Soc. Exp. Biol. Med. 175, 522-529.

Thomas, J. P., Bachowski, G. J., and Girotti, A. W. (1986) Biochim. Biophys. Acta 884, 448-461.

Thor, H., Smith, M. T., Hartzell, P., Bellomo, G.,Jewell, S. A., and Orrenius, S. (1982) J. Biol. Chem. 257, 12419-12425.

Thornalley, P. J. and Vasak, M. (1985) Biochim. Biophys. Acta. 827, 36-44.

Waalkes, M. P., Hjelle, J. J. and Klaassen, C. D. (1984) Toxicol. Appl. Pharmacol. 74, 230-236.

Wormser, U. and Calp, D. (1988) Toxicology, 53, 323-329.

Yasukochi, Y. and Masters, B. S. (1976) J. Biol. Chem. 251, 5337-5344.

Metallothionein III
K. T. Suzuki, N. Imura and M. Kimura (eds)

NUCLEAR AND CYTOPLASMIC LOCALIZATION OF METALLOTHIONEIN IN HUMAN LIVER DURING DEVELOPMENT AND IN TUMOR CELLS.

M. George Cherian

Department of Pathology The University of Western Ontario London, Ontario, Canada N6A 5C1

SUMMARY: Metallothioneins (MT) are present in high levels in fetal mammalian livers, and also in rat fetal neonatal liver bound to zinc and copper. They are also present in two major isoforms in low basal levels in various organs of adults in several species. Although a number of functions have been proposed for MT, their major biological roles may be in the storage of zinc and copper during rapid growth and development and also in the detoxification of certain toxic metals. In adult liver MT is mainly localized in the cytoplasm while it is localized also in the hepatocyte nuclei in human fetal liver and fetal/neonatal rat liver, as determined by immunohistochemical staining with a specific MT antibody. Because of its high expression in fetal development, the potential role of MT in human tumors was investigated. The cellular localization of MT was demonstrated in various human tumors such as thyroid tumors, testicular germ cell carcinoma, bladder transitional cell carcinoma and salivary gland tumors. In most of these tumor tissues, MT was found in high levels in nucleus and cytoplasm in both benign and malignant tumors, although the proliferating edge of the malignant tumors showed most intense MT staining. The expression of MT is not universal to all tumor growth and its presence may depend on various factors such as the type of tumor, its cellular origin, morphological heterogeneity or its stage of growth. The localization of MT is mainly found in epithelial cells in both normal tissues and tumors. Human testicular seminomas, which are well differentiated, showed little expression of MT irrespective of the staging as compared to less differentiated embryonal carcinomas. Preliminary studies with DNA flow cytometry on adenocarcinoma of breast and colon, using a fluorescent antibody to MT showed that MT was present in the nucleus of the epithelial cells only in the S-phase of the cell cycle. Thus the nuclear presence of MT in tumor cells is indicative of their proliferative activity and MT staining may be used as a marker of the proliferation index which is considered as a prognostic indicator of certain tumors.

INTRODUCTION

The cellular requirement of essential metals, their homeostatic control and biological and toxicological effects are controlled by a number of specific metal binding proteins. These metalloproteins are involved in the storage, transport and biological properties of many metals. Therefore it is important to study the type of binding of metals to proteins and their intracellular localization in order to understand the biological functions and mechanism of action of metals. The role of proteins such as ceruloplasmin, transferrin and ferritin in the metabolism of copper and iron is extensively studied in various laboratories (Freiden, 1986 ; Sarkar, 1983 ; Young and Aisen, 1982 ; Linder and Munro, 1973 ; Cousins, 1973). Because of their high affinity for metals, metallothioneins (MT) may play a physiological role in the absorption, storage and metabolism of important trace metals such as zinc and copper as well as in detoxification of certain toxic metals such as cadmium and mercury. This brief review will focus some of the recent work from our laboratory on the changes in intracellular localization of MT during mammalian development in the liver and also its expression in certain human tumors.

MT IN FETAL LIVER;

Although MT was first isolated and characterized as a cadmium binding protein from horse kidney (Kagi and Vallee, 1960), several studies (Bremner et al., 1977 ; Webb and Cain, 1982 ; Panemangalore et al., 1983) have shown its presence in many tissues as a zinc and copper binding protein. Most adult mammalian tissues contain low basal levels of MT which may vary with age and type of tissues. The presence of high levels of intracellular MT during gestation and the early postnatal period, bound to zinc and copper in mammalian liver, is well documented (Wong and Klaassen, 1979 ; Gallant and Cherian,

1986). The estimation of MT and metals in human liver and kidney at different age groups showed (Chung et al., 1986) high levels of zinc and copper bound to MT in human fetal livers (14 to 16 gestational weeks). There is no detectable cadmium in any human tissue at birth and most of MT is bound to zinc and copper. In contrast to high hepatic MT levels in the fetus, the concentration of MT in the kidney is low (Fig. 1). The few samples analyzed from newborn children and those under the age of twenty showed low MT levels both in liver and kidney. The binding of zinc and copper to MT and its high expression in human fetal liver suggest that MT may serve as a temporary reservoir for these metals during fetal development. The requirement for zinc is universal during mammalian growth and development and MT may serve as a storage protein for zinc. Since DNA and RNA polymerases and a number of other enzymes may require zinc, the homeostatic control of zinc is essential for catalysis of metabolic processes, nucleic acid metabolism and protein synthesis.

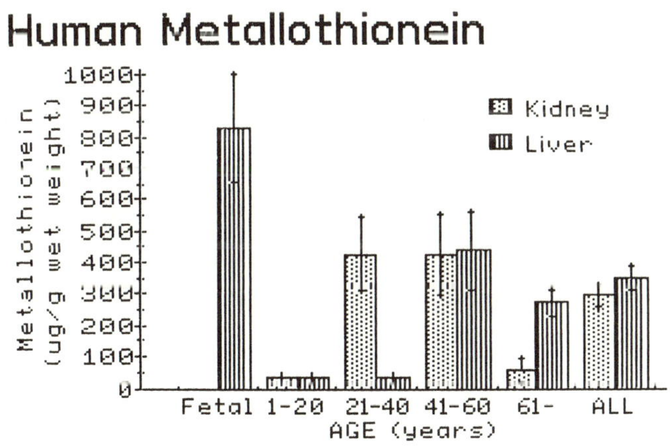

FIG 1. Metallothionein levels in human liver and kidney at different age groups. MT was estimated by Cd-saturation method and details of the data are presented in Chung et al (1986).

Similar to human fetal liver, high level of MT was detected in fetal and neonatal rat liver. In Sprague-Dawley rats, while MT was barely detectable in liver on the 15th day of gestation, a dramatic increase in hepatic MT concentration was observed on day 18 and it was continued until birth (Fig. 2). Both zinc and

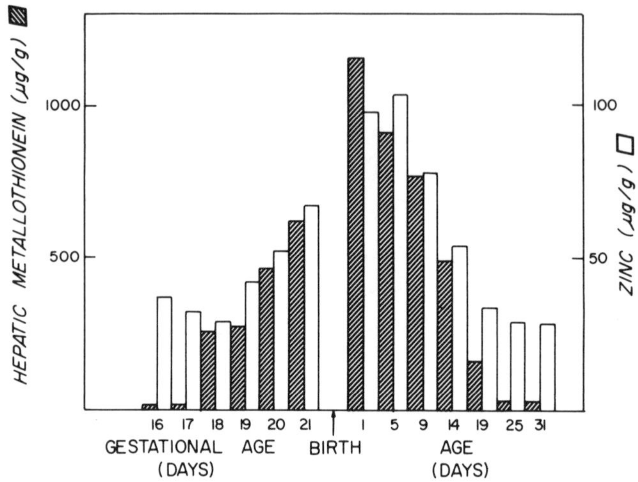

FIG. 2. Hepatic zinc and metallothionein levels in rats during perinatal and postnatal development. The data is adapted from Panemangalore et. al. (1983) and Templeton et. al. (1985).

MT concentrations in liver were highest soon after birth and they remained high for about 14 days of age. As MT content in the liver increased on day 18 of gestation, the Zn/MT molar ratio in liver decreased and then remained constant at about eleven in late gestation and neonatal periods. The results suggest that about 65 to 70% of total zinc pool in the rat liver during this period of rapid growth and development is bound to MT. The high hepatic concentration of both zinc and MT in newborn rats decreased at a linear logarithmic rate as the pups grew older and attained low near adult levels on weaning (Panemangalore et al., 1983 ; Templeton et al., 1985).

Developmentally regulated expression of MT also has been studied in cell culture using embryonal carcinoma cells induced to differentiate to primitive visceral endoderm by retinoic acid treatment (Andrews et al., 1984). These results suggest that the increased synthesis of MT in hepatocytes during fetal life may be due to the high requirement for essential metals such as zinc which is necessary for various metabolic processes in embryonic development, cellular differentiation and cell proliferation.

CELLULAR LOCALIZATION OF MT;

In addition to the high expression of MT in mammalian liver during early development, there is a marked difference in the cellular localization of MT as determined by immunohistochemical staining with a specific MT antibody. We have demonstrated that although MT is localized mainly in the cytoplasm in adult liver (Fig. 3), it can also be detected in hepatocyte nuclei (Fig. 4)

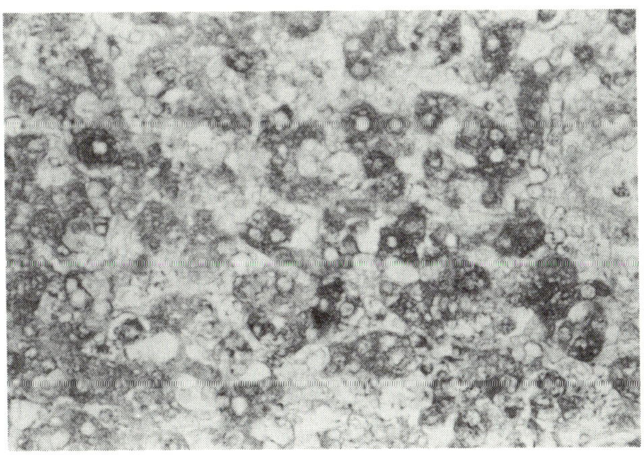

FIG. 3. Cellular localization of metallothionein in adult human liver. MT was localized by immunohistochemistry using a specific polyclonal antibody and peroxidase-antiperoxidase method as described in Nartey et. al. (1987A). The staining for MT was mainly in the cytoplasm and most of the nuclei did not stain. No counter stain.

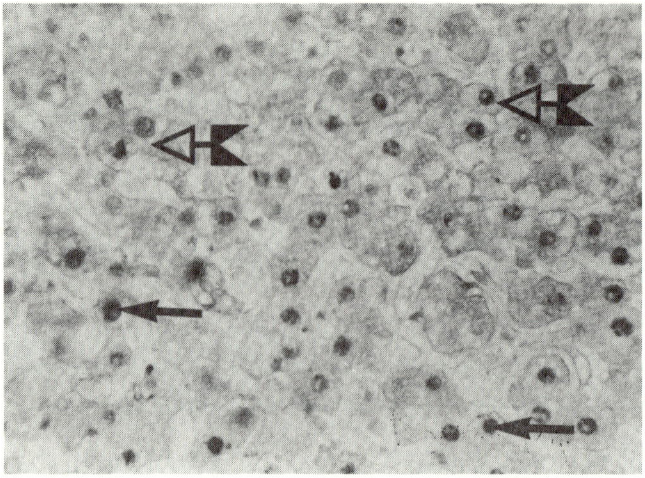

FIG. 4. Cellular localization of MT in human fetal liver. Fourteen week old fetal liver was stained with a specific antibody to MT as described by Nartey et. al. (1987A). Cytoplasmic (open arrowhead) and nuclear (closed arrow) staining for MT are shown. No counter stain.

in human fetal liver and fetal/neonatal rat liver (Nartey et al., 1987a ; Andrews et al., 1987). The high expression of MT and its mRNA in fetal rat liver is dependent on the maternal zinc status (Andrews et al., 1987). The hepatic MT level in newborn rats is markedly decreased when the pregnant rats were fed with zinc-deficient diet at late gestation period (Gallant and Cherian, 1986). The neonates born from zinc-deficient mothers were able to induce MT synthesis following injection of zinc salts, suggesting that the accumulation of zinc in the fetal liver during late gestation plays an important role in the induction of MT. Although there are distinct differences in the nuclear-cytoplasmic presence of endogenous MT in hepatocyte depending on the developmental stage, the significance of these changes in intracellular localization of MT is not yet understood. These changes in developmental and intracellular localization profiles and gene expression of MT suggest that MT could be considered as an oncodevelopmental gene product.

MT IN HUMAN TUMORS;

There is increasing evidence that certain proteins such as α-fetoproteins whose normal expressions are restricted to the embryonal and fetal periods, are frequently re-expressed in certain tumours. The re-expression of these proteins in tumors may be closely related to cellular differentiation and some of these proteins are used as tumor markers in certain neoplasia. In a report published in 1987, we were able to demonstrate the distinct presence of MT in 31 of 34 human thyroid tumors in archival paraffin-embedded tissue samples by immunohistochemical localization while only less than 20 per cent of normal thyroid glands showed a diffused distribution of MT (Nartey et al., 1987b). The positive staining for MT was present in both nucleus and cytoplasm of thyroid adenomas and carcinomas. In subsequent studies (Table I), we have demonstrated the presence

Table I
Localization of Metallothionein in human tumors

Organ	Tumors	Reference
Thyroid	Adenoma Follicular adenocarcinoma Papillary adenocarcinoma	Nartey et al (1987B)
Testes	Seminoma Embryonal Carcinoma Teratocarcinoma Choriocarcinoma	Chin et al (1993) Kontazoglou et al (1989)
Urinary bladder	Transitional cell carcinoma	Bahnson et al (1991)
Salivary gland	Pleomorphic adenoma Adenoid cystic carcinoma	Chauvin et al (1992)
Breast	Adenocarcinoma	Haile-Meskel et al (1992)
Colorectal	Adenocarcinoma	Haile-Meskel et al (1992)

Archival paraffin-embedded tissue sections fixed in formalin were stained for MT using a specific polyclonal antibody and peroxidase-antiperoxidase method as described in detail by Nartey et. al. (1987B).

of MT in various human tumors including transitional cell carcinoma of urinary bladder (Bahnson et al., 1991), testicular embryonal carcinoma (Kontozoglou et al., 1989), other types of testicular tumors (Chin et al., 1993), pleomorphic adenoma and adenoid cystic carcinoma of salivary gland (Chauvin et al., 1992), and adenocarcinoma of breast and colorectum (Haile-Meskel et al., 1992). These studies show that MT is expressed in both benign adenomas and malignant carcinomas and can be detected in nucleus and cytoplasm of tumor cells. However, the staining for MT is not universal to all tumor growth; but it may depend on various factors which may be related to the type of tumor, its cellular origin, morphological heterogeneity or its stage of growth. The development of subpopulation of microregions of heterogeneous cells with a variety of growth and functional properties within tumors (Sutherland, 1988) creates additional obstacles to study the role of expression of certain specific proteins such as MT.

The immunohistochemical staining of MT in various tumors showed its presence mainly in epithelial cells, especially on the proliferating edge of the tumor (Kontozoglou et al., 1989). There was a distinct difference between MT staining in seminomas and non-seminomas in human testicular tumors (Table II). Pure seminomas showed little or no staining for MT, irrespective of the clinical stage while non-seminomas stained heavily for MT, especially in the more advanced clinical stages (Chin et al., 1993). In general, seminomas are well differentiated tumors and are very sensitive to both radiation and chemotherapy (Logothetis et al., 1987). However, it is unknown whether there is any direct relationship between the radiosensitivity or chemosensitivity of seminomas with the absence of MT or differences in subcellular localization. The staining of MT in pleomorphic adenomas and adenoid cystic carcinoma of human salivary gland showed its presence mainly in myoepithelial cells with intense staining in the proliferating edge (Chauvin et al., 1992). These tumors are known to be relatively radioresistant.

Table II
Distribution of Metallothionein staining and clinical staging of germ cell tumors.

Seminoma (N=10)

MT Staining	STAGE I	II	III
0	1	6	1
+	2	0	0
++	0	0	0
+++	0	0	0

Non-Seminoma (N=22)

MT Staining	STAGE I	IIA	IIB	III
0	1	2		0 0
+	3	0		0 0
++	0	6		0 0
+++	2	2		3 3

0 - No staining, + - weak, ++ - moderate and +++ - strong MT staining. Non-seminomas are embryonal carcinoma, teratocarcinoma and choriocarcinoma. Adapted from Chin et al (1993).

The differences in MT localizationin embryonal carcinoma cells are closely related to their well-defined histomorphological patterns (Kontozoglou et al., 1989). This study showed that tumors with a typical glandular-papillary pattern, corresponding to the epiblastic or ectodermal pattern were strongly positive for MT, while tumors growing in solid masses, designated embryoblastic were negative or weakly positive for MT at the periphery. These results suggest that the variability of MT staining in human tumors may reflect their morphological heterogeneity which may also correlate with the degree of differentiation and maturation of the tumor cells. Thus, the tumors with solid pattern which resembles the better differentiated seminomas may have less cellular metabolic activity and less requirement for zinc than other morphological patterns of embryonal carcinomas which therefore may have little MT. These differences in growth patterns within the tumors may explain the different expression of MT in various tumors. The potential role of MT in tumors and their drug resistant pattern are discussed in recent articles (Cherian et al., 1993 ; Satoh et al., 1993).

SIGNIFICANCE OF NUCLEAR MT;

The significance of the presence of MT in nucleus of hepatocytes during development (Panemangalore et al., 1983 ; Nartey et al., 1987A) or partial hepatectomy (Tohyama et al., 1991) and also in certain tumor cells (Nartey et al., 1987B ; Kontozoglou et al., 1989 ; Chin et al., 1993) is not yet understood but may be related to the interactions of metal ions like zinc with various nuclear constituents during cell cycle. The metals which bind to MT can also bind to histones, nucleolar RNA and nuclear acidic proteins and may modify gene expression and phosphorylation of regulatory proteins. The presence of MT in the cell nucleus can effectively donate zinc to several enzymes or chelate zinc from transcription factors such as zinc finger proteins (Zeng et al., 1991). These protein-metal interactions in the nucleus can modify various cellular processes, including gene transcription and enzymatic activity.

The subcellular localization of MT in primary cultured adult rat hepatocytes stimulated by epidermal growth factor and insulin showed that there was a transient presence of MT in the nucleus during the early S-phase of cell cycle (Tsujikawa et al., 1991). Metallothionein also has been localized in hair follicles in normal skin and in the basal layer of hyperplastic epidermis, indicating its association with proliferation of epidermal keratinocytes (Karasawa et al., 1991). In a recent DNA flow cytometry study (Haile-Meskel et al., 1992), we have demonstrated the localization of MT in the nucleus of both diploid and aneuploid cells during the S-phase and not during G-1 phase of cell cycle of adenocarcinomas of breast and colon. In a previous study (Panemangalore et al., 1983), the attempts to demonstrate the presence of MT in cell nuclei isolated by biochemical techniques from newborn rat were not successful because of the leaching out of low molecular weight MT through the nuclear pores. In the DNA flow cytometry study, the cell nuclei were isolated from archival tumor tissue, fixed in formalin by the method of Hedley et al (1985). The polymerized form of MT in the nucleus of formalin fixed tissues cannot leach out, and therefore intact nuclei containing MT can be isolated. Thus the physiological role of MT is closely related to its nuclear presence in proliferating phase of the cell cycle where MT may have a significant role to supply zinc to enzymes involved in the synthesis of nucleic acid and protein or to chelate zinc from transcriptional factors. The mechanism of transfer of MT from cytoplasm to nuclei is not well understood. The site of synthesis of MT has been shown to be on the free polysomes and not on the polysomes attached to the endoplasmic reticulum where most of the secretory and membrane proteins are synthesized. In addition, newly synthesized MT does not contain any signal peptide, which is characterstic of most of the secretory proteins.

It is well established that the cell kinetic status of malignant tumors has prognostic implications (Merkel and McGuire, 1990). An active tumor growth implies cell proliferation and loss of differentiation which may depend on the growth fraction, cell cycle time and cell loss. The abnormal ploidy and elevated proliferative activity are associated with poor clinical outcome in locoregional breast, non-small cell lung, ovarian and colorectal cancers. Therefore, for diagnostic and therapeutic purposes, it is of interest to characterize tumors with expression of different antigens related to the cell cycle. The results of preliminary DNA flow cytometry studies of breast and colon carcinoma to localize MT in cellular nucleus suggest that it may provide a good indication of the proliferating fraction of cell cycle in tumors (Haile-Meskel, 1992, 1993). Further studies are required to demonstrate whether MT labelling index in flow cytometry can be used for prognosis in histopathological evaluation of epithelial malignancies.

REFERENCES;

Andrews, G.K., Adamson, E.D., and Gedamu, L. (1984) Dev. Biol. 103, 294-303.

Andrews, G.K., Gallant, K.R., and Cherian, M.G. (1987) Eur. J. Biochem. 166, 527-531.

Bahnson, R.R., Banner, B.F., Ernstoff, M.S., Lazo, J.S., Cherian, M.G., Banerjee, D., and Chin, J.L. (1991) J. Urology 146, 1518-1520.

Bremner, I., Williams, R.B., and Young, B.W. (1977) Brit. J. Nutr. 38, 87-92.

Chauvin, P.J., Cherian, M.G., Wysocki, G.P., and Pringle, G.A. (1992) Toxicologist 12: 379.

Cherian, M.G., Hunag, P.C., Klaassen, C.D., Liu, Y.P., Longfellow, D.G., and Waalkes, M.P. (1993) Cancer Res. 53, 922-925.

Chin, J.L., Banerjee, D., Kadhim, S.A., Kontozoglou, T.E., Chauvin, P.J., and Cherian, M.G. (1993) Cancer (In Press).

Chung, J., Nartey, N.O., and Cherian, M.G. (1986) Arch. Environ. Hlth. 41, 319-323.

Cousins, R.J. (1985) Physiol. Rev. 65, 238-309.

Freiden, E. (1986) Clin. Physiol. Biochem. 4, 11-19.

Gallant, K.R., and Cherian, M.G. (1986) Biochem. Cell. Biol. 64, 8-12.

Haile-Meskel, H., Cherian, M.G., Martinez, G., Veinot, L., and Frei, J.V. (1992) Proc. Cand. Assoc. Pathology, Toronto, Ontario.

Haile-Meskel, H., Cherian, M.G., Martinez, V.J., Veinot, L.A., and Frei, J.V. (1993) Mod. Pathol. (In Press).

Hedley, D.W., Friedlander, M.L., and Taylor, I.W. (1985) Cytometry 6: 327-333.

Kagi, J.H.R., and Vallee, B.L. (1960) J. Biol. Chem. 235, 3460-3465.

Karasawa, M., Nishimura, H., Tohyama, C., Hashiba, H., and Kuroki, T. (1991) J. Invest. Dermatology, 97, 97-100.

Kontozoglou, T.E., Banerjee, D., and Cherian, M.G. (1989) Virchows. Archiv. A. Pathol. Anat. 415, 545-549.

Linder, M.C., and Munro, H.N. (1973) Enzyme 15, 111-138.

Logothetis, C.J., Samuels, M.L., Ogden, S.L., Dexeus, F.H., and Chong, C.D.K. (1987) J. Urology. 138, 789-794.

Merkel, D.E., and McGuire, W.L. (1990) Cancer 65, 1194-1205.

Nartey, N.O., Banerjee, D., and Cherian, M.G. (1987A) Pathology 19, 233-238.

Nartey, N., Cherian, M.G. and Banerjee, D. (1987B) Am. J. Pathol. 129, 177-182.

Panemangalore, M., Banerjee, D., Onosaka, S., and Cherian, M.G. (1983) Dev. Biol. 97, 95-102.

Sarkar, B., Ed. (1983) Raven Press, New York.

Satoh, M., Kloth, D.M., Kadhim, S.A., Chin, J.L., Naganuma, A., Imura, N., and Cherian, M.G. (1993) Cancer Res. 53, 1829-1832.

Sutherland, R.M. (1988) Science 240, 177-240.

Templeton, D.M., Banerjee, D., and Cherian, M.G. (1985) Can. J. Biochem. Cell Biol. 63, 16-22.

Tohyama, C., Ghaffar, A., Nishimura, H., and Nishimura, N. (1991) In: Metallothionein in Biology and Medicine, (C.D. Klaassen and K.T. Suzuki, Eds.), CRC Press, Boca Raton, Florida, pp. 249-263.
Tsujikawa, K., Imai, T., Kakutani, M., Kayamoni, Y., Mimura, T., Otaki, N., Kimura, M., Fukuyama, R., and Shimizu, N. (1991) FEBS Lett. 283, 239-242.
Webb, M., and Cain, K. (1982) Biochem. Pharmacol. 31, 137-142.
Wong, K.L., and Klaassen, C.D. (1979) J. Biol. Chem. 254, 12399-12403.
Young, P.S., and Aisen, P. (1982) In: The liver biology and pathobiology, (I.M. Arias, H. Popper, D. Schachter and D.A. Shafritz, Eds.), Raven Press, New York, pp. 393-404.
Zeng, J., Vallee, B.L., and Kagi, J.H.R. (1991) Proc. Natl. Acad. Sci. 88: 9984-9988.

Metallothionein III
K. T. Suzuki, N. Imura and M. Kimura (eds)

^{125}I-METALLOTHIONEIN BINDING TO SPERMATOZOA

Katsuyuki Nakajima [1], Takanori Suzuki[2], Keiji Suzuki [3], and Hidetoshi Yamanaka [2]

1)Japan Immunoresearch Laboratories Co., Ltd, Takasaki, 370 Gunma.
2)Department of Urology, School of Medicine, Gunma University, Maebashi, 371 Gunma.
3)Department of Pathology, Medical Care and Technology, Gunma University, Maebashi, 371 Gunma.

SUMMARY :

The localization of metallothionein (MT) in the prostate and prostatic fluid in rat and man was studied using immuno-histochemical and RIA methods. Binding of ^{125}I-MT to spermatozoa was shown to be specific and temperature dependent, requiring 4-6 hrs to reach saturation. Anti MT antiserum and formalin fixation strongly inhibited the ^{125}I-MT binding to spermatozoa, but sonicated or isolated nuclei still retained binding capacity. Sucrose gradient analysis revealed that the bound MT was localized in the nuclear fraction. Inhibition of induced nuclear chromatin decondensation in the spermatozoa by Zn as well as MT was observed. This inhibition of nuclear chromatin decondensation is important for the viability of the spermatozoa, and thus represents a novel function of MT.

INTRODUCTION :

The biological significance of MT has been reported and discussed (Webb and Cain.,1982, Thornalley an Vasak., 1985; Miura et al.,1988; Riordan and Vallee.,1992) . But the presence of an MT receptor, or receptor related mechanism for MT incorporation has not been reported. To study the physiological role of MT we have selected a system to study the relationship between the high concentration of MT in the prostate and its possible effects on spermatozoa.

The localization of MT in the lateral, dosal and ventral lobes of rat prostate under physiological and Cd-treated conditions has been reported (Waalkes et al., 1982; Nolan and Shaikh., 1986; Bataineh et al., 1986; Umeyama et al., 1987; Suzuki et al., 1991). The secreted materials in these lobes were shown to be strongly stained positive for MT (Nishimura et al., 1990), and it followed that prostatic fluid contained large amounts of zinc (Mawson and Fisher., 1952; Anderson and Fair.,1976) and MT (Suzuki et al., 1992). These results suggested that MT in the secreted prostatic fluid has some biological function for the spermatozoa. Normal prostatic fluid is known to be of importance in maintaining the motility and viability of the spermatozoa (Kvist., 1980). Zinc has been thought to be the crucial factor for these effects.

We have used ^{125}I-MT, prepared by reacting rat MT-II with Tyr-MBS (M-maleimidobenzoyl-N-hydroxy succimide) and radioiodinated by the chloramine T method (Nakajima et al.,1992a), for the binding studies of MT to rat and human spematozoa.

MATERIALS AND METHODS :

Preparation of spermatozoa : Adult male rats (Wistar, age 12-15 weeks) were purchased from Imai Experimental Animals (Saitama, Japan). After rats were anesthesized by ether, the abdominal wall was incised in the midline and testis, epididymis and spermatic cord were removed from the scrotum. The tail portion of epididymis was pressed by finger and the spermatozoa were taken out from the spermatic cord. The spermatozoa were immediately washed by sperm-Ringer solution and then centrifuged at 1,500 rpm (Hitachi, Tokyo) for 5 min. After the seminal plasma was removed, the spermatozoa were washed twice and kept in an ice cold sperm-Ringer solution. Human ejaculated spermatozoa were prepared with the same procedure for this study.

Reagents : ^{125}I-Tyr-MT and anti-MT-I antibody were prepared, and RIA and immunohistochemical staining by the ABC method were performed according to the method described by Nakajima et al. (1992b). Rabbit MT-II and Noridate were purchased from Sigma Chem. Co., (St. Louis, MO). ^{65}Zn in 0.5M HCl solution was purchased from Dai-ichi Chemical Co., (Tokyo). ^{125}I-Insulin was purchased as an Insulin RIA Kit from Dainabott (Tokyo).

Assay procedure of ^{125}I-MT binding to spermatozoa : MT binding assays were performed at 0°C, room temperature (R.T) and 37°C for 0-6 hr in the presence of 4% normal rat serum, with gentle shaking. The number of spermatozoa used for the assay was 1-5 $\times$ 10^5 / tube in 100ul of sperm-Ringer solution and 100ul of ^{125}I-MT (approximately 10,000-100,000 cpm) were added to the reaction mixture. Competition curves were obtained with ^{125}I-MT (50nM) in the presence of increasing amounts of unlabelled rabbit MT-II (0.063-1.0mg/ml) for 4 hr at R.T. The spermatozoa were fixed by 10% buffered formalin (10% formalin in PBS) for 30 min and used for the assay after washing by sperm-Ringer solution. Sonication was performed for 5 min (Sonifier Model 200, Branson Co., Ltd. CA) at R.T. The inhibition of ^{125}I-MT binding to spermatozoa by anti MT antiserum was performed by adding the diluted antiserum (100 to 3,200 fold dilution) to the ^{125}I-MT solution and reacted for 10 min at R.T before mixing with the spermatozoa. The reaction was carried out at R.T for 4 hr. ^{65}Zn diluted by sperm-Ringer solution (45,000cpm) was used in the same way with ^{125}I-MT. All reactions were stopped by adding 2 ml of ice-cold sperm-Ringer solution and the tubes were centrifuged immediately at 3,000 rpm (CD-100R, Tomy Seiko Co., Ltd., Tokyo) for 10 min at 4°C. Supernatants were removed and the radioactivity of ^{125}I and ^{65}Zn was counted by an automatic gamma counter (ARC-600, Aloka, Tokyo).

Sucrose Gradient Analysis: Fractionation of the spermatozoa by sucrose gradient (10-70%) centrifugation was performed after incubation of the ^{125}I-MT with spermatozoa at 37°C for 4 hr, sonication for 5 min with Noridate (0.5%) treatment in 0.1 ml of PBS with 20% rat serum, and centrifugation at 10,000 rpm for 2 hr at 20°C

Nuclear Chromatin Decondesation: Semen samples from 7 healthy donors were used for the study of MT and Zn effects on human sperm nuclear chromatin decondensation. Spermatozoa were prepared as described above. One aliquot was exposed to 1% sodium dodecyl sulfate (SDS) in 0.05M borate buffer, pH 9.0 and the other exposed to 1% SDS containing 1.2 mM of Zn. MT was added to the two groups at the final concentration of 0, 0.75, 7.5, and 75 μ g/ml. Chromatin decondensation was studied after 60 min incubation at 37°C. The reaction was stopped by adding an equal amount of 2.5% glutaraldelyde in 0.05M borate buffer, pH 9.0. The degree of nuclear swelling was assessed under phase contrast microscopy (×400) in a randomly chosen optical field.

RESULTS :

The localization of MT in the ventral, lateral and dorsal lobes of prostate from adult rat and man was studied both by immuno-histological and RIA analysis. MT positive immunostaining was rarely seen in the ventral lobe of the prostate as compared with strong immunoreactivity in the lateral lobe. The dorsal lobe of prostate demonstrated MT immunoreactivity not only in the cytoplasm and nucleus of epithelial cells, but also in apocrine secretions (Fig. 1).

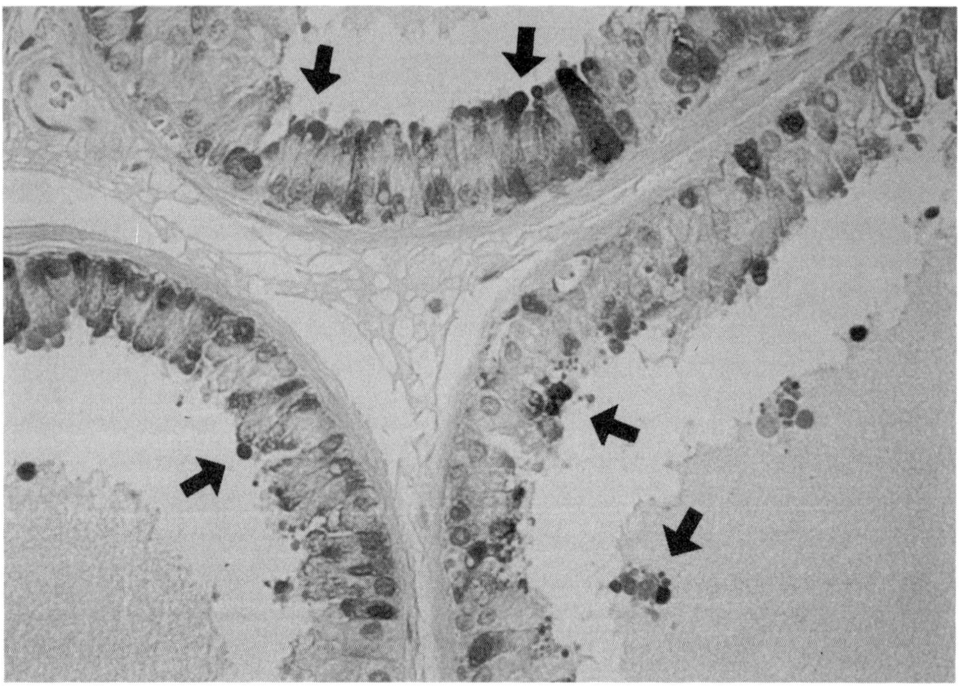

Fig. 1 The cytoplasm and nuclei of epithelial cells and the apocrine secretions are stained positive for MT by ABC method. Rat dosal prostate (×200)

The MT concentration in the ventral, lateral, and dorsal lobes was determined before and after $CdCl_2$ administration. Cd treatment induced significant amount of MT in all lobes of prostate to the same degree as in liver and kidney (Table 1).

The gel filtration chromatography of rat prostate extract showed the highest Zn content in the MT fraction, suggesting that the major Zn binding protein in the prostate is MT (Fig. 2).

Table 1 Changes of metallothionein (MT) concentration in the prostatic ventral lobe, lateral lobe, dorsal lobe, kidney and liver by $CdCl_2$ administration

Group	MT concentration (ug/g wet tissue)				
	Prostate			Kidney	Liver
	Ventral	Lateral	Dorsal		
1	0.3±0.1	30.4±24.0	5.2±0.9	25.0± 6.4	2.0± 1.5
2	3.3±2.6*a	103.5±82.1*a	6.0±2.1	25.7±13.5	10.2± 33.9*a
3	9.0±5.0*a	130.4±81.8*a	16.7±8.0*a	56.7±25.2*a	197.9± 64.3*a
4	20.0±9.9*b	64.8±60.5	21.9±7.7*b	199.6±87.9*b	505.5±186.5*b

Group 1 was injected with saline as control ; group 2 was injected with $CdCl_2$ (0.3 mg/kg) ; group 3 was injected with CdCl (0.9 mg/kg) ; group 4 was injected with $CdCl_2$ (2.7 mg/kg). All rats were injected one time subcutaneously. The values are means±S.D..

*a significantly different from group 1 ($p<0.05$).

*b significantly different from groups 1,2 and 3 ($p<0.05$).

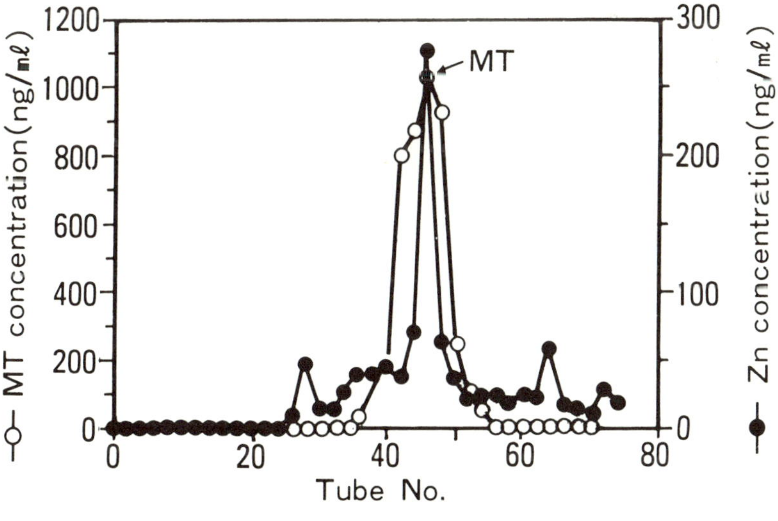

Fig. 2 Gel-filtration chromatography (Superose 12 HR 10/30) of MT and Zn extracted from a rat prostate. Elution buffer, 5mM Tris-HCl, pH 8.6

The concentration of MT in human prostatic fluid has been determined by RIA and shown to be as follows : 11.5±5.7 μg/g protein (normal control, n=6) ; 16.9 ±3.0 μg/g protein (benign prostatic hyperplasia, n=6). The level of MT in the prostatic fluid was shown to be approximately 60-100 times that of normal serum MT levels (less than 10ng/ml). Gel-filtration chromatography of prostatic fluid revealed high molecular weight complexes, indicating the possibility of MT binding to higher molecular weight (M.W.) proteins, or MT multimers.

MT binding to spermatozoa was studied using ^{125}I-MT, which had been radioiodinated by the chloramine T method after chemical modification of the rat MT-II. Tyrosine residues were introduced to six or seven cysteine residues per MT molecule prior to iodination. The effect of temperature (0°C, R.T. and 37°C) and incubation time on ^{125}I-MT binding to rat spermatozoa showed that maximum binding was observed at 37°C for 4 hr, while lower temperature showed decreased binding. The binding reaction was temperature and time dependent (Fig. 3).

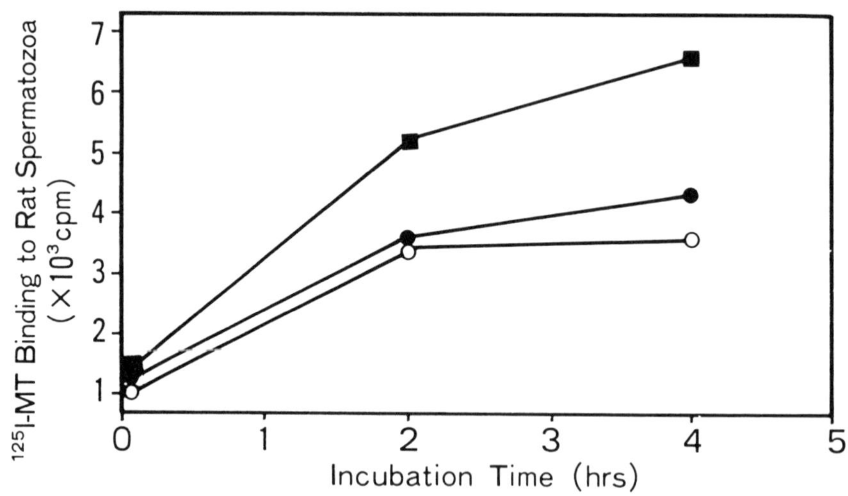

Fig. 3 Effect of incubation time and temperature on ^{125}I-MT binding to rat spermatozoa. Rat spermatozoa (5×10^5/tube) in 100 μl of sperm-Ringer solution and 100 μl of ^{125}I-MT (30,000 cpm) were incubated at 0°C(○-○), room temperature (●-●) and 37°C (■-■) shaking gently.

Affinity constants of the binding reaction by Scatchard plot analysis indicated that ^{125}I-MT bound to the spermatozoa in a bimodal fashion, with both a high affinity (Kd1=2×10^{-9}M) and a low affinity form (Kd2=2×10^{-8}M) (Fig. 4).

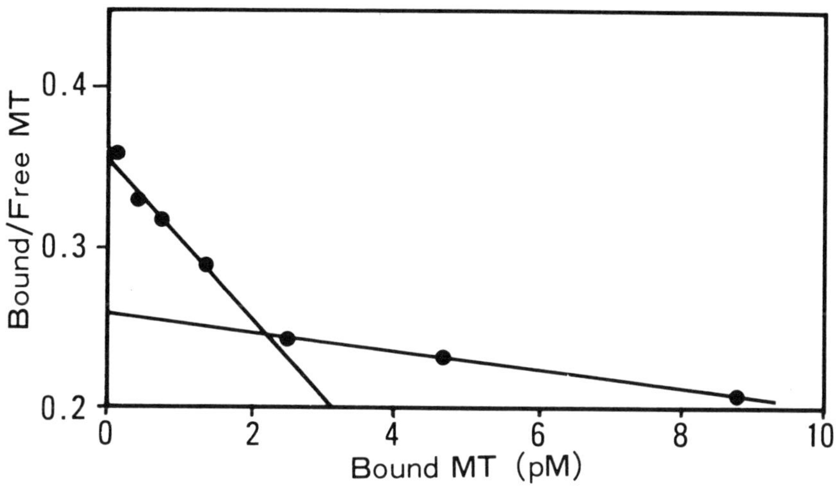

Fig. 4 Scatchard analysis of ^{125}I-MT binding to rat spermatozoa. Rat spermatozoa (5×10^{5}/tube) were incubated in the presence of 1.6～50 nM ^{125}I-MT at 37 ℃ for 4 hr. Affinity (Kd) was shown as high (Kd1=2×10^{-9}M) and low (Kd2=1×10^{-8}M) sites.

Inhibition of rat MT binding was studied using unlabelled rabbit MT-II, showing partial inhibition of ^{125}I-MT binding to rat spermatozoa in 4 hr incubation at R.T. (Fig. 5).

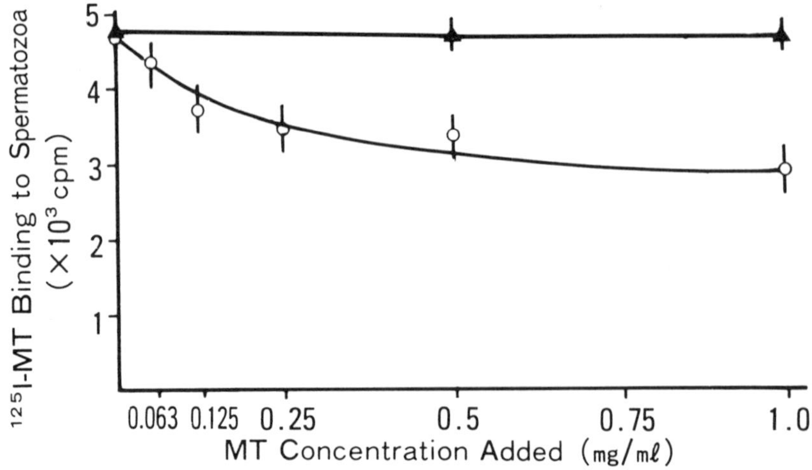

Fig. 5 Competitive effect of MT on ^{125}I-MT binding to rat spermatozoa. Rat spermatozoa (5×10^5/tube) with ^{125}I-MT (50nM) were incubated in the presence of 0.063-1.0 mg/ml unlabelled rabbit MT-II at R.T. for 4 hr. Data are means of triplicat observations of three indipendent experiments conducted. Control,▲-▲ ; MT added, ○-○.

Ten per cent formalin fixation of rat spermatozoa for 30 min at R.T. before incubation with ^{125}I-MT resulted in nearly complete inhibition of MT binding (Fig. 6).

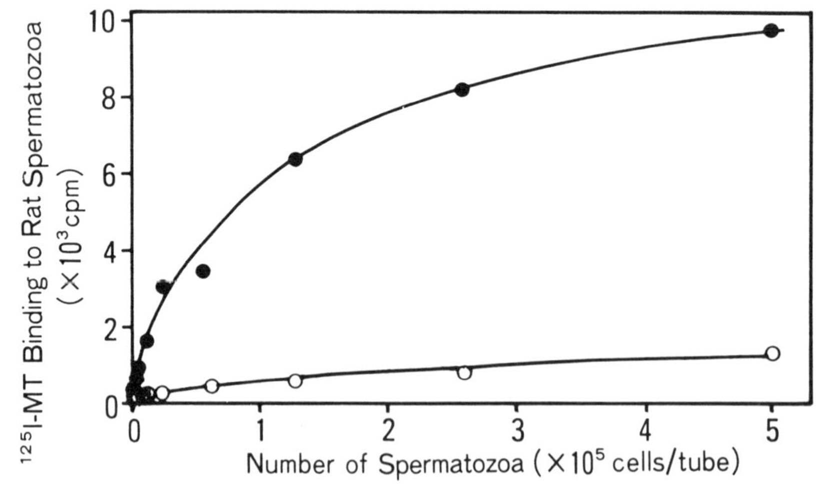

Fig. 6 Effect of formalin fixation on ^{125}I-MT binding to rat spermatozoa. Rat spermatozoa were fixed by 10% buffered formalin for 30 mis before use. Fixed (○-○) and not fixed (●-●) rat spermatozoa ($1 \sim 5 \times 10^5$/tube) were incubated with ^{125}I-MT at R.T. for 4 hr.

Sonication of rat spermaotzoa for 5 min before incubation with MT produced the same result as that with viable spermatozoa, indicating that binding does not require viable cells (Fig. 7).

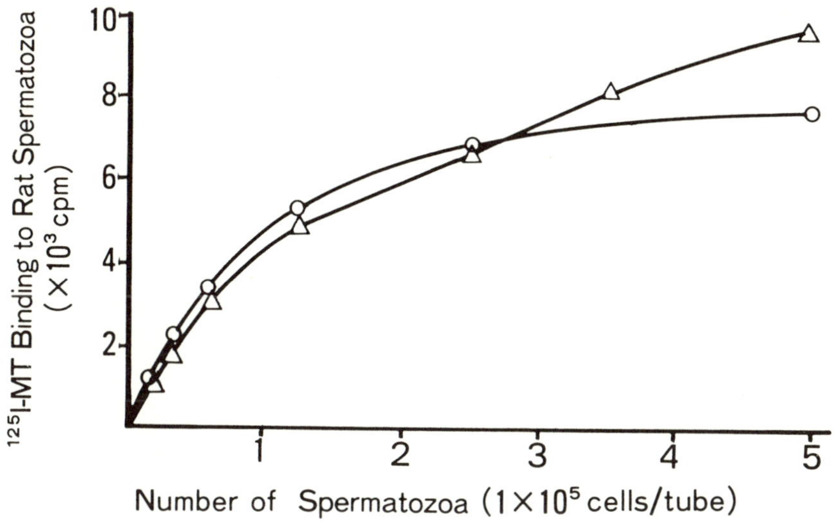

Fig. 7 Effect of sonication on ^{125}I-MT binding to rat spermatozoa. Rat spermatozoa were sonicated for 5 min and confirmed that the head and tail portion were separated by micromopic sbvervation. Sonicated (○-○) and viable rat spermatozoa (△-△) (1∼5× 10^5/tube) were incubated with ^{125}I-MT at R.T. for 4 hr.

Anti MT antiserum that we have raised and characterized (recognition epitope : NH_2 terminal 1-7) was used in a binding inhibition study. Anti-MT antiserum in the reaction mixture (diluted 1:100) almost completely inhibited the reaction when antiserum was added either before, or right after ^{125}I-MT and spermatozoa (5×105 / tube) were mixed (Fig. 8).

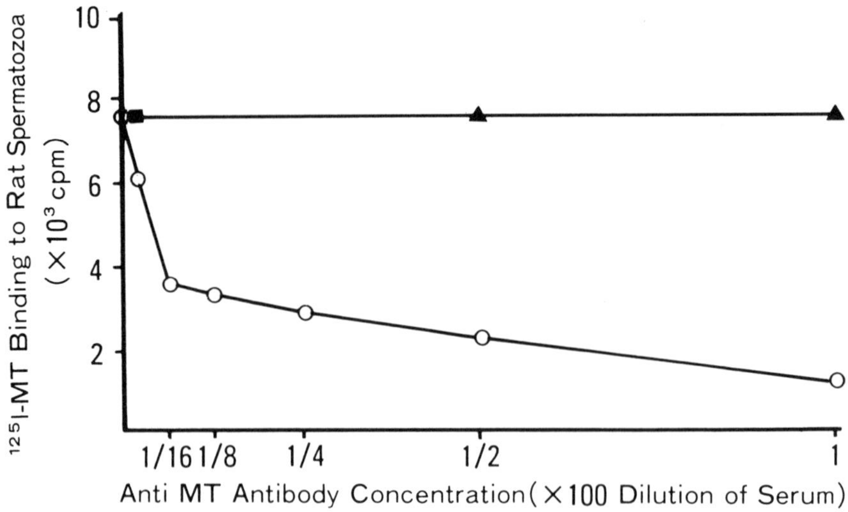

Fig. 8 Effect of anti-MT antiserum on ^{125}I-MT binding to rat spermatozoa. Anti-MT antiserum diluted from 100 to 3,200 fold was incubated with ^{125}I-MT for 10 min at R.T. before mixing with spermatozoa. Spermatozoa (viable and sonicated) were incubated with ^{125}I-MT, already reacted with diluted antiserum at R.T. for 4 hr.(antiserum treated, ○-○ ; control ▲- ▲)

Inhibition of ^{125}I-MT binding to rat spermatozoa by anti MT antiserum was observed using sonicated spermatozoa as well. The specificity of ^{125}I-MT binding was studied using rat lymphocytes, which did not bind ^{125}I-MT at all; in addition ^{125}I-Insulin did not bind to rat spermatozoa under the same conditions. ^{65}Zn ion was used for the detection of ^{65}Zn binding to rat spermatozoa, and ^{65}Zn bound at relatively high levels to rat spermatozoa when incubated at 37°C for 4 hr. This result was the same when fixed with formalin or not (Fig. 9), suggesting that ^{65}Zn binding to the spermatozoa was non-specific.

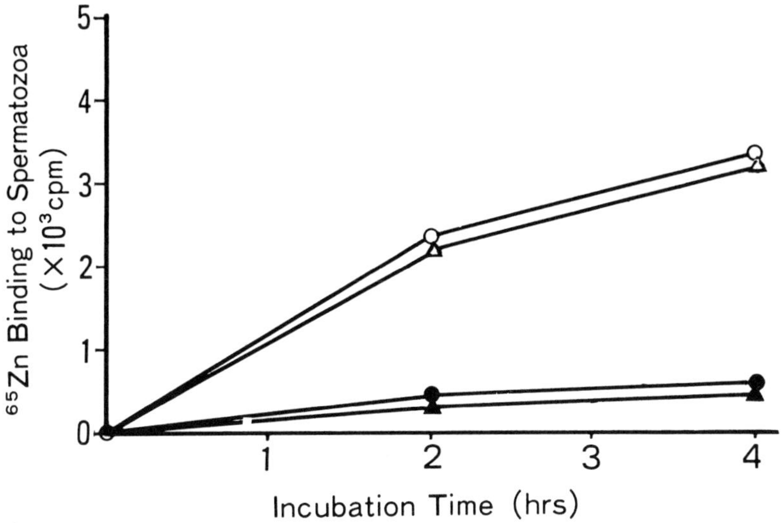

Fig. 9 ^{65}Zn binding to viable and fixed rat spermatozoa at 0°C and 37°C. Rat spermatozoa (5× 10^5/tube) with or without 10% buffered folmalin sixed were incubated with ^{65}Zn (45,000cpm) at 0°C and 37°C for 4 hr. Viable spermatozoa at 37°C, △-△ ; fixed spermatozoa at 37°C; ○-○ ; viable spermatozoa at 0°C, ▲-▲ ; fixed spermatozoa at 0°C, ●-●

The localization of ^{125}I-MT bound to spermatozoa (4 hr, 37°C incubation) was studied by fractionating the rat and human spermatozoa, first by sonication and followed by further digestion of the acrosomal membrane by Noridate (0.5%) or sodium dodesyl sulfate (SDS) (2%). Fig. 10 shows that the highest concentration of ^{125}I-MT bound was associated with the heaviest sucrose density fraction, > 60%, in rat and human spermatozoa.

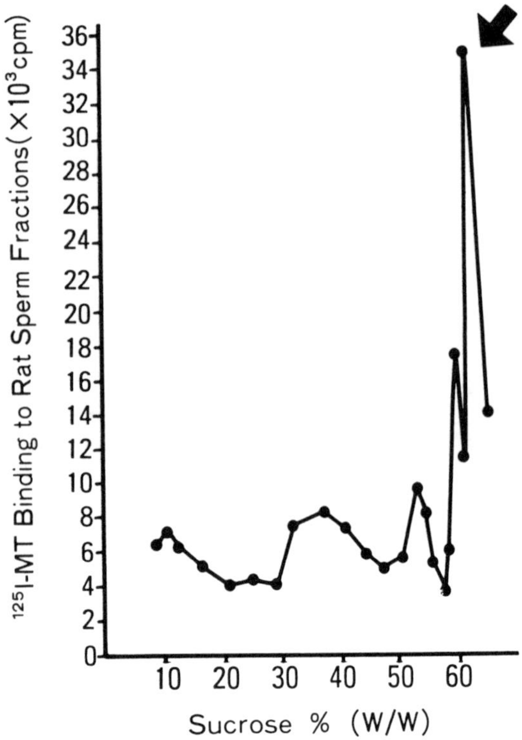

Fig.10 ^{125}I-MT binding profile to rat sperm fractions by sucrose gradient centrifugation. Rat spermatozoa (5×10^6/ml) incubated with ^{125}I-MT at 37°C for 4 hr and washed with ice-cole sperm-Ringer solution were sonicated for 5 min with 0.5% Noridate in 0.1 ml of PBS with 20 % rat serum. Then applied on sucrose gradient solution (10-70 % w/w) and centrifuged at 10,000 r.p.m for 2 hr at 20°C. (The arrow indicates the peak containing only nuclei)

These results indicated that ^{125}I-MT did not mainly bind to cytoplasmic or mitochondrial fractions, but to the head or head and tail complex fractions.

Zinc has been reported to inhibit the sperm nuclear chromatin decondensation (NCD) induced by membrane disintegration. One per cent of SDS was added to the human spermotozoa from 7 normal volunteers (1×10^6 / tube in 1 ml sperm-Ringer solution) and incubated at 37°C for 60 min to induce NCD. Fifty per cent of spermatozoa had reached an advanced stage of NCD, but when 1.2 mM of Zn was added to the spermatozoa in SDS buffer solution, very strong inhibition (60%) was observed. Equivalent levels of MT as found in prostatic fluid (3-10 μg / ml) showed a similar inhibitory effect on NCD (Fig. 11).

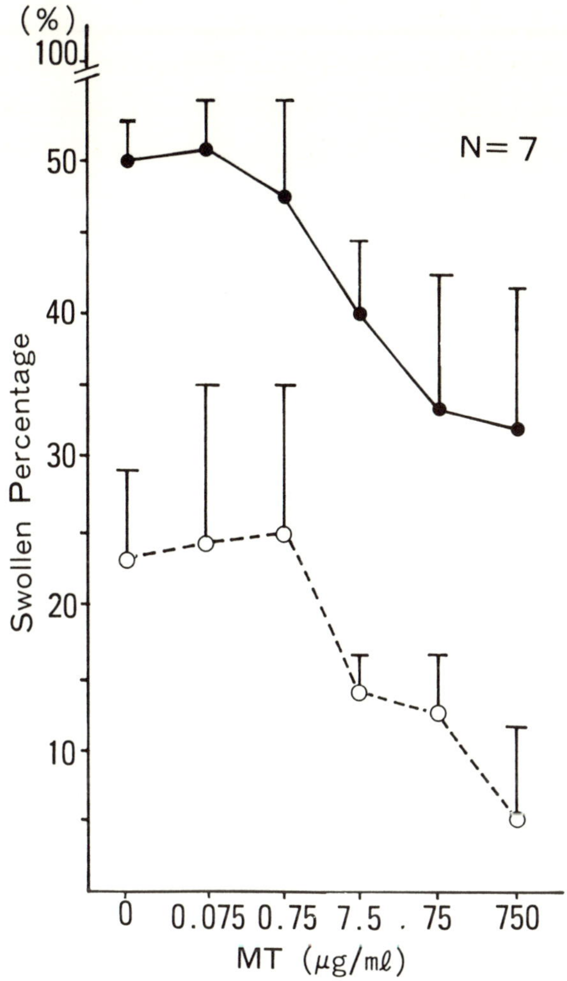

Fig.11 Effect of MT and Zn on human nuclear chromatin decondensation Human ejaculated spermatozoa (1×10^6/tube) were exposed to 1 % SDS in 0.05 M borate buffer (pH.9.0) in the presence of MT (0, 0.75, 7.5, 75 μ g/ml) for 60 min at 37°C,(●-●). The other exposed to 1% SDS containing 1.2 mM of Zn and MT (○-○). The reaction was stopped by adding as 2.5% glutaldehyde in 0.05 M borate buffer, pH 9.0. The degree of nuclear swelling was assenced ander phase contrast microscopy (× 400) counted over 200 cells.

DISCUSSION :

We have investigated the binding of MT to spermatozoa in this study with special attention to two major factors. One is species specificity: as the labelled MT-II was prepared from rat liver, and as the antibody to MT was raised against rat liver MT-II, we have done our basic studies using rat spermatozoa. The other factor was the radiolabelling of MT. Since Bolton-Hunter labelling of MT may alter the NH_2-terminal region, and since MT does not contain tyrosine or histidine for iodination, we used a sulfhydryl linked method (Tyr-MBS). Specific radioactivity obtained by this method was shown to be about 10 fold higher than that with the Bolton-Hunter method (Bolton and Hunter, 1973; Vander Mallie and Garvey, 1979; Tohyama and Shaikh, 1981; Mehra and Bremner, 1983). Also this ^{125}I-Tyr-MT continued to bind well with the MT antibody, which recognizes the NH_2-terminal end of MT (Nakajima et al.,1992a). The binding site of MT is, of course, not known, but this region (NH_2-terminal, 1-7) may be a ligand for a receptor because the amino acid sequence of this region is well conserved among most of the vertebrates (Kojima et al., 1979).

The presence of MT in the cytoplasm and nucleus is well known (Panemangalore et al.,1983), but few have reported that MT is secreted from the cells or organs. Nishimura et al (1990) reported that MT positive apocrine secretions were observed in the dorsal lobe of rat prostate. The analysis of prostatic fluid showed high MT content, about 100 times more concentrated than serum MT levels (Suzuki et al.,1992). Also Zn in the prostate seemed to be mainly bound to MT according to our gel filtration chromatography (Fig. 2). Arver (1982) and Arver and Eliasson (1982) reported the presence of low molecular weight Zn, such as Zn citrate, in prostatic fluid, but it did not seem to be the dominant form of Zn in the prostate. Therfore we believe that the MT found in the prostatic fluid was originally secreted from the prostate gland.

Prostatic fluid has been thought to be of importance in maintaining spermatozoal motility and viability (Huacuja et al.,1973; Lindohlmer, 1974). Espccially for thc capacitation of spcrmatozoa, contact with prostatic fluid seems to be essential and prevents nuclear choromatin decondensation (Kvist and Eliasson,1980; Blazak and Overstreet,1982; Bjorndahl and Kvist,1985). Nuclear chromatin decondensation has been thought to be affected by Zn contained in prostatic fluid. These observations prompted us to study the interaction between MT and spermatozoa. One component of MT binding to spermatozoa was shown to be high affinity, with formalin fixation of the spermatozoa and anti-MT antibody both inhibiting the binding activity completely. In addition, unlabelled rabbit MT-II showed partial inhibition of the labelled rat MT-II binding. As a specificity control, ^{125}I-insulin, a similar molecular weight peptide, did not bind to spermatozoa under the same conditions. The binding of MT to other cells such as lymphocytes and cancer cell lines was also negative. As the acrosomal membrane, which surrounds the head of spermatozoa, is known to adsorb various glycoproteins (Voglmayr,1986) during the maturation process, we had to know if MT binding was of this kind. Sucrose gradient analysis revealed that MT bound to spermatozoa is mainly localized in the nucleus, and not in the acrosomal membrane. Therefore, we suggest that the spermatozoa absorbs MT in a specific manner, possibly by an MT receptor or some other binding site on the surface membrane of the head portion.

This specific uptake of MT raises the question of function, one of which could be Zn transport to the nucleus. The presence of MT (and thus Zn) in the nucleus may serve to prevent nuclear chromatin decondensation in order to keep the spermatozoa viable. The relation of Zn-citrate and MT for this biological function is unknown, but the effective dose of each for inhibition of NCD seems to reflect the concentration levels found in prostatic fluid (Zn ; 1-2 mM and MT ; 10-20 μ M). Kvist (1980) reported that the depletion of spermatozoal Zn by EDTA enhanced NCD. Even though this SDS-induccd NCD may not cxactly mimic the naturally occuring process, inhibition of NCD in this assay by Zn or MT indicates that they may be important for the storage and capacitation of spermatozoa.

These results suggest that MT secreted from the prostate binds to the sperm nucleus, where it may act to prevent nuclear chromatin decondensation until the spermatozoon fuses with the ooplasm.

REFERENCES

Anderson, R.U. and Fair, W.R.(1976) Invest. Urol. 14, 137-140.

Arver, S. (1982) Acta, Physiol. Scand. 116, 67-73.

Arver, S and Eliasson, R.(1982) Acta Physiol. Scand. 115, 217-224.

Bataineh, Z. M., Heidger, P.M., Thompson, S. N., and Timms, B. G. (1986) Prostate 9, 397-410.

Bjorndahl, L., and Kvist, U. (1985) Acta Physiol. Scand. 124, 189-194.

Blazak, W. F., and Overstreet, J.W. (1982) Gamete Res. 5, 153-160.

Bolton, A. E., and Hunter, W. M. (1973) Biochem. J. 133, 529-538.

Huacuja, L., Sosa, A., Delgado, N. M., and Rosado A. (1973) Life Sci 13, 1383-1394.

Kojima, Y., Berger, C., and Kagi, J. H. R in "Metallothionein" (Kagi, J. H. R. and Nordberg, M. eds) (1979) p.153 Birkhauser, Basel.

Kvist. U. (1980) Acta Physiol. Scand. 109. 00-20.

Kvist, U and Eliasson, R. (1980) Int. J. Androl, 3, 143-152.

Lindholmer, C. (1974) Andrologia 6, 7-16.

Mawson, C, A., and Fisher, M. I. (1952) Can. J. Med. Sci., 30, 336-339.

Mehra, R. K and Bremner, I (1983) Biochem. J. 213, 459-465.

Miura, T., Tsujikawa, K., Yasuda, N., Nakajima, H., Haruyama, M., Ohmura, T and Okabe, M. (1988) Biochem. Biophys. Res. Commun. 151. 725-729.

Nakajima, K., Suzuki, K., Otaki N. and Kimura, M. (1992 a) Meth. Enzymol. 205, 387-395.

Nakajima, K., Suzuki K, Otaki N. and kimura, M. (1992 b) Meth. Enzymol. 205, 174-184.

Nishimura, H., Nishimura, N., and Tohyama. C (1990) J. Histochem Cytochem, 38, 927-990.

Nolan, C. V., and Shaikh, Z. A., (1986) Toxicol. Appl. Pharmacol 85, 135-142.

Panemangalore M., Banerjee D., Onosaka S., and Cherian M. G. (1983) Dev. Biol. 97, 95-102.

Riordan, J. F and Vallee, B. L. (1992)
Meth. Enzymol. Vol.205, 1-610.

Suzuki. T., Umeyama, T., Ohma C., Yamanaka H., Suzuki, K, Nakajima, K. and Kimura M., (1991) Prostate 19, 35-42.

Suzuki, T., Yamanaka, H., Tamura, Y., Nakajima, K., Knatani, K. Kimura, M. and Otaki, N. (1992) Tohoku J. Exp. Med., 166, 251-257.

Thornalley, P. J and Vasak M., (1985) Biochim. Biophys. Acta 827, 36-45.

Tohyama, C and Shaikh, Z. A (1981) Fundam. Appl. Toxicol. 1, 1-7.

Tsujikawa, K, Imai, T., Kakutani, M., Kayamori, Y., Miura, T., Okaki, N., Kimura, M. Fukuyama, R and Shimizu. N (1991) FEBS Lett. 283, 239-242.

Umeyama, T., Saruki, K., Imai, K., Yamanaka, H., Suzuki, K., Ikei, N., Kodaira, T., Nakajima, K., Saitoh, H. and Kimura, M. (1987) Prostate, 10, 257-264.

Vander Mallie, R. J and Garvey, J.S (1979)
J. Biol. Chem. 254, 8416-8421.

Voglmayr, J, K. (1986) Growth and Differ. Suppl., 28, 7-8.

Waalkes, M. P., Donovan, M. P. and Thomas J. A. (1982)
Prostate 3, 23-25.

Webb, M and Cain, K, (1982) Biochem. J. 31, 137-142.

Metallothionein III
K. T. Suzuki, N. Imura and M. Kimura (eds)

Degradation of Metallothionein

Curtis D. Klaassen, Supratim Choudhuri, James M. McKim, Jr., Lois D. Lehman-McKeeman, and William C. Kershaw

Department of Pharmacology, Toxicology, and Therapeutics, University of Kansas Medical Center, Kansas City, Kansas, USA

The objectives of the present investigation were to: 1) locate the subcellular site(s) of MT degradation, 2) identify proteases that catalyze the hydrolysis of MT and 3) obtain a better understanding of the processes that regulate the degradation of MT. To meet the initial objective, the abilities of selected liver subcellular fractions to degrade apo-MT *in vitro* were compared. The rates of apo-MT degradation in homogenates, cytosol and lysosomes obtained from naive rats were 0.78 ± 1.16, 0.13 ± 0.03, and 50.2 ± 4.3 nmol/mg protein/min, respectively. These maximal activities were noted at pH 5.5 (homogenates and lysosomes) and pH 7.2 (cytosol). Homogenate promoted hydrolysis of apo-MT at a pH similar to that present within lysosomes (pH 5.5) which was 3-times more extensive than that observed at pH 7.2. These data suggest that lysosomes are a major site of MT degradation *in vivo*. To address the second objective, apo-MT was incubated with lysosomal extracts plus cathepsin inhibitors, or with purified cathepsins. Leupeptin (inhibitor of cathepsin B and L), E-64 (cathepsin C), and pepstatin (cathepsin D) inhibited the degradation of apo-MT by 80, 50 and 20%, respectively. Apo-MT degradation rates in the presence of cathepsin B, C, or D were 12,000 ± 1500, 350 ± 33, or 276 ± 15 pmol MT degraded/Unit cathepsin/min, respectively. These data suggest that cathepsin(s) B and/or L play a critical role in the catabolism of MT. Finally, a series of *in vitro* and *in vivo* studies were conducted to evaluate the relationship between MT metal composition and the regulation of MT degradation. The rates of lysosomal extract-promoted hydrolysis of apo-MT, Cd-MT or Zn-MT were 50,200 ± 4300, 35.0 ± 5.2, and 20.5 ± 2.8 pmol/mg protein/min, respectively. In the presence of cathepsin B, the hydrolysis rates of these same substrates were 12,000 ± 1500, 13.0 ± 1.6, and 2.0 ± 0.3 pmol/mg protein/min. The exceptional stabilizing effect of metals against MT degradation was also demonstrated in other *in vitro* experiments in which lysosomal extract- or cathepsin B-induced degradation of apo-MT was carried out in the presence of increasing amounts of Zn bound to MT. Independent of the source of catalytic activity, the rate of MT degradation was inversely related to the amount of Zn bound to MT. Almost complete cessation of proteolytic activity occurred at $\geq$ 5 molar equivalents of Zn per mole MT.

The influence of metal composition on the stability of MT was further examined *in vivo*. Pulse-labeled MT-I and MT-II were isolated from the livers of Zn-, Cd- or ethanol-treated rats at various time points after ^{35}S-cysteine injection. Pulse-labeled hepatic iso-MTs were also isolated from livers of naive adult and neonatal rats. Zn, Cu, and Cd compositions of radiolabeled MTs were analyzed in selected experimental groups. Two noteworthy findings with respect to the metal stabilizing phenomenon were noted. First, iso-MTs of neonates and adults treated with Zn or ethanol had similar total metal contents and were composed principally of Zn (5.5 to 6.6 g atoms Zn/mol MT). However, the half-lives of these isoproteins were remarkable different (9 to 73 hr). Second, MT-II of neonates, which was composed entirely of Zn, was more stable than iso-MTs containing Cd; a metal having a strong stabilizing effect vis-a-vis its relatively high binding affinity. These *in vivo* observations do not preclude the role of metal composition in the regulation of MT degradation as demonstrated by the *in vitro* studies. Rather, these pulse-labeling experiments indicate that additional factors unrelated to the structure of MT also influences its degradation.

The role of MT synthesis in the homeostasis of metals is well defined. In contrast, the degradation of the protein, which may cause the release of metals and influence metal toxicity, is poorly understood. Moreover, because of its unique biochemical properties, the regulation of MT degradation may be quite different from that of many other proteins.

The principle objectives of the current research activities are to locate the subcellular site (or sites) in which the proteolytic breakdown of hepatic MT occurs, identify the proteases that catalyze the catabolism of MT, and obtain a more thorough understanding of the chemical and biological processes that regulate the degradation of this important metal-binding protein.

The initial goal was to identify cellular compartments or organelles capable of degrading MT (Choudhuri *et al.,* 1992). To this end, the relative abilities of rat liver homogenates, cytosol and lysosomes to degrade metal-free MT (*i.e.* apo-MT) were assessed *in vitro*. Apo-MT was chosen as the substrate because the metal-bound form of the protein (*i.e.* CdMT and ZnMT) was found to be completely resistant to degradation for at least 16 hr in the most rigorous of incubation conditions (*i.e.* lysosomal extracts at pH 5.5).

Subcellular fractions of male rat liver were prepared using a differential centrifugation followed by phase-separation technique described by Osada *et al.,* (1990). The substrate (*i.e.* ^{35}S-radiolabeled apo-MT-I) incubated with these cellular fractions, was prepared according to Sendelbach and Klaassen (1988), from livers of rats that received subacute exposure to inducing agents (Zn or Cd), and [^{35}S]cysteine. Metals associated with the ^{35}S-apo-MT moiety were removed using a dialysis membrane against an aqueous 0.1 M HCl solution. This final step was omitted for subsequent experiments in which Zn- or CdMT were used as substrates.

Proteolytic reactions were conducted *in vitro* at pH 5.5 or 7.2. These slightly acidic and neutral milieus were selected as they are representative of pHs found in proposed subcellular compartments of protein degradation, namely lysosomes and cytosol. The ratio of apo-MT-I to homogenate or cytosolic protein used was 1:1 (50 μg protein of apo-MT, homogenate or cytosol), whereas the ratio for lysosomal protein was 80:1 (50 μg apo-MT and 0.625 μg lysosomal protein). Incubates with a final volume of 100 μl were maintained at 37°C, and the reaction sampled at 0, 0.5, 1.0 and 2.0 hr after introduction of all components to the digestion mixture. Samples were rapidly placed in denaturing buffer, frozen in liquid nitrogen and stored at -80°C. The reaction was quantified by measuring the disappearance of radiolabeled substrate. Undegraded apo-MT was separated from hydrolysis products by sodium dodecyl sulfate, polyacrylamide gel electrophoresis (SDS-PAGE), as described by Hames (1981), and stained with Coumassie brilliant blue, as employed by Otsuka *et al.*, (1988). MT bands in the gels were excised, solubilized, mixed with scintillation solution and analyzed for radioactivity using a scintillation spectrometer. The radioactivities of undegraded substrate bands that remained after proteolytic reactions were subtracted from pre-reaction substrate radioactivity (*i.e.* the time zero band), and the resulting cpm values were converted to amount MT degraded according to the original specific activity of the radiolabeled substrate.

The rates of apo-MT degradation in homogenates, cytosol and lysosomes were 0.78 ± 0.16, 0.13 ± 0.03 and 50.2 ± 4.3 nmol/mg protein/min, respectively. These maximal hydrolytic activities were observed at pH 5.5 for homogenates and lysosomes, and at pH 7.2 for cytosol. The greatest degradative activity clearly resides in lysosomes having 65- and 390-times greater proteolytic activity than those detected in homogenates and cytosol, respectively. These observations are consistent with previous studies that indicate lysosomes as the intracellular site of MT degradation *in vivo* (Feldman *et al.*, 1978; Meulenaar *et al.*, 1985; Chen and Failla, 1989; Steinebach and Wolterbeek, 1992).

Additional evidence that confirms lysosomes as the major subcellular site of MT degradation was obtained by examining the relationship between pH, and extent of MT degradation in various intracellular compartments. Figure 1 (top panel) illustrates the degradation of apo-MT in the presence of rat liver homogenate or cytosol with each extract buffered at pH 5.5 or 7.2. These data demonstrate that homogenate-promoted degradation of MT is about 3-times more extensive in acidic conditions than in a neutral environment.

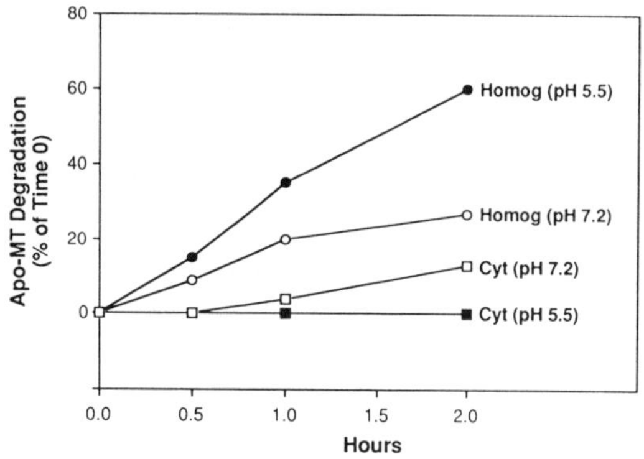

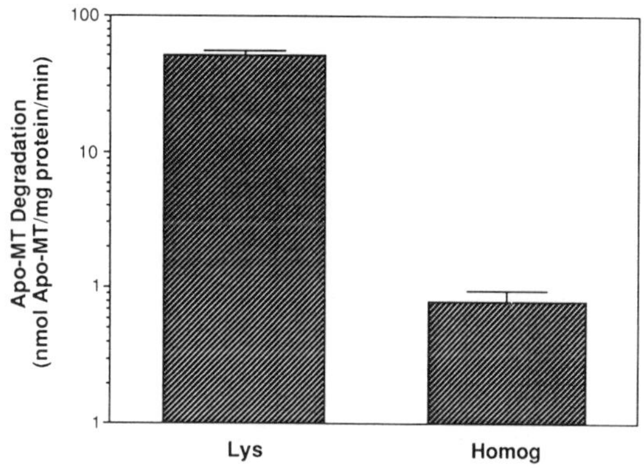

FIG. 1. (Top panel) Effect of pH on homogenate- and cytosol-promoted degradation of MT in vitro. Apo-MT was incubated separately with either rat liver homogenate or cytosol at two different pHs, 5.5 and 7.2. Samples were collected at 0.5, 1.0, and 2.0 hr and extent of substrate catabolism evaluated as described in the text. (Bottom panel) Comparison of homogenate and lysosomes to degrade apoMT-I at pH 5.5. Separate incubations of apoMT with homogenes and lysosomal functions were carried out at 37°C and pH 5.5 for 2 hr.

In contrast, apo-MT was essentially left intact after 2 hr of incubation with cytosol at pH 5.5, whereas the rate of cytosol-promoted catabolism at pH 7.2 was slow (0.13 ± 0.03 nmol/mg protein/min) and limited (13% of initial substrate mass converted to hydrolysis products). The bottom panel of Figure 1 illustrates a comparison of homogenate and cytosol to degrade apo-MT-I at pH 5.5. The relative effectiveness of homogenate at pH 5.5 to degrade MT is consistent with the "lysosomal theory of MT degradation", in as much as, lysosomal proteases have maximal activity at acidic pH conditions near hydrogen ion concentrations (pH 3.6 to 5.0) typically found in this compartment (Gordon, 1973). The finding that cytosol at pH 7.2 has degradative activity, albeit minimal, suggests that neutral soluble proteases, may also degrade MT. Other investigators have also reported that MT is a substrate for these cytosolic enzymes (Feldman *et al.,* 1978).

Incubations with lysosomal extracts and apo-MT as described above were also conducted in the presence of compounds that inhibit the lysosomal cathepsins. These acid proteolytic enzymes are abundant in lysosomes of mammalian tissues, and they have broad substrate specificity. They mediate a critical catabolic pathway in liver for a wide array of exogenous and endogenous proteins (Barret, 1977; Kunio *et al.,* 1985; Marzella *et al.,* 1987). Digestion mixtures contained 10 μM of leupeptin (inhibitor of cathepsins B and L), E-64 (trans-epoxysuccinyl-L-leucylamido-4-guanidino butane; an inhibitor of cathepsin C) or pepstatin (inhibits cathepsin D).

Figure 2 illustrates the relative abilities of cathepsin inhibitors to decrease the extent of apo-MT degradation *in vitro*. Leupeptin-induced inhibition of cathepsin(s) B and/or L activities caused the most notable decrease in the hydrolysis of apo-MT (80% decrease in degradation value). The inhibition of cathepsin C by E-64 or cathepsin D by pepstatin had moderate (50% inhibition) and minimal (20%) inhibitory effects on apo-MT degradation, respectively. These data indicate that cathepsin(s) B and/or L are most important in the lysosomal degradation of MT. The moderate inhibitory activity of E-64 suggests a role for cathepsin C in MT degradation as well, although this conjecture must be viewed with caution in light of the recent finding that E-64 has a slight inhibitory effect on cathepsins B and L (Funabiki *et al.,* 1990). The minimal effect of pepstatin seems to preclude the involvement of cathepsin D in MT turnover *in vivo*. This is consistent with the specificity of cathepsin D for hydrophobic amino acid residues (Kirschke and Barrett, 1987) and the absence of such regions in the tertiary structure of MT (Furey *et al.,* 1986).

The inhibitor studies presented in Figure 2 suggested the identity of the cathepsins with the greatest activity towards MT. In order to further characterize the identity of these problems, similar experiments were performed using purified cathepsins B, C and D (McKim *et al.*, 1992). The composition of incubates was similar to that employed with crude liver extracts except for their substitution with cathepsins B (6.0 mUnits), C (90 mUnits), or D (75 mUnits). Substrate/product separation and radiochemical analyses were identical to those noted above. The degradation rates of apo-MT in the presence of cathepsin B, C, or D were 12,500 ± 1500, 350 ± 33, or 276 ± 15 pmol MT degraded/Unit cathepsin/min, respectively. Clearly, cathepsin B has the greatest specific activity towards the metal-free form of MT. For example, cathepsin B degraded apo-MT 36- and 45-times faster than cathepsins C and D, respectively. These data further support our conclusion that cathepsin B, a cysteinyl-proteinase, plays an important role in the degradation of MT.

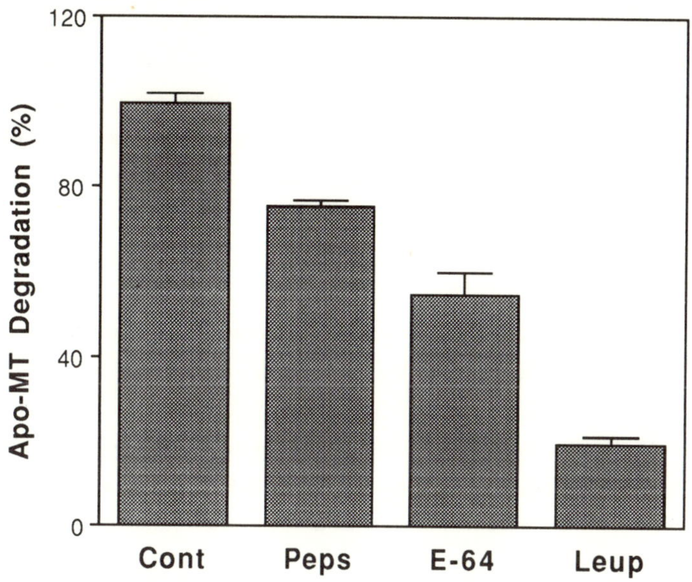

FIG. 2. Effect of cathepsin inhibitors on lysosomal enzyme mediated degradation of MT in vitro. Apo-MT (50 μg protein) was incubated with lysosomal extract (0.625 μg protein) prepared from rat liver. Proteolytic reactions occurred in the presence of the leupeptin, E-64, or pepstatin (10 μM each inhibitor). Methods employed to measure the extent of MT degradation are described in the text.

It is generally accepted that the susceptibility of MT to proteolytic breakdown is regulated by the metals contained in the structure of MT. This hypothesis is based predominately on three observations. First, apo-MT is hydrolyzed rapidly in the presence of rat liver lysosomal extracts whereas Zn-, Cu-, or Cd-containing MTs are substantially more resistant to hydrolysis (Feldman *et al.*, 1978; Bremmner and Mehra, 1983). Second, CdMT is more resistant than ZnMT to degradation in rat liver *in vivo* (Anderson *et al.*, 1978; Cain and Holt, 1979; Cain and Griffiths, 1984), which is consistant with the fact that Cd has a higher binding affinity for MT than does Zn *in vitro* (Waalkes and Klaassen, 1984). Taken together these observations suggest that metals stabilize

MT with respect to intracellular degradation and that the extent of stabilization is inversely related to the dissociation constants of bound metals. The mechanism of metal-promoted stabilization may be related to the high degree of order that metals confer to the tertiary structure of MT (*i.e.* a prolate ellipsoid). The ordered structure changes to a random coil configuration as metals are released from MT (Vasak *et al.*, 1980). In this context, it is noteworthy that proteins with a random configuration are more susceptible to degradation (Schimke, 1970).

To define further the role of various enzymes in MT degradation, Choudhuri *et al.*, (1992) and McKim *et al.*, (1992) studied the susceptibility of fully saturated Zn_7-MT-I (approximately 7 g atoms Zn per mole MT-I) or Cd_7-MT-I to degradation with rat liver lysosomes or purified cathepsins in a manner described previously.

A photograph (Figure 3) of a representative SDS-PAGE isolation of Cd_7MT hydrolysis products after 6 to 16 hr of incubation with lysosomal extracts demonstrates the resistance of CdMT to lysosomal degradation. In a separate experiment, the relative ability of Zn or Cd to stabilize MT against hydrolysis were quantified by conducting *in vitro* incubations (pH 5.5), in which either apo-MT-I, Cd_7MT-I, or Zn_7MT-I were utilized as substrates. The rates of degradation for apo-MT, CdMT and ZnMT were 50,200 ± 4300, 35.0 ± 5.2 and 20.5 ± 2.8 pmol/mg lysosomal protein/min, respectively. Relative to the extremely rapid breakdown of apo-MT, the rates of ZnMT and CdMT degradation were negligible.

Note that CdMT was degraded 1.5-times faster than ZnMT. This is surprising as previous work suggested that the half-life of hepatic MT in Cd-treated rats to be three to four times longer than those measured in Zn-treated rats (Feldman and Cousins, 1976; Anderson *et al.*, 1978; Feldman

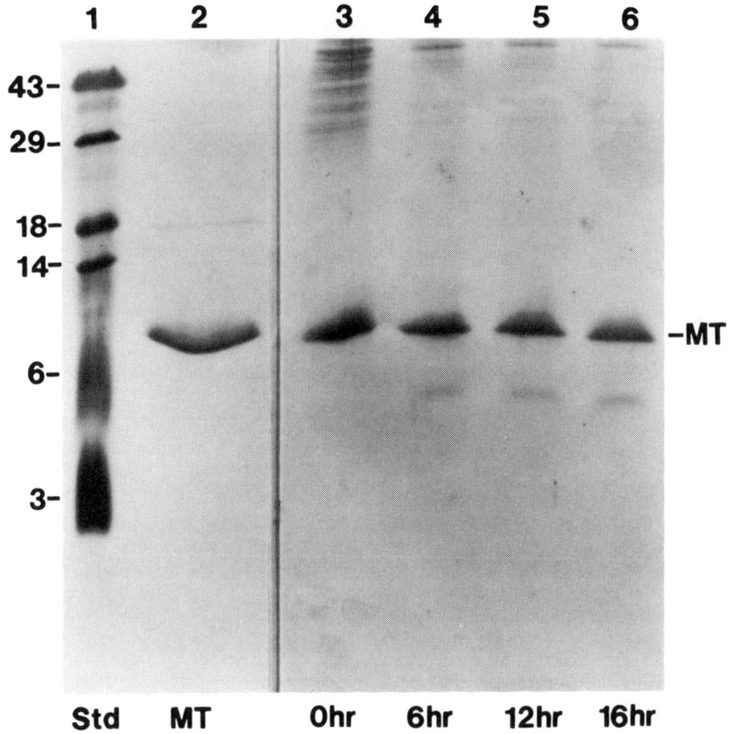

FIG. 3. SDS-PAGE analysis of CdMT following degradation by lysosomal enzymes in vitro. Lane 1 contains molecular weight markers (x 10^{-3}). Lane 2 contains MT standard (20 μg) and is located at approximately 7000 Da. Lanes 3 to 6 contains CdMT samples taken from incubation mixtures at 0, 6, 12 and 16 hr after the initiation of proteolytic reactions.

et al., 1978; Caino and Holt, 1979; Cain and Griffiths, 1984; Meulenaar *et al.*, 1985; Dunn *et al.*, 1987; Lehman-McKeeman *et al.*, 1988; Kershaw and Klaassen, 1992). This apparent anomaly between *in vitro* and *in vivo* stabilities will be discussed later.

The phenomenon of metal related protection against degradation was also demonstrated in another fashion. Apo-MT was placed in 0.5 M sodium acetate buffer (pH 5.5) with increasing amounts of Zn and the extent of degradation assessed. As the data presented in Figure 4 shows, the rate of MT degradation decreases linearly from about 36 nmol MT/mg lysosomal protein/min to about 7 nmol MT/mg lysosomal protein/min, as the protein undergoes a step-wise transition

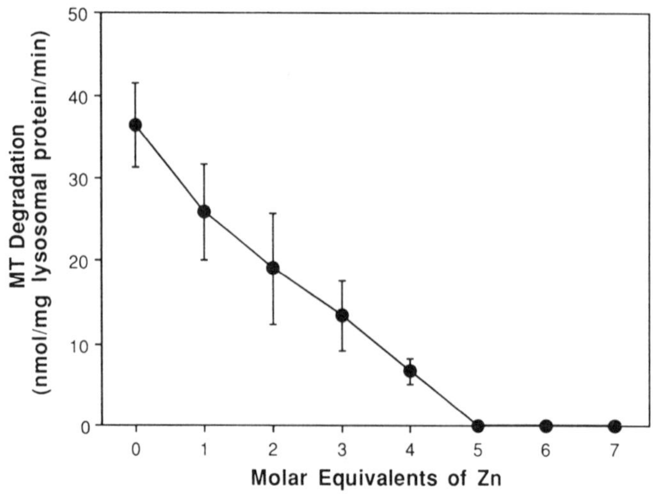

FIG. 4. Effect of MT metal composition on lysosomal extract promoted degradation of MT in vitro. Apo-MT (50 μg) was reconstituted with 0 to 7 molar equivalents of Zn and incubated with lysosomal protein (0.625 μg) at 37°C. Assessment of MT degradation was preformed as described in the text.

from a metal-free form to a protein that contains 4 molar equivalents of Zn. Degradation under the employed *in vitro* conditions was negligible at 5 or more molar equivalents of Zn. The incremental addition of Zn into incubation mixtures containing apo-MT-1 and cathepsin B were employed to further assess metal dependent stabilization of MT. These data (Figure 5) are strikingly similar to those observed with lysosomes (Figure 4). Cathepsin-mediated degradation of MT decreased in a linear fashion from approximately 11 to 3 nmol MT/Unit cathepsin B/min as molar equivalents of Zn increased from 0 to 4. Near complete inhibition of catabolism occurred at 5 or more molar equivalents of Zn.

A similar series of proteolytic reactions were carried out in which lysosomal extracts were replaced with cathepsin B as the enzymatic source of catalysis. Cathepsin B was chosen as it was determined previously to be a possible endogenous acid protease that degrades MT *in vivo*. As depicted in the SDS-PAGE analysis (Figure 6), the ability of metals to stabilize MT against proteolytic breakdown was also seen under these experimental conditions. Specifically, Zn_7MT and Cd_7MT in the presence of cathepsin B were stable *in vitro* for at least 34 hr and for 12 hr, respectively.

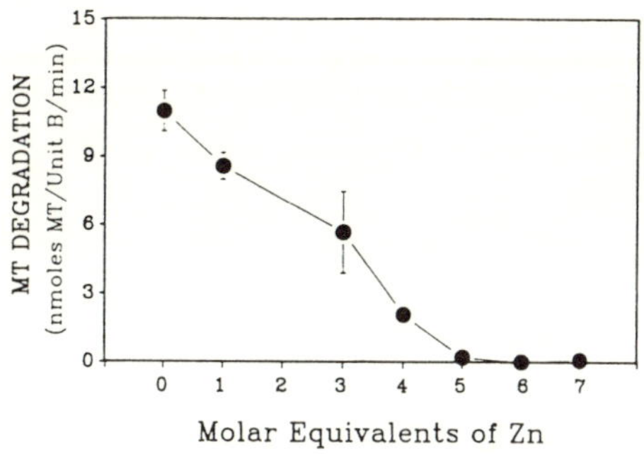

FIG. 5. Effect of MT metal composition on cathepsin B promoted degradation of MT in vitro. Apo-MT (50 μg) was reconstituted with 0 to 7 molar equivalents of Zn and incubated with cathepsin B (6 mUnits) at 37°C. Assessment of MT degradation was performed as described in the text.

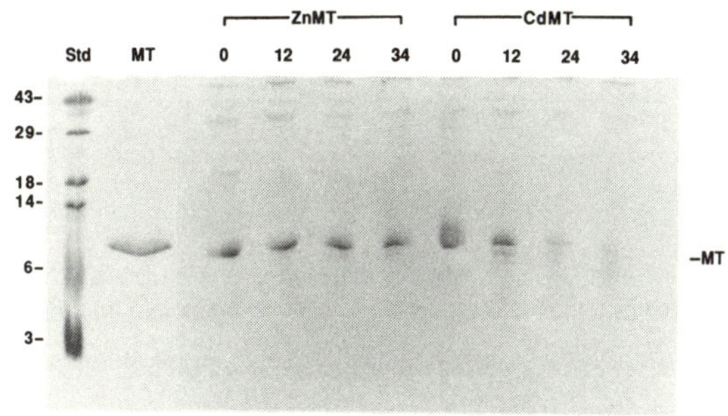

FIG. 6. SDS-PAGE analysis of ZnMT and CdMT following degradation by cathepsin B in vitro. Two representative gels are shown, one with purified ZnMT (20 μg) and molecular weight markers (x 10^{-3}), and another gel depicting the degradation of ZnMT and CdMT by cathepsin B, at 0, 12, 24, and 34 hr. MT protein bands are indicated at approximately 7000 Da. Cathepsin B and an unidentified protein can be seen at about 29,000 and 43,000 Da, respectively.

Quantitative assessments of the relative susceptibilities of metal-free and metal-containing MTs were also conducted. The hydrolysis of apo-MT was measured at selected times during a 15-min period, whereas the hydrolysis of Zn_7MT-I or Cd_7MT-I was measured 0, 12 and 24 hr following the initiation of hydrolysis. These reactions were linear over the indicted incubation times and were used to calculate the relative rates of degradation for each form of MT. The rates of degradation for apo-MT, CdMT and ZnMT were 12,500 ± 1500, 13.0 ± 1.6, and 2.0 ± 0.3 pmol MT degraded/Unit cathepsin B/min, respectively. Clearly, the presence of metals imparts considerable resistance against cathepsin-induced digestion of MT as Cd- and Zn-containing MTs were degraded nearly 1,000- and 6,000-times slower, respectively, than the metal-free form. As noted with reactions catalyzed by lysosomal extracts, CdMT was unexpectedly less stable than ZnMT in the presence of cathepsin B.

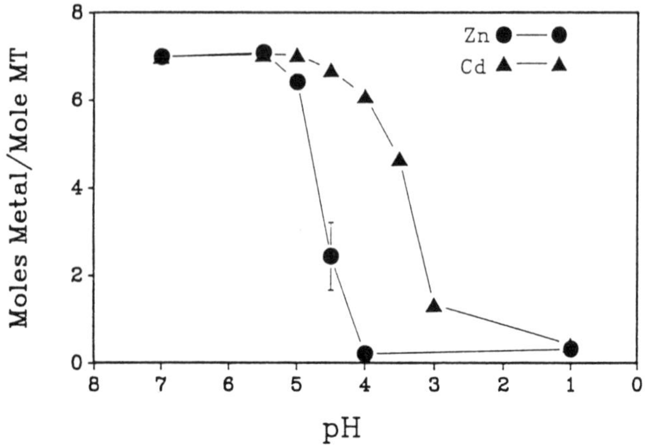

FIG. 7. Displacement of Zn and Cd from MT at various pHs. Aliquots of either Zn_7MT-2 or Cd_7MT-2 were placed in dialysis chambers and dialyzed into 4 liters of a sodium acetate buffer at the pHs shown for 24 hr.

The previous studies indicate that metals must be released from MT before degradation can occur to a significant extent. The displacement of metals by hydrogen ions may provide an explanation by which metals are removed from MT *in vivo*. Figure 7 shows Zn or Cd loss from MT at various pHs. At a pH of 5 or more, both metals remained bound to each respective protein (*i.e.* Zn_7MT-II and Cd_7MT-II). The metal content of ZnMT decreases markedly as the pH decreases from 5.0 to 4.5. ZnMT is completely void of metals at pH 4 and below. In contrast, Cd remained bound to MT at pH of 4.5 with complete removal of metals only occurring at a pH of 1. These data show that considerably more hydrogen ions are needed to displace Cd compared to those required for Zn displacement. This is consistent with the rank order of binding affinities for these metals (*i.e.* Cd > Zn).

Under the experimental conditions employed in these studies (*i.e.* pH 5.5), CdMT was approximately 2 to 6-times more susceptible to degradation than ZnMT, depending on whether lysosomal extracts or cathepsin B were used as the enzyme source. As mentioned previously, numerous investigators have shown the reverse rank order of stability *in vivo.* Specifically, hepatic ZnMT is degraded faster than CdMT in rats treated with each respective metal. It must be pointed out that *in vitro* data presented herein are not necessarily in contradiction with the *in vivo* observations. Lysosomes presumably accumulate MT *in vivo* as suggested by indirect evidence provided by intra-hepatocellular disposition studies with Zn- and Cu-containing MTs of human and animal origin (Riordan and Richards, 1980; Johnson *et al.,* 1981; Mehra and Bremner, 1984; Sasa *et al.,* 1986). Once inside the lysosomal compartment, MT is in an acidic environment (pH 3.6 - 5.0) that strongly favors complete release of Zn from MT (Figure 7), followed immediately by rapid digestion of the metal-free form of the protein. In comparison, the upper limit of the hydrogen ion concentration attainable in lysosomes is associated with only partial removal of Cd from the protein. Consequently, the degradation of CdMT would be expected to be slower than that of ZnMT, which is consistent with the numerous reports concerning the relatively protracted half-life of CdMT *in vivo*.

In order to further examine the influence of metal composition on the proteolytic breakdown of MT, the degradation and metal contents of MT-I and MT-II in rat liver were examined during various conditions of induction, such as neonatal development and acute exposures to Zn (1 mmol/kg, *sc*), ethanol (109 mmol/kg, *po*) or Cd (10 μmol/kg, *sc*). This was accomplished by measuring the disappearance of pulse-labeled MT-I and MT-II in liver (Kershaw and Klaassen, 1992). Injections of ^{35}S-cysteine 24 hr after parturition, or 6 hr after inducer pretreatment were employed to radiolabel each isoprotein. High dosages of non-radiolabeled cysteine (1 mmol/kg, *po*) were administered 4 hr after ^{35}S-cysteine and every 24 hr thereafter to minimize reutilization of radiolabeled amino acid. Iso-MTs contained in hepatic cytosol were isolated by gel-filtration followed by anion exchange and HPLC chromatography for the assessment of the metal content and specific radiological activity. Zn, Cu, and Cd compositions of MT-I and MT-II in each experimental group were measured by atomic absorption spectrometry. Chemical and radiological concentrations of iso-MTs were measured by the HPLC-atomic absorption spectrophotometric method of Lehman and Klaassen (1986) and by liquid scintillation spectrometry, respectively.

The metal compositions and half-lives of hepatic iso-MTs in neonates, and adults treated with Cd, Zn, or ethanol are presented in Table 1. The validity of the hypothesis that MT degradation is regulated by the amount and/or type of metals associated with the protein can be assessed by comparing the half-life values of these isoproteins with their respective metal contents. There are two noteworthy findings in the context of the metal composition hypothesis. First, iso-MTs of neonates and Zn- or ethanol-treated rats had similar metal contents, being composed principally of Zn (*i.e.* 5.5 to 6.6 g atoms Zn/mol MT), yet the half-lives of these isoproteins were quite divergent (*i.e.* 9 to 73 hr). Second, an iso-thionein composed entirely of Zn (*i.e.* MT-II of neonates) was more stable than MTs that contained Cd (*i.e.* MTs of Cd-treated rats). Curiously, the metal composition hypothesis is not compatible with these results. The hypothesis predicts that MTs with markedly different stabilities would have qualitative and/or quantitative differences with respect to their metal contents. However, a consistent and causal relationship between Zn content and half-life was not observed among the experimental groups. The hypothesis also predicts that isothioneins associated with tightly bound metals (*i.e.* Cd) would be more stable than those associated with metals having lower binding affinities (*i.e.* Zn). Consequently, the remarkable stability of Zn-containing MT-II of neonates relative to MTs of Cd-treated rats would be unexpected.

These *in vivo* observations appear inconsistent with those of *in vitro* experiments that clearly demonstrate the release of metals from MT to be a prerequisite for degradation. It should be emphasized that these *in vivo* findings do not eliminate the fact that metals stabilize MT, but rather, they imply that additional factors unrelated to the structure of MT influences its degradation. On a theoretical basis, it seems reasonable to speculate that inducers of MT have either a direct or indirect effect on processes involved in the catabolism of MT. There is empirical evidence to support this supposition. For example, the administration of Cd disrupts hepatic heterolysosome formation and inhibits lysosomal protein degradation (Mego and Cain, 1975) and Zn stabilizes hepatic lysosomal membranes *in vitro* (Chvapil *et al.,* 1972). Other factors, such as the availability of MT to the degradation machinery may also determine its accumulation and half-life. For example, it has been reported that MT is localized mainly in the nucleus during development (Nartey *et al.*m 1987; Tsujikawa *et al.*, 1991). In theory, such nuclear localization of MT will reduce its availability to the cytosolic or lysosomal degradation machinery, thereby increasing its accumulation and half-life. This can also explain why neonatal rat liver has high MT level (Wong and Klaassen, 1979) and extended half-life (Table 1).

TABLE 1
Metal Contents and Half-Lives of MT-1 and MT-2[a]

Treatment	iso-MT	Metal content (g atoms metal/mol MT)				Half-life (hr)
		Zn	Cu	Cd	Total	
Neonate	MT-1	5.5	0.1	ND[b]	5.6	49
	MT-2	5.9	ND	ND	5.9	73
Cd-treated	MT-1	3.2	0.2	3.2	6.6	58
	MT-2	3.4	0.1	2.8	6.3	61
Zn-treated	MT-1	6.6	ND	ND	6.6	21
	MT-2	5.8	0.3	ND	6.1	33
Ethanol-treated	MT-1	6.0	ND	ND	6.0	9
	MT-2	5.8	ND	ND	5.8	9

[a] Metal contents and half-lives of hepatic isometallothioneins were determined as described under Materials and Methods. Metal contents represent mean values of three determinations made upon a single pooled sample prepared from six adult rats (Cd-, Zn-, or ethanol-treated) or 76 nontreated neonates.

[b] Not detected. The lowest detectable specific metal content was approximately 0.1 g atom metal/mol MT.

References

Anderson, R.D. Winter, W.P., Maher, J.J., and Berstein,, I.A. (1978). Turnover of metallothioneins in rat liver. *Biochem. J.* **174,** 327-338.

Barrett, A.J. (1977). In *Proteinases of Mammalian Cells and Tissues* (A.J. Barrett, Ed.) pp. 209--248. Elsevier, North-Holland, Amsterdam.

Bremner, I., and Mehra, R.K. (1983). Metallothionein: Some aspects of its structure and function with special regard to its involvement in copper and zinc metabolism. *Chem. Sci.* **21,** 117-121.

Cain, K., and Holt, D.E. (1979). Metallothionein degradation: Metal composition as a controlling factor. *Chem.-Biol. Interact.* **28,** 91-106.

Cain, K., and Griffiths, B.L. (1984). A comparison of isometallothionein synthesis in rat liver after partial hepatectomy and parental zinc administration. *Biochem. J.* **217,** 85-92.

Chen, M.L., and Failla, M.L. (1989). Degradation of zinc-metallothionein in monolayer cultures of rat hepatocytes. *Proc. Soc. Exp. Biol. Med.* **191,** 130-138.

Choudhuri, S., McKim, J.M., Jr. and Klaassen, C.D. (1992). Role of hepatic lysosomes in the degradation of metallothionein. *Toxicol. Appl. Pharmacol.* **115,** 64-71.

Chvapil, M., Ryan, J.N., and Brada, Z. (1972). Effects of selected chelating agents and metals on the stability of liver lysosomes. *Biochem. Pharmacol.* **21,** 1097-1105.

Dunn, M.A., Blalock, T.L., and Cousins, R.J., (1987). Metallothionein. *Proc. Soc. Exp. Biol. Med.* **185,** 107-119.

Feldman, S.L., and Cousins, R.J. (1976). Degradation of hepatic zinc-thionein after parenteral zinc administration. *Biochem. J.* **160,** 583-588.

Feldman, S.L., Failla, M.L., and Cousins, R.J. (1978). Degradation of rat liver metallothionein *in vitro*. *Biochim. Biophys. Acta*. **544,** 638-646.

Funabiki, R., Yagasaki, K., Shirakawa, C., Sugita, H., and Ishiura, S. (1990). Activity measurement of lysosomal cysteine proteinases, cathepsin B, H and L, in some crude tissue extracts and their relation to the fractional rate of protein degradation. *Int. J. Biochem.* **22,** 1303-1306.

Furey, W.F., Robbins, A.H., Clancy, L.L., Winge, D.R., Wang, B.C. and Stout, C.D. (1986). Crystal structure of Cd, Zn metallothionein, *Science* **231,** 704-710.

Gordon, A.H., (1973). In *Lysosomes in Biology and Pathology* (J.T. Dingle, Ed.) pp. 89-137. Elsevier, North-Holland, Amsterdam.

Johnson, G.F., Morell, A.G., Stockert, R.J., and Sternlieb, I. (1981). Hepatic lysosomal copper protein in dogs with a inherited copper toxicosis. *Hepatology* **1,** 243-248.

Kershaw, W.C. and Klaassen, C.D. (1992). Degradation and metal composition of hepatic isometallothionein in rats. *Toxicol. Appl. Pharmacol.* **112,** 24-31.

Kirschke, H., and Barrett, A.J. (1987). Chemistry of lysosomal proteases. In *Lysosomes: Their role in protein Breakdown* (H. Glauman and F.J. Ballrads, Eds.), pp. 193-238. Academic Press, New York.

Kunio II, Hizawa, K., Kominami, E., Bando, Y., and Katunuma, N. (1985). Different immunolocalizations of cathepsins B, H, and L in the liver. *J. Histochem. Cytochem.* **33,** 1173-1175.

Lehman, L.D., and Klaassen, C.D. (1986). Separation and quantification of metallothioneins by high performance liquid chromatography coupled with atomic absorption spectrophotometry. *Anal. Biochem.* **153,** 305-314.

Lehman-McKeeman, L.D., Andrews, G.K., and Klaassen, C.D. (1988). Mechanisms of regulation of rat hepatic metallothionein-I and metallothionein-II levels following administration of zinc. *Toxicol. Appl. Pharmacol.* **92,** 1-9.

Marzella, L., and Glaumann, H. (1987). Autophagy, microautophagy, and crinophagy as mechanisms for protein degradation. In *Lysosomes: Their Role in Protein Breakdown* (H. Glaumann and F.J. Ballard, Eds.) pp. 319-367. Academic Press, New York.

McKim, J.M., Jr., Choudhuri, S., Klaassen, C.D. (1992). In Vitro degradation of apo-, zinc- and cadmium-metallothionein by cathepsins B, C, and D. *Toxicol. Appl. Pharmacol.* **116,** 117-124.

Mehra, R.K., and Bremner, I. (1984). Species differences in the occurrence of copper-metallothionein in the particulate fractions of the liver of copper loaded animals. *Biochem J.* **219,** 539-546.

Mego, J.L., and Cain, J. (1975). An effect of cadmium on heterolysosome formation and function in mice. *Biochem. Pharmacol.* **24,** 1227-1232.

Meulenaar, H., Van Den Hamer, C.J.A., Van Den Ingh, T.S.G.A.M., and Stolk, T.M.W. (1985). The effect of pH and lysosomal extract on rat liver metallothionein (Cd-thionein). *Biol. Trace Elem. Res.* **7,** 49.54.

Nartey, N.O., Banerjee, D., and Cherian, M.G. (1987). Immunohistochemical localization of metallothionein in cell nucleus and cytoplasm of fetal human liver and kidney and its changes during development. *Pathology* **19,** 233-238.

Osada, J., Aylagas, H., Sanchez-Prieto, J., Sanchez-Vegazo, I., and Palacios-Alaiz, E. (1990). Isolation of rat liver lysosomes by a single two-phase partition on dextran/polyethylene glycol. *Anal. Biochem.* **185,** 249-253.

Riordan, J.R., and Richards, V. (1980). Human fetal liver contains both zinc and copper-rich forms of metallothionein. *J. Biol. Chem.* **255,** 5380-5383.

Sasa, A., Hayashi, H., Yagi, A., Ohguchi, S., Kidokoro, R., Sato, Y., and Sakamoto, N. (1986). A copper-sulfur complex in the liver of a patient with Wilson's disease. *Gastroenterol. Jpn.* **21,** 633-637.

Schimke, R.T. (1970). Regulation of protein degradation in mammalian tissues. In: *Mammalian Protein Metabolism* (ed. H.N. Munro), pp. 177-228. Academic Press, New York.

Sendelbach, L.E. and Klaassen, L.E. (1988). Kidney synthesizes less metallothionein than liver in response to cadmium chloride and cadmium metallothionein. *Toxicol. Appl. Pharmacol.* **92,** 95-102.

Steineback, O.M. and Wolterbeek, B.T. (1992). Metallothionein biodegradation in rat hepatoma cells: A compartmental analysis aided ^{35}S-radiotracer study. *Biochemica et Biophysica Acta* **1116,** 155-165.

Tsujikawa, K., Imai, T., Kakutani, M., Kayamori, Y., Mimura, T., Otaki, N., Kimura, M., Fukuyama, R., and Shimizu, N. (1991). Localization of metallothionein in nuclei of growing primary cultured adult rat hepatocytes. *FEBS Lett.* **283,** 239-242.

Vasak, M., Galdes, A., Hill, H.A.O., Kagi, J.H.R., Bremner, I., and Young, B.W. (1980). Investigation of the structure of metallothioneins by proton nuclear magnetic resonance spectroscopy. *Biochemistry* **19,** 416-425.

Waalkes, M.P., and Klaassen, C.D. (1984). Relative *in vitro* affinity of hepatic metallothionein for metals. *Fundam. Appl. Toxicol.* **5,** 473-477.

Wong, K.-L., and Klaassen, C.D. (1979). Isolation and characterization of metallothionein which is highly concentrated in newborn rat liver. *J. Biol. Chem.* **254,** 12399-12403.

Metallothionein III
K. T. Suzuki, N. Imura and M. Kimura (eds)

γ-GLUTAMYLCYSTEINYL ISOPEPTIDES: STRUCTURE, SYNTHESIS AND METAL-BINDING COMPLEXES

W.E. Rauser

Department of Botany, University of Guelph, Guelph, Ontario, N1G 2W1 Canada

Summary: Class III metallothioneins (MTs) from certain yeasts and algae through to flowering plants were first identified as (γ-glutamyl-cysteinyl)$_n$-glycines or (γEC)$_n$Gs. The range of γEC isopeptides now includes those lacking Gly to those where the Gly is replaced by β-Ala, Ser or Glu. A γEC dipeptidyl transpeptidase synthesizes (γEC)$_n$G from glutathione, a related activity polymerizes γEC into (γEC)$_n$. Various γEC isopeptides, metal and acid-labile sulfide form metal-binding complexes. The exact compositions vary with organisms and the metal, cadmium or copper. The γEC isopeptides provide Cd-tetrathiolate coordination. The binding of low quantities of sulfide is uncertain, in high sulfide complexes a CdS crystallite can occur.

The contents of γEC isopeptides rise and fall with fluctuating supplies of Cu and zinc suggestive of a role in metal homeostasis. Copper and Zn(γEC)$_n$G complexes activate apoforms of metal-requiring enzymes and the γEC isopeptides can protect Cd-sensitive enzymes. About 62% of intracellular Cd is sequestered by γEC isopeptides in cultured plant cells and up to 59% in maize roots. The speciation of the remaining cellular Cd is unknown. Complexation of Cd by γEC isopeptides is essential for survival of Cd-adapted yeasts and cultured plant cells. The role of such complexes in metal tolerance of intact plants is uncertain.

Genes related to archetypal MT occur in flowering plants. The gene product expressed in transformed E. coli binds Cu, Cd, or Zn, however, the region linking the two Cys-rich domains is vulnerable to proteolytic attack in vivo. Antibodies to rat MT react in certain cells of seeds and seedlings. Perhaps the putative archetypal plant MT functions in regenerative parts of plants whereas mature cells in the bulk of the plant respond to metal by synthesis of γEC isopeptides.

INTRODUCTION

Organisms react to excess metals from the environment in various ways including the accumulation of thiol groups able to chelate intracellular metal. In plants, algae and certain yeasts the major thiols for this purpose are the class III metallothioneins (MTs). These are atypical, nontranslationally synthesized thiolate polypeptides ranging in length from 5 to 23 amino acids. The primary structure of such peptides was first shown to be $(\gamma\text{-Glu-Cys})_n\text{-Gly}$ or $(\gamma EC)_nG$ where n=2 or 3 in the fission yeast Schizosaccharomyces pombe (Kondo et al., 1984). Independent studies provided the same primary structure for the n=3 to 7 peptides from cultured cells of the plant Rauvolfia serpentina as reported by Dr. E. Grill at the Second International Meeting on Metallothioneins in 1985 (Grill, 1987). Mixtures of the peptides form complexes with metal. The purpose of this essay is to summarize developments and highlight current information in a selective rather than comprehensive manner. Several reviews on the subject are available (Rauser, 1990; Reddy and Prasad, 1990; Robinson, 1990; Steffens, 1990).

STRUCTURE

PRIMARY STRUCTURE: The major sources of peptides for structural analyses have been Cd-binding complexes from S. pombe and various plants particularly cultured cells. The structure of $(\gamma EC)_nG$ peptides was confirmed by chemical synthesis of the penta- and heptapeptides (Kondo et al., 1984, 1985) and the nanopeptide (Grill et al., 1985). The recurrent γ-carboxyamide linkage prevented exclusive use of Edman protein sequencing methodology. Steffens et al. (1986) confirmed the sequence of the hepta- and nanopeptides from tomato (Lycopersicon esculentum) by mass spectrometry. Enzymic digestions provided the sequence information of peptides from Datura innoxia, the γ-glutamyl linkages were identified by ^{13}C-NMR spectroscopy (Jackson et al., 1987).

Once the primary structure of $(\gamma EC)_nG$ peptides was established for fission yeast, three dicotyledonous plants and the alga Chlorella fusca

(Gekeler et al., 1988), occurrence of the peptides in other species was tested. Using Cd as the inducing metal, all algae (Gekeler et al., 1988), bryophytes, pteridophytes, gymnosperms and angiosperms tested were found to contain $(\gamma EC)_nG$ peptides (Gekeler et al., 1989). The yeast *Candida glabrata* produced MT in the presence of Cu and $(\gamma EC)_nG$ with Cd (Mehra et al., 1988). Other microbes producing MT could form $(\gamma EC)_nG$ peptides (Kneer et al., 1992). Both an MT-free mutant and a wild type strain of *Saccharomyces cerevisiae* exposed to Cd produced $(\gamma EC)_2G$ as verified by mass spectrometry. A small amount of $(\gamma EC)_2G$ occurred in *Neurospora crassa*. The n=2 to 6 peptides were also detected in six species of Basidiomycetes and four species of *Schizosaccharomyces* in addition to *S. pombe*. The proportion of cellular Cd bound by MT and $(\gamma EC)_nG$ peptides was not evaluated.

The $(\gamma EC)_nG$ peptides discussed so far are related to glutathione (γECG) because the γEC portion is repeated 2 to 11 times and the Gly is carboxy-terminal (Grill et al., 1987). Amongst certain legumes, however, glutathione is partly replaced by the homologue homoglutathione (γ-Glu-Cys-β-Ala or γECβA) depending on whether roots, leaves or seeds are analyzed (Klapheck, 1988). Roots of soybean (*Glycine max*) contained abundant homoglutathione and on exposure to Cd produced $(\gamma EC)_n\beta A$ peptides (Grill et al., 1986). The primary structure of this series of peptides was established by a cyclic regime of γ-glutamyltranspeptidase treatment and Edman degradation with analysis of the liberated amino acids, β-Ala being identified by mass spectrometry. Of the 43 species of legumes challenged with Cd, 7 contained $(\gamma EC)_nG$ peptides, 13 had $(\gamma EC)_n\beta A$ peptides and 23 species had both series reflecting the relative abundance of both glutathione and homoglutathione.

The tripeptide γ-Glu-Cys-Ser (γECS), a second homologue of glutathione, was identified in several graminaceous species (Klapheck et al., 1992a). The abundance of this tripeptide ranged from being equal to that of glutathione in rice (*Oryza sativa*), to 3.5-fold less than glutathione in *Agrostis tenuis*, to not being detectable in *Saccharum*, *Sorghum* and _Zea_ species. Plants of rice and wheat (_Triticum aestivum_), able to form both γECG and γECs, produced both the $(\gamma EC)_{2\text{-}3}G$ and the $(\gamma EC)_{2\text{-}3}S$ series of peptides on exposure to Cd, rice responded similarly to excess Pb and Zn (Klapheck et al., 1992b). Two peptides containing Glu, Cys and Ser also occurred in the Cd-binding complex of _Agrostis gigantea_ (Rauser et al., 1988) resembling the situation in rice.

Maize (_Zea mays_) roots exposed to Cd produced a range of thiols some of which participated in forming Cd-binding complex (Fig. 1). Several peptides were purified as the thionitrobenzoate derivatives and their primary structure determined by mass spectrometry. Peptide 6 was the familiar $(\gamma EC)_2G$ related to glutathione. Peptide 8 was $(\gamma EC)_2E$, a pentapeptide not based on glutathione but rather on the third homologue, the tripeptide γ-Glu-Cys-Glu (γECE) found as peak 1 in Figure 1A. The γ linkage was inferred from parallels in fragmentation patterns with γECG, γEC and $(\gamma EC)_2G$. Of the peptides induced by Cd in maize roots, peptide 6 or $(\gamma EC)_2G$ was most abundant in acid extracts up to 9 hr but was then surpassed by peptide 7, the single most abundant peptide in Cd-binding complexes from day 1 through 7 (Fig. 1B). Structural analysis showed that peptide 7 was $(\gamma EC)_2$. This desGly form of $(\gamma EC)_2G$ occurred in lesser quantities to $(\gamma EC)_2G$ in the Cd complex of _C. glabrata_ (Mehra et al., 1988), the n=3 and 4 desGly variants occurred in _S. pombe_ (Mehra and Winge, 1988). It is possible that these peptides are products of incomplete biosynthesis or of catabolism. The unusual accumulation of γEC in Cd-exposed roots of maize after the level of glutathione had fallen (Rauser et al., 1990; Meuwly and Rauser, 1992; Rüegsegger and Brunold, 1992) could be interpreted as availability of the precursor for biosynthesis of $(\gamma EC)_n$. Structural identification of the other thiolate peptides in Figure 1 is in progress.

NOMENCLATURE: The peptides related to glutathione have been given the trivial names cadystins (Kondo et al., 1984), phytochelatins (Grill et al., 1985), γ-glutamyl peptides (Mehra and Winge, 1988) and poly(γ-

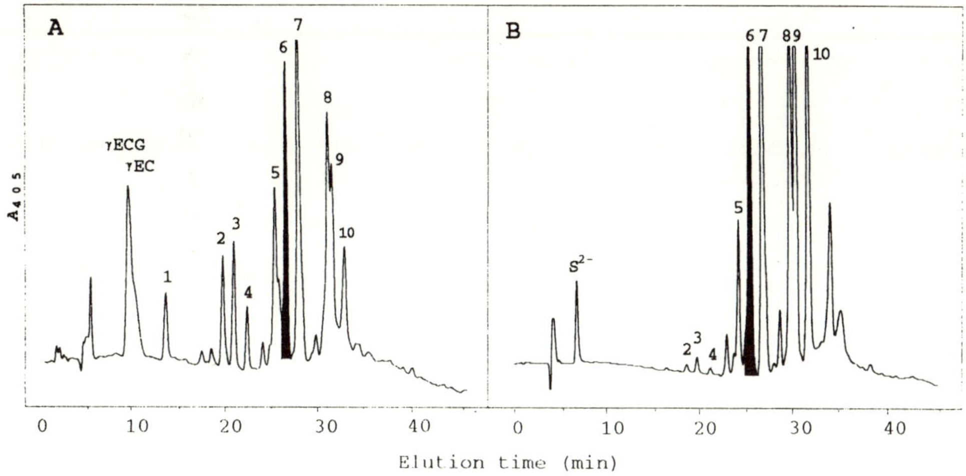

FIG. 1. Thiol profiles from roots of maize exposed to 3 μM $CdSO_4$ for seven days. Acid-soluble thiols in crude extract equivalent to 0.26 g fresh weight (A) and in a Cd-binding complex equivalent to 0.62 g fresh weight (B) were separated by reverse-phase HPLC. These groups of thiols were eluted from a C18 column by a linear gradient of 0-20% acetonitrile in 0.1% trifluoroacetic acid and detected by post-column reaction with Ellman's reagent. The amount of thiol shown for peptides 1, 6, 7, and 8 in the acid extract (A) was 4.34, 20.72, 33.87 and 20.92 nmol, respectively, and for peptides 6, 7 and 8 from the Cd-binding complex (B) 21.90, 33.81 and 25.83 nmol, respectively.

glutamylcysteinyl)glycines or $(\gamma EC)_nG$ (Jackson et al., 1987). The polypeptides related to homoglutathione have been designated homophytochelatins (Grill et al., 1986). Cadystins hold priority, all names have some drawbacks (Rauser, 1990; Steffens, 1990). From the preceding primary structures of the compounds involved in Cd-binding complexes two fundamental features arise: γEC pairs occur in multiples and the γ-carboxyamide linkage requires nontranslational biosynthesis. Both aspects are captured in the proposed trivial name γ-glutamylcysteinyl isopeptide (γ-Glu-Cys isopeptide; γEC isopeptide). This name accomodates the broad range of peptides now known (i.e. the series of desGly, Gly, β-Ala, Ser and Glu isopeptides), specifies the amino acid providing the functional thiolate and applies irrespective of the phylogenetic classification of the organism producing them.

INDUCTION BY METALS: The appearance of γEC isopeptides depends on furnishing a suitable metal. For *Rauvolfia serpentina* cell cultures Grill et al. (1987) showed that excess Ag, Au, Bi, Cu, Cd, Hg, Ni, Pb, Sb, Sn, Te, W and Zn, and the anions arsenate and selenate, induced various γEC isopeptides. No such response was found with Al, Ca, Co, Cs, K, Mg, Mn, Mo, Na or Va. The γEC isopeptides were also found in the algae *Scenedesmus acutiformis* and *Chlorella fusca* on exposure to excess Ag, Cd, Cu, Hg, Pb and Zn (Gekeler et al., 1988). Responses to a wide variety of metals have received little independent verification. Cadmium is effective in all organisms tested. The γEC isopeptides were induced in fission yeast by Cd and Cu (Reese et al., 1988) and in tobacco cells by Cd (Reese and Wagner, 1987), however, in both instances Zn was without effect.

BIOSYNTHESIS: The $(\gamma EC)_nG$ isopeptides have been considered most extensively because of their relationship to glutathione. The pool of glutathione in *R. serpentina* cells declined dramatically on adding Cd, $(\gamma EC)_2G$ accumulated rapidly and levelled off by 90 min (Grill et al., 1987). The $(\gamma EC)_3G$ isopeptide began accumulating at 30 min, the n=4 and 5 isopeptides did so successively later as if the shorter isopeptide was the precursor to the longer one. This conclusion was confirmed with Cd-exposed *Datura innoxia* cells (Robinson et al., 1988; Berger et al., 1989). Incorporation of γEC residues into $(\gamma EC)_nG$ isopeptides in the first 3-6 hr of Cd exposure equalled the consumption of glutathione (Grill et al., 1987; Berger et al., 1989). Reducing the glutathione available by administering buthionine sulfoximine, an inhibitor of γEC synthetase (EC6.3.2.2) reduced $(\gamma EC)_nG$ synthesis (Steffens et al., 1986; Grill et al., 1987; Scheller et al., 1987) and decreased the amount of Cd-binding complex (Reese and Wagner, 1987). Mutants of *S. pombe* with low activities of γEC synthetase or glutathione synthetase (EC6.3.2.3) were unable to synthesize Cd-binding complexes (Mutoh and Hayashi, 1988).

Direct evidence of glutathione involvement in the biosynthesis of $(\gamma EC)_nG$ isopeptides was shown _in vitro_ with an enzyme preparation from cell cultures of _Silene cucubalus_ (Grill et al., 1989; Loeffler et al. 1989). The γEC moiety from glutathione as a donor molecule was initially transferred to an acceptor molecule of glutathione to form $(\gamma EC)_2G$ and Gly. As $(\gamma EC)_2G$ accumulated and competed with glutathione as an acceptor molecule the γEC moieties from glutathione donor molecules were transferred to the pentapeptide to form $(\gamma EC)_3G$ and Gly. In the absence of glutathione the enzyme transferred γEC from $(\gamma EC)_2G$ as donor molecule to $(\gamma EC)_2G$ as acceptor molecule to form $(\gamma EC)_3G$ and glutathione. The enzyme was named γ-glutamyl-cysteinyl dipeptidyl transpeptidase and the general reaction cartalyzed was

$$(\gamma EC)_nG + (\gamma EC)_nG \longrightarrow (\gamma EC)_{n-1}G + (\gamma EC)_{n+1}G$$

where n=1, 2, 3, etc. (Grill et al., 1989). The kinetics observed _in vitro_ mimicked the appearance of $(\gamma EC)_nG$ isopeptides described _in vivo_ for _R. serpentina_ cells (Grill et al., 1987).

The enzyme had a molecular weight of 25,000 under denaturing conditions (Grill et al., 1989). In the non-denatured state activity was evident at M_r 95,000 early in the purification process and at 50,000 in later stages, as if the tetrameric form occurred _in vivo_ yet the dimer retained activity. The pH optimum was 7.9 and the K_M for glutathione was 6.7 mM with 100 μM Cd. Glutathione modified at its sulfur atom with the monobromobimane adduct was an even better substrate (K_M = 1.6 mM) indicating that a metal-glutathione complex was not essential for peptide cleavage. The enzyme was constitutive since it occurred in cells not previously exposed to Cd. The level of enzyme activity did not change in cells cultured with up to 500 μM Cd for 8 days (Loeffler et al., 1989).

Catalysis by γEC dipeptidyl transpeptidase depended absolutely on the presence of Cd and, in decreasing order of effectiveness, on Ag, Bi, Pb, Zn, Cu, Hg and Au (Grill et al., 1989). These were the metals that induced the appearance of $(\gamma EC)_nG$ isopeptides in vivo in plant cells and tissues. No enzyme activity was detected with Al, Ca, Fe, Mg, Mn, Na and K, non-inducing elements in vivo. Transfer of γEC moieties by the Cd-activated enzyme ceased when sufficient $(\gamma EC)_nG$ had formed to complex Cd at a thiol:Cd ratio close to 2:1 (Loeffler et al., 1989). The Cd-activated enzyme was stopped prematurely when free Cd was chelated by addition of NaEDTA or apo$(\gamma EC)_nG$. It was suggested that the dipeptidyl transpeptidase was self-regulated in that free Cd activated the enzyme and once all free Cd was chelated by the products catalysis ceased.

Hayashi et al. (1991) confirmed the γEC dipeptidyl transpeptidase reaction using radiolabelled substrates and an enzyme preparation from S. pombe. They discovered that with a low level of glutathione (12 μM rather than 1 mM) γEC from glutathione was transferred to γEC, $(\gamma EC)_2$ and $(\gamma EC)_3$ to produce the respective $(\gamma EC)_{n+1}$ polymer. In the absence of glutathione γEC was polymerized directly into $(\gamma EC)_2$ and $(\gamma EC)_3$. These data provide the first indication that the desGly variants of $(\gamma EC)_nG$ isopeptides can be products of synthesis and not necessarily products of degradation as suggested previously (Gekeler et al., 1989).

One major difference is evident: the enzyme preparation from S. pombe catalyzed the various reactions with or without Cd (Hayashi et al., 1991) whereas catalysis by the preparation from Silene cucubalus depended absolutely on the presence of Cd (Grill et al., 1989). Either the dipeptidyl transpeptidase from the two organisms differs markedly or the results are the manifestations of the enzyme preparations. Impure enzymes were used by both groups, a crude extract concentrated by phenyl Sepharose chromatography in the case of Silene (Grill et al., 1989) and the ammonium sulfate precipitate at 75% saturation from S. pombe

(Hayashi et al., 1991). The status of metal activation is preferably evaluated with highly purified enzymes. It remains unclear whether the various transpeptidase reactions apparent with the S. pombe preparation are catalyzed by a single or separate enzymes. Studies on isopeptide formation using γECβA, γECS and γECE as substrates are also required.

METAL-BINDING COMPLEXES

The major putative function of γEC isopeptides is metal binding. Assessment of such function *in vivo* is currently limited to isolating and characterizing metal-binding complexes. The most rigorous data are for Cd complexes in fission yeast and a variety of cultured plant cells, roots and shoots. There is some information on Cu complexes and minimal evidence for Zn- and Pb-complexes (Grill, 1989). Whether any of the other metals stated to induce γEC isopeptides actually form complexes is unknown.

STRUCTURE: The complexes to be understood comprise a mixture of γEC isopeptides, metal, acid-labile sulfide and perhaps sulfite. A complex from cells of *R. serpentina* contained in μmol: glutathione 3.9, $(\gamma EC)_2G$ 1.7, $(\gamma EC)_3G$ 71.0, $(\gamma EC)_4G$ 22.9, $(\gamma EC)_5G$ 3.9, $(\gamma EC)_6G$ 1.1, sulfide 1.0, Cd 89.4 for a Cd:SCys ratio of 1:3.8 (Strasdeit et al., 1991). This description probably reflects a composite or aggregate of several complexes rather than a stoichiometric composition of an individual complex. Both lyophilized material and that in solution were examined by extended X-ray absorption fine structure. The immediate environment of the Cd was four sulfur atoms at a Cd-S bond length of 2.52 ± 0.02 Å. It was proposed that $(\gamma EC)_nG$ isopeptides formed discrete $Cd(SCys)_4$ units. The carboxylate groups in the isopeptides were non-coordinating and oriented toward the surface. The latter explains the high negative charge so characteristic of Cd-binding complexes and their tendency to increase in M_r at low ionic strength (Grill et al., 1987; Reese et al., 1988). In this complex sulfide was not a significant ligand.

S. pombe exposed to 1 mM Cu produced a complex with $(\gamma EC)_3G$ and $(\gamma EC)_4G$ isopeptides binding Cu and acid-labile sulfide was absent (Reese et al., 1988). In parallel experiments with Cd the yeast produced two types of complexes with $(\gamma EC)_{3-4}G$ isopeptides: a low M_r complex with low acid-labile sulfide content and a larger M_r complex with high acid-labile sulfide content and a higher Cd:S ratio. In C. glabrata exposed to Cd the complex was formed by $(\gamma EC)_2G$ and $(\gamma EC)_2$ isopeptides and variable amounts of acid-labile sulfide (Mehra et al., 1988). In both studies the stoichiometry of peptide units and Cd bound were undefined because of uncertainties in M_r of the complex. The strength of Cd binding was enhanced in the presence of acid-labile sulfide (Reese et al., 1988) and with increasing length of $(\gamma EC)_nG$ isopeptide (Hayashi et al., 1988).

Anion exchange chromatography was used to separate three Cd-binding complexes from S. pombe (Plocke and Kägi, 1992). Complex I contained equimolar amounts of glutathione, $(\gamma EC)_2G$ and Cd, acid-labile sulfide was absent, Cd was in a mixed-ligand environment. The stoichiometry for complex II was

$$[(\gamma EC)_2G]_6\cdot[(\gamma EC)_3G]_2\cdot[S^{2-}]_1\cdot[Cd^{2+}]_6$$

and for complex III $[(\gamma EC)_2G]_2\cdot[(\gamma EC)_3G]_3\cdot[(\gamma EC)_4G]_1\cdot[S^{2-}]_1\cdot[Cd^{2+}]_6$.

Spectra from UV and magnetic and non-magnetic circular dichroism indicated Cd-tetrathiolate coordination in complexes II and III, binding of the single sulfide was uncertain.

Electron micrographs and X-ray diffraction of high sulfide complexes from C. glabrata and S. pombe showed particles 16-18 Å in diameter with properties of CdS quantum semiconductor crystallites (Dameron et al., 1989a). The 8 Å coherence length was interpreted as corresponding to a short network of about four identical Cd-S bonds. It was deduced that the thiolates in $(\gamma EC)_2G$ and $(\gamma EC)_2$ isopeptides were surface ligands to a core of CdS units. This arrangement led to the prominent UV transition around 305-318 nm in sulfide-containing complexes. In a rich nutrient broth C. glabrata exposed to Cd produced CdS crystallite coated with a mixture of both glutathione and $(\gamma EC)_2$ whereas in a synthetic medium the coating peptides were $(\gamma EC)_2G$ and $(\gamma EC)_2$ (Dameron et al., 1989b). Reese et al. (1991) showed that roots of tomato, like yeasts, produced a population of low and high sulfide-containing complexes, the latter showing UV transitions typical of small core clusters of CdS crystallites. In the high sulfide complex $(\gamma EC)_4G$ predominated whereas $(\gamma EC)_3G$ and $(\gamma EC)_4G$ predominated in the low sulfide complex.

CELLULAR LOCATION: The intracellular location of Cd-binding complexes demands that both Cd and γEC isopeptides occur in the same compartment. This was found for isolated vacuoles from leaves of tobacco (Nicotiana rustica) (Vögeli-Lange and Wagner, 1989). Within experimental error purified vacuoles contained all Cd and all $(\gamma EC)_{2\text{-}3}G$ isopeptides found in the isolated protoplasts. Glutathione was absent from the vacuoles yet occurred in protoplasts. Since γEC dipeptidyl transpeptidase was readily solubilized (Grill et al., 1989) it was likely that γEC isopeptides were synthesized extravacuolarly where glutathione was available. This situation requires transport across the vacuolar membrane of Cd and γEC isopeptides individually or as a complex. Direct evaluation of these possibilities is required. A Cd sensitive mutant of S. pombe was deficient in accumulating the high M_r sulfide-containing Cd-binding complex (Ortiz et al., 1992). The gene *hmt1* complementing the mutant shared sequence identity with the family of (ATP-binding cassette)-type transport proteins. The HMT1 polypeptide occurred in the vacuolar membrane making it a contender for transporting γEC isopeptides into the vacuole.

METAL HOMEOSTASIS: Copper and Zn are essential micronutrients which in excess can interfere with sulfhydryl groups in proteins. Temporary binding of Cu and Zn through γEC isopeptides would lower the content of free metal in cells, these complexes would be a supply of Cu and Zn to apoenzymes as they are formed. Such a homeostatic role was suggested for various cultured plant cells transferred from a spent medium to a fresh one (Grill et al., 1988). New supplies of Cu and Zn in micronutrient concentrations in fresh medium caused the appearance of $(\gamma EC)_nG$

isopeptides, these began to disappear with a half-time of about 2.5 days once the medium was depleted of Cu and Zn. No direct evidence of Cu- and Zn-$(\gamma EC)_nG$ complexes was provided. Thumann et al. (1991) reconstituted separate Cu and Zn complexes with individual $(\gamma EC)_nG$ isopeptides. The apoform of diamino oxidase from spinach was reactivated by $Cu(\gamma EC)_2G$ nearly as effectively as by an equal amount of Cu from $CuSO_4$. The $Cu(\gamma EC)_4G$ and $Cu(\gamma EC)_5G$ complexes were less efficient in activating the apoenzyme perhaps due to stronger binding of metal by the longer isopeptides. Apocarbonic anhydrase from erythrocytes was reactivated by $Zn(\gamma EC)_2G$ and $Zn(\gamma EC)_7G$, however, reactivation was 30% and 60% slower respectively than with an equal amount of Zn from $ZnSO_4$. Five metal-sensitive plant enzymes showed full activity when Cd was presented as $Cd(\gamma EC)_nG$ complex rather than as free metal, suggesting a protective role of the metal-binding γEC isopeptides (Kneer and Zenk, 1992). The roles of metal donor and of enzyme protector attributed to γEC isopeptides need to be compared to the relative effectivenes of the organic acids prevalent in plants.

<u>METAL CHELATION AND METAL TOLERANCE</u>: The importance of γEC isopeptides in binding intracellular metal can be evaluated by estimating the amount of cellular metal occurring as γEC isopeptide complex. This approach requires the selective removal of extracellular Cd, customarily through desorption of cells or tissues in a solution of divalent cation, followed by a quantitative partitioning of the remaining Cd into that bound by cell debris, that occurring as Cd-γEC isopeptide complex and residual soluble Cd (complexes with other chelators, free metal). From this viewpoint, reported values based solely on Cd partitioning during gel filtration are overestimated. Two studies following the most inclusive precepts are available. When cells of <u>R</u>. <u>serpentina</u> cultured in 100 μM Cd for 3 days were washed in Ca and Mg solution to exchange extracellular Cd, 36% of the radioactivity was associated with pelleted cell debris and 64% was buffer-soluble (Kneer and Zenk, 1992). Even though

nearly all the soluble Cd was complexed by γEC isopeptides, this form represented 62% of the Cd in washed cells. A comparable situation was found in maize. Roots exposed to 3 μM Cd for 1 to 7 days were desorbed in Ca solution and the root material was extracted with buffer six times. Of the Cd in such washed roots 92-93% was consistently solubilized. The proportion of root Cd complexed by γEC isopeptides increased with time from 19% on day 1 to 33%, 50%, 51% and 59% on days 2, 4, 5 and 7 respectively (Rauser, unpublished data). The lower proportion of complexed Cd in maize roots during the first 2 days of exposure was associated with decreased rates of root elongation whereas the 50-59% values from day 4 onward were associated with a recovered rate of root elongation (Meuwly and Rauser, 1992). The appreciable amounts of soluble Cd not complexed by γEC isopeptides indicate that other agents, perhaps organic acids (Krotz et al., 1989; Wang et al., 1991), may be of importance in cellular reactions to excess Cd.

The ability to form Cd-γEC isopeptide complexes is essential for adaptation of yeast and cultured plant cells to excess Cd. Inhibition of γEC isopeptide synthesis by reducing γEC synthesis with buthionine sulfoximine inhibited growth of cells in otherwise non-toxic concentrations of Cd (Steffens et al., 1986; Huang et al., 1987). Formation of high sulfide-containing γEC isopeptide complexes in particular was essential for survival of S. pombe in excess Cd (Mutoh and Hayashi, 1988; Ortiz et al., 1992). Overproduction of γEC isopeptides by cultured cells of Datura innoxia was not responsible for Cd tolerance, only rapid production of Cd-γEC isopeptide complex correlated with tolerance (Delhaize et al., 1988). According to Schat and Kalff (1992) the amounts of γEC isopeptide in root apecies of Silene cucubalus did not correlate with Cu tolerance, rather they reflected metal-imposed strain. This study is exemplary in that comparisons of cellular constituents are based on the target tissue of roots showing the same relative growth inhibition rather than on growing roots of the metal-tolerant ecotype and partly or totally inhibited roots of the metal-sensitive ecotype. Schat and Kalff, however, did not measure the amount of Cu-γEC isopeptide complex which, in the case of Cd, was correlated with tolerance (Delhaize et al., 1988). Ernst et al. (1992) used the few available imperfect experiments to argue that γEC isopeptides were not involved in metal tolerance. There is still room for critical assessments of the specific role of metal-sulfide-γEC isopeptide complexes in differential metal tolerance, preferably for a variety of plant species and metals.

ARCHETYPAL METALLOTHIONEIN IN PLANTS

Identification of γEC isopeptides in plants arose from the search for archetypal MT. The challenge to isolate and characterize archetypal MT in plants has been eclipsed by the fact that a considerable portion of metal, notably Cd, is complexed by γEC isopeptides.

Mature wheat embryos contained a Cys-rich Zn-binding protein with amino acid sequence homology to rat liver MT and crab MT (Lane et al., 1987). The protein disappeared in seedlings. Investigations of Cu-regulated genes in Cu-tolerant *Mimulus guttatus* uncovered two genes similar to those of MTs (de Miranda et al., 1990). The strong homology was due to two sequences coding for 14 or 15 amino acids including 6 Cys in two separate domains near either end of the molecule of 72 amino acid length. Similar genes were found in pea (Evans et al., 1990), barley (Okumura et al., 1991), soybean (Kawashima et al., 1991) and maize (de Framond, 1991). These specific gene products remain to be isolated from plants. The $PsMT_A$ gene from pea was fused to glutathione-S-transferase and the recombinant fusion protein expressed in *E. coli* (Tommey et al., 1991). After growth in 500 μM Cu, Cd, or Zn the *E. coli* produced fusion protein binding each of the metals. The pH of half-dissociation of Cu, Cd and Zn from the fusion protein was 1.45, 3.95 and 5.75, respectively, values similar to those for equine renal MT. The $PsMT_A$ gene was re-isolated from pea with a PCR strategy and expressed via a plasmid in *E. coli* (Kille et al., 1991). Ion exchange chromatography and gel electrophoresis showed at least three pools of protein all with the same total amino analysis, as if the linker region was susceptible to proteolytic attack in vivo. If this was the case for the recombinant molecule exposed to the normal proteolytic enzymes of E. coli, Kille et al. suggested that a similar set of circumstances inside plant cells would explain previous difficulties in characterising archetypal MTs in plant extracts.

Antibodies to rat MT reacted with parts of soybean (Chongpraditnun et al., 1991; Nakajima et al., 1991). In hydrated seeds immunohistochemical reaction occurred in cells of the root cap and the epidermis of cotyledons, in seedlings reaction also occurred in root cap cells and in the shoot apex below and to the side of the tunica-corpus. Perhaps the putative archetypal plant MTs are involved particularly in highly regenerative parts of plants whereas the mature cells in the bulk of the tissues respond rapidly, certainly to Cd, with the synthesis of γEC isopeptides.

Acknowledgement

Work in the authors laboratory has been supported by grants from the Natural Sciences and Engineering Research Council of Canada.

REFERENCES

Berger, J.M., Jackson, P.J., Robinson, N.J., Lujan, L.D. and Delhaize, E. (1989) Plant Cell Rep. 7, 632-635.

Chongpraditnun, P., Suzuki, K., Kawaharada, U., Nakajima, K. and Chino, M. (1991) Water Air Soil Pollut. 57-58, 545-553.

Dameron, C.T., Reese, R.N., Mehra, R.K., Kortan, A.R., Carroll, P.J., Steigerwald, M.L., Brus, L.E. and Winge, D.R. (1989a) Nature 338, 596-540.

Dameron, C.T., Smith, B.R. and Winge, D.R. (1989b) J. Biol. Chem. 264, 17355-17360.

de Framond, A.J. (1991) FEBS Lett. 290, 103-106.

de Miranda, J.R., Thomas, M.A., Thurman, D.A. and Tomsett, A.B. (1990) FEBS Lett. 260, 277-280.

Delhaize, E., Jackson, P.J., Lujan, L.D. and Robinson, N.J. (1989) Plant Physiol. 89, 700-706.

Ernst, W.H.O., Verkleij, J.A.C. and Schat, H. (1992) Acta Bot. Neerl. 41, 229-248.

Evans, I.M., Gatehouse, L.N., Gatehouse, J.A., Robinson, N.J. and Croy, R.R.D. (1990) FEBS Lett. 262, 29-32.

Gekeler, W., Grill, E., Winnacker, E.-L. and Zenk, M.H. (1988) Arch. Microbiol. 150, 197-202.

Gekeler, W., Grill, E., Winnacker, E.-L. and Zenk, M.H. (1989) Z. Naturforsch. 44C, 361-369.

Grill, E. (1987) Experientia Suppl. 52, 317-322.

Grill, E. (1989) Metal Ion Homeostasis: Molecular Biology and Chemistry, eds. D.H. Hamer and D.R. Winge, pp. 283-300. Alan R. Liss, Inc., New York.

Grill, E., Winnacker, E.-L. and Zenk, M.H. (1985) Science 230, 674-676.

Grill, E., Winnacker, E.-L. and Zenk, M.H. (1986) FEBS Lett. 205, 47-50.

Grill, E., Winnacker, E.-L. and Zenk, M.H. (1987) Proc. Natl. Acad. Sci. USA 84, 439-443.

Grill, E., Thumann, J., Winnacker, E.-L. and Zenk, M.H. (1988) Plant Cell Rep. 7, 375-378.

Grill, E., Löffler, S., Winnacker, E.-L. and Zenk, M.H. (1989) Proc. Natl. Acad. Sci. USA 86, 6838-6842.

Hayashi, Y., Nakagawa, C.W., Uyakul, D., Imai, K., Isobe, M. and Goto, T. (1988) Biochem. Cell Biol. 66, 288-295.

Hayashi, Y., Nakagawa, C.W., Mutoh, N., Isobe, M. and Goto, T. (1991) Biochem. Cell Biol. 69, 115-121.

Huang, B., Hatch, E. and Goldsbrough, P.B. (1987) Plant Sci. 52, 211-221.

Jackson, P.J., Unkefer, C.J., Doolen, J.A., Watt, K. and Robinson, N.J. (1987) Proc. Natl. Acad. Sci. USA 84, 6619-6623.

Kawashima, I., Inokuchi, Y., Chino, M., Kimura, M. and Shimizu, N. (1991) Plant Cell Physiol. 32, 913-916.

Kille, P., Winge, D.R., Harwood, J.L. and Kay, J. (1991) FEBS Lett. 295, 171-175.

Klapheck, S. (1988) Physiol. Plant. 74, 727-732.

Klapheck, S., Chrost, B., Starke, J. and Zimmermann, H. (1992a) Bot. Acta 105, 174-179.

Klapheck, S., Chrost, B. and Zimmermann, H. (1992b) 2nd Workshop Sulfur Nutrition and Sulfur Assimilation in Higher Plants, April 21-25, Grainau, Germany.

Kneer, R. and Zenk, M.H. (1992) Phytochem. 31, 2663-2667.

Kneer, R., Kutchan, T.M., Hochberger, A. and Zenk, M.H. (1992) Arch. Microbiol. 157, 305-310.

Kondo, N., Imai, K., Isobe, M., Goto, T., Murasugi, A., Wada-Nakawa, C. and Hayashi, Y. (1984) Tetrahedron Lett. 25, 3869-3872.

Kondo, N., Isobe, M., Imai, K. and Goto, T. (1985) Agric. Biol. Chem. 49, 71-83.

Krotz, R.M., Evangelou, B.P. and Wagner, G.J. (1989) Plant Physiol. 91, 780-787.

Lane, B.L., Kajioka, R. and Kennedy, T. (1987) Biochem. Cell Biol. 65, 1001-1005.

Loeffler, S., Hochberger, A., Grill, E., Winnacker, E.-L. and Zenk, M.H. (1989) FEBS Lett. 258, 42-46.

Mehra, R.K. and Winge, D.R. (1988) Arch. Biochem. Biophys. 265, 381-389.

Mehra, R.K., Tarbet, E.B., Gray, W.R. and Winge, D.R. (1988) Proc. Natl. Acad. Sci. USA 85, 8815-8819.

Meuwly, P. and Rauser, W.E. (1992) Plant Physiol. 99, 8-15.

Mutoh, N. and Hayashi, Y. (1988) Biochem. Biophys. Res. Commun. 151, 32-39.

Nakajima, K., Suzuki, K., Otaki, N. and Kimura, M. (1991) Methods Enzymol. 205, 174-189.

Okumura, N., Nishizawa, N.-K., Umehara, Y. and Mori, S. (1991) Plant Mol. Biol. 17, 531-533.

Ortiz, D.F., Kreppel, L., Speiser, D.M., Scheel, G., McDonald, G. and Ow, D.A. (1992) EMBO J. 11, 3491-3499.

Plocke, D.J. and Kägi, J.H.R. (1992) Eur. J. Biochem. 207, 201-205.

Rauser, W.E. (1990) Annu. Rev. Biochem. 59, 61-86.

Rauser, W.E., Quesnel, A.A., Lam, J.S. and Southam, G.G. (1988) Plant Sci. 57, 37-43.

Rauser, W.E., Schupp, R. and Rennenberg, H. (1990) Plant Physiol. 97, 128-138.

Reddy, G.N. and Prasad, M.N.V. (1990) Environ. Exptl. Bot. 30, 251-264.

Reese, R.N. and Wagner, G.J. (1987) Plant Physiol. 84, 574-577.

Reese, R.N., Mehra, R.K., Tarbet, E.B. and Winge, D.R. (1988) J. Biol. Chem. 263, 4186-4192.

Reese, R.N., White, C.A. and Winge, D.R. (1991) Plant Physiol. 98, 225-229.

Robinson, N.J. (1990) Heavy Metal Tolerance in Plants, ed. A.J. Shaw, pp. 195-214. CRC Press, Boca Raton.

Robinson, N.J., Ratliff, R.L., Anderson, P.J., Delhaize, E., Berger, J.M. and Jackson, P.J. (1988) Plant Sci. 56, 197-204.

Rüegsegger, A. and Brunold, C. (1992) Plant Physiol. 99, 428-433.

Schat, H. and Kalff, M.M.A. (1992) Plant Physiol. 99, 1475-1480.

Scheller, H.V., Huang, B., Hatch, E. and Goldsbrough, P.B. (1987) Plant Physiol. 85, 1031-1035.

Steffens, J.C. (1990) Annu. Rev. Plant Physiol. Plant Mol. Biol. 41, 553-575.

Steffens, J.C., Hunt, D.F. and Williams, B.G. (1986) J. Biol. Chem. 261, 13879-13882.

Strasdeit, H., Duhme, A.-K., Kneer, R., Zenk, M.H., Hermes, C. and Nolting, H.-F. (1991) J. Chem. Soc., Chem. Commun. 16, 1129-1130.

Thumann, J., Grill, E., Winnacker, E.-L. and Zenk, M.H. (1991) FEBS Lett. 284, 66-69.

Tommey, A.M., Shi, J., Lindsay, W.P., Urwin, P.E. and Robinson, N.J. (1991) FEBS Lett. 292, 48-52.

Vögeli-Lange, R. and Wagner, G.J. (1989) Plant Physiol. 92, 1086-1093.

Wang, J., Evangelou, B.P., Nielsen, M.T. and Wagner, G.J. (1991) Plant Physiol. 97, 1154-1160.

Metallothionein III
K. T. Suzuki, N. Imura and M. Kimura (eds)

MEDICAL IMPLICATIONS OF METALLOTHIONEIN

M.P. Waalkes

National Cancer Institute, Frederick Cancer Research and Development Center, Frederick, MD, USA

SUMMARY: This report summarizes and discusses the data and issues brought forth during the Symposium entitled *Medical Implications*. Several areas in which metallothionein (MT) may play an important role in human health and disease as well as in therapeutics were addressed. One of the primary areas in which MT may have a critical role is in the chemotherapy of certain cancers, both in the development of tolerance to chemotherapeutics and as an adjunct to reduce toxic side effects. In particular, it appears that induction of MT in the kidney can reduce the renal toxicity of cisplatin (CDDP), a primary limiting factor in its therapeutic usefulness. Furthermore it is clear that MT gene expression may be a primary cause of tumor cell resistance to electrophilic antineoplastic agents, including CDDP. In this regard a variety of human and animal tumor cell lines were shown to express large quantities of MT, again emphasizing its potential importance in tumor cell resistance to electrophilic chemotherapeutics. New data in mouse models of chemically induced liver and lung cancer indicate that cadmium can markedly reduce tumor incidence and multiplicity and that this correlates with poor expression of MT in these particular tumor cells, creating a potential chemotherapeutic utility for cadmium and a rational basis for its usage. A new form of MT, called growth-inhibitory factor (GIF), has been isolated specifically from the brain and its down regulation in certain nervous cells may be an important factor in the pathogenesis of Alzheimer's disease. As the medical implications of MT continue to be explored, other important roles in the diagnosis and treatment of disease will no doubt continue to be discovered.

INTRODUCTION:

Much of the previous research effort on metallothionein (MT) has concerned its role in either metal homeostasis or toxicity and the molecular control mechanisms of its genetic expression. Recently another area has emerged in MT research which concerns its medical implications, including the roles of MT in disease states and their therapy. An important aspect of this, with clear human relevance, is the role of MT in cancer cell pathobiology and drug resistance, as well as in adjunctive therapy to reduce drug toxicity. Roles for MT in other diseases, potentially including debilitating neurological diseases such as Alzheimer's disease, have also been discovered and creates exciting new avenues for further study. Other areas of importance of MT in the diagnosis of disease and its medical management will no doubt continue to be uncovered. This manuscript will review and discuss these issues as they were brought forth during the Symposium entitled *Medical Implications* and elsewhere during the meeting. Interested readers are directed at the individual papers in this volume for more in-depth coverage of the reference materials available on this topic.

ROLE OF MT IN CANCER CHEMOTHERAPY:

It is well known that the synthesis of MT is induced by various toxic metals and it is thought that this induction provides protection from toxicity by allowing sequestration of the metal. Therefore, it was initially suspected that the toxicity of metallic chemotherapeutics, such as CDDP, could be reduced by MT as well. In fact evidence from cell culture in many, but not all, cases indicates elevated MT can indeed prevent CDDP-induced cytotoxicity. Efforts in rodents demonstrated that the administration of a bismuth salt, specifically bismuth subnitrate (BSN), could induce the synthesis of MT in the kidney and thereby reduce the renal toxicity of CDDP (Naganuma et al., 1993), the clear limiting factor in its efficacy. What is most exciting about these studies is the observation that BSN does not compromise the chemotherapeutic activity of CDDP in mice, because MT was not induced by bismuth within the tumors (Naganuma et al., 1993). These findings in laboratory animals clearly set the stage for exploring the clinical use of BSN as an adjunctive therapy.

In this regard two recent efforts investigated the clinical utility of the induction of MT by BSN in controlling CDDP toxicity in cancer patients (Saijo et al., 1993; Kondo et al., 1993). Saijo et al. (1993), in a study of several patients with renal or pulmonary cancer, gave BSN daily for five days prior to the resection of tumors. Five patients with lung cancer received no BSN and served as controls. MT was markedly increased by BSN treatment in normal renal tissue compared to renal tumors. Lung MT levels were not modified by BSN either in normal or tumorous areas compared to controls. Based on the rodent studies (Naganuma et al., 1993) the enhancement of MT synthesis in the normal human kidney, but not in renal or pulmonary tumors, would indicate a potential to reduced CDDP toxicity with BSN treatments without compromising therapeutic effectiveness. In fact, treatment of advanced lung cancer patients undergoing CDDP therapy with BSN resulted in a reduction in the CDDP-nephrotoxicity, as indicated by decreased urinary β_2-microglobulin and N-acetyl-D-glucosaminidase. These results are very encouraging and it would seem quite possible to increase CDDP efficacy by increasing dosage while having the same level of toxic side effects.

Kondo et al. (1993) also studied the effects of BSN on CDDP toxicity and efficacy in humans bearing various malignancies, including renal, bladder, ureteral and testicular carcinoma. After initial studies in healthy human volunteers indicating that administration of citrate, a possible bismuth chelator, markedly enhanced oral bismuth absorption, this was employed as well in an attempt to further enhance the induction of MT synthesis by BSN. In renal cancer patients treated with BSN and citrate and then undergoing nephrectomy, MT was markedly increased by the treatments in normal renal tissue but not in the renal tumors, similar to the findings of Saijo et al. (1993). This appears to be due to a rather remarkable differential uptake of bismuth, as levels in normal renal tissue were between 360 to 1045 times higher than those found in renal tumor tissue. In clinical trials of adjunct treatments with BSN and citrate in patients with urogenital tumors undergoing multidrug chemotherapy, of which CDDP was a component, several indicators of renal dysfunction were significantly decreased by the adjunctive treatments. Again these results are very encouraging and indicate that BSN given with citrate could be an effective adjunct in chemotherapy of urogenital tumors using CDDP.

It is noteworthy that unlike many other chemotherapeutics used in combination with CDDP, which are frequently bone marrow suppressive, CDDP has a relatively low myelotoxicity. This makes a reduction in the limiting nephrotoxicity of CDDP of even greater value. The results presented in this symposium indicate that this is a real clinical possibility with BSN treatments.

<u>THE ROLE OF METALLOTHIONEIN IN CISPLATIN RESISTANCE</u>:

Chemotherapeutics that interact with DNA, such as CDDP, are a group of widely used and effective anticancer agents for solid tumors. However, both acquired and indigenous resistance to these agents can markedly reduce efficacy. MT has been implicated in many cases in acquired cellular resistance to CDDP, though the mechanisms of resistance are clearly multifactorial and include such things as reduced cellular accumulation of CDDP and increased DNA repair. Although controversy exists as to this point, it appears that acquired CDDP resistance is due, at least in some cases, to MT. This was discussed during the symposium by Lazo et al. (1993), Petering et al (1993) and Saijo et al. (1993). Other relevant information on this topic was presented by Kitazawa et al.

(1992), Koropatnick (1992), Satoh et al. (1992), Takeda et al. (1992) and Toyoda et al. (1992) during other sessions.

Lazo et al. (1993) found that various human cells lines that are made CDDP resistant, including human carcinoma cells, clearly overexpressed MT as a function of resistance. Immunohistochemical studies indicated that the subcellular localization of MT was not altered by acquisition of cellular resistance as resistant SSC25/CP cells showed similar distributions of MT compared to parental SSC25 cells. Of the multiple isoforms of human MT gene the one most commonly overexpressed in the CDDP resistant cells was MT IIA. After exposure with the pyrimidine analog 5-azacytidine, which is known to induced DNA hypomethylation, MT IIA gene expression was increased in the parental SCC25 cell line but not in the resistant SCC25/CP cell line (Lazo et al., 1993). Restriction enzyme analysis of the methylation status of the 5' regulatory region and genomic sequences of human MT IIA gene did not, however, detect differences between parental and resistant cells. This indicates methylation differences may exist between as yet to be defined cellular *trans* activating factors controlling human MT IIA expression in resistant and parental cells (Lazo et al., 1993). Thus the methylation status of the MT gene, specifically, the 5' regulatory region and genomic sequences, may be unaltered with acquired resistance and the status of activating factors may play a more important role. The molecular basis of MT-dependent resistance to CDDP clearly deserves more research effort.

Saijo et al. (1993) found that in five CDDP resistant lung cancer cell lines several forms of resistance apparently occurred. In three cell lines the main cause of CDDP resistance was the reduced accumulation of the drug and the expression of MT gene was, in some cases, actually decreased compared with the parental cell lines. On the other hand, in two resistant lines no reduction occurred in CDDP accumulation, though the formation of DNA inter-strand crosslinks was significantly decreased. In these lines the levels of MT expression was inversely correlated with sensitivity to CDDP. Similarly, Kitazawa et al. (1992) showed a positive correlation between MT expression in several clones of a mouse colon adenocarcinoma cell line and resistance to CDDP.

Petering et al (1993) discussed their findings that several animal and human tumors and cell lines had high levels of MT which may be important in relation to the zinc requirements of growing tumors. Their chemical studies also support the reactivity of the thiolate groups in MT with and CDDP (Petering et al, 1993). NMR studies imply that reaction of Cd_7-MT with CDDP stoichiometrically and cooperatively destroys the metal clusters, suggesting possible direct sequestration. In light of previous findings of the role of MT in CDDP resistance, the presence and/or induction of MT in tumors and in normal liver as a results of tumor-induced stress suggests that the antitumor efficacy may be significantly reduced by presence of MT (Petering et al., 1993). The work of Takeda et al. (1992) also indicates that MT is elevated in liver in tumor-bearing animals and this could possibly be of diagnostic value. It is important to note that effective inducers of MT, like dexamethasone and prednisone, are often components of chemotherapeutic regimes (Petering et al., 1993) which might likewise modify efficacy of drugs such as CDDP.

Apometallothionein is a hypothetical metal free form of MT and has never been convincingly shown to exist in vivo. The assertion that large amounts of apometallothionein are present in certain tumor cells (Petering et al., 1993), based on the difference between gel-filtration elution profiles of cadmium saturated and non-saturated samples, requires more rigorous substantiation. Gel-filtration separation is based on molecular weight and does not precisely define the molecular species that is binding to the cadmium upon saturation. It is obvious that other proteins that could bind cadmium when added in vitro exist in this molecular weight range, and such "non-specific" binding would appear as "apometallothionein". Supersaturation with cadmium is also another problem in that it would overestimate the number of "free" cadmium binding sites in a sample of MT, thus showing the presence of "apometallothionein". Many of the metal saturation type assays used for MT analysis, such as the Cadmium-hemoglobin assay (Onosaka et al., 1978) have similar problems, although in these assays specificity is enhanced by partial purification of MT, such as through heat treatment. The question of the existence of apometallothionein is very important and deserves very careful and critical analysis but it will be very difficult, if not impossible, with current technology to convincingly shown its existence under physiological conditions.

Recent gene transfer studies using "sense" and "antisense" gene constructs capable of up- and down-regulating MT expression, as discussed by Koropatnick (1992), indicate cells overexpressing MT had little resistance to CDDP in vitro. In this regard the mechanism by which MT overexpression induces resistance to CDDP induced in cells by continuous exposure to CDDP, is as yet undefined. Direct scavenging of CDDP by MT is a possibility and it has been shown in vitro that CDDP displaces zinc from MT. However, in light of the transfection (Koropatnick, 1992) studies it would appear that direct sequestration of CDDP by MT is not the primary cause of resistance. Thus unknown factors in addition to overexpression of MT may participate in the acquisition of CDDP the resistance phenotype. MT may interact with other cellular constituents, such as elements of DNA repair systems, or modification of homeostatic roles of MT may have a secondary effect on resistance (Koropatnick, 1992). However, Toyoda et al. (1992) found a lack of correlation with MT synthesis in HeLa cell transformants containing multiple copies of the mouse MT-I cDNA in their genomes and overexpressing MT with in vitro resistance to CDDP. In sharp contrast, a clear correlation existed between MT expression and in vivo resistance to CDDP of tumors resulting from inoculation with these transformants, establishing a major discrepancy between in vitro and in vivo results (Toyoda et al., 1992). Thus, the precise role of MT in acquired resistance to CDDP is as yet undefined, but may well depend on the system in question. This should be a prime area for future investigation.

Regardless of the precise mechanism if the presence of MT contributes to CDDP resistance then blocking its synthesis in tumor cells should enhance CDDP efficacy. Satoh et al. (1992) investigated such a possibility by using proparglycine (PPG), a cystathionase inhibitor. In mice inoculated with mouse bladder tumor, zinc treatments markedly enhanced MT in the tumor and reduced the antitumor activity of CDDP. When PPG was given along with zinc, MT induction in tumors was inhibited and development of CDDP resistance diminished. PPG similarly affected MT synthesis induced by zinc in transplanted human bladder tumor cells and fibrosarcoma cells but not in colon adenocarcinoma cells (Satoh et al., 1992). This indicates PPG could be an effective adjunct to reduce acquired resistance to CDDP, at least in some tumors.

In summary, numerous studies indicate resistance to CDDP and other electrophilic antineoplastic agents is due to several distinct mechanisms, or combinations thereof, including decreased drug accumulation, enhanced DNA repair, and increased cellular thiol pools. MT represents a major protein thiol pool within most normal cells and at least some tumors cells and probably plays a role in acquired resistance to anticancer drugs in many cases. The precise mechanisms by which such cells overexpress MT or how MT modifies CDDP toxicity are unknown and will require further research. The possibility exists that synthesis of MT could be suppressed specifically in tumor cells thereby enhancing chemotherapeutic efficacy of drugs such as CDDP.

ROLE OF MT IN THE TUMOR SUPPRESSIVE EFFECTS OF CADMIUM:

In a series of experiments originally designed to determine the carcinogenic and tumor promoting efficacy of cadmium the metal's ability to initiate or promote tumors in an established mouse liver and lung model system was studied (Waalkes et al., 1993). In sharp contrast to what was expected, when N-nitrosodiethylamine (NDEA) was given as an initiator followed by cadmium in drinking water, the NDEA-induced elevations of liver and lung tumors were prevented, rather than increased, by cadmium. NDEA-induced liver or lung tumors were inhibited by cadmium even when the metal was given months after NDEA. In initiation studies, when cadmium was given to mice before a defined promoting regimen of oral sodium barbital (BB), elevations of liver tumor multiplicity over control induced by BB were prevented by cadmium. Evidence of a deficiency of MT was seen immunohistologically in NDEA-induced hepatic and pulmonary tumors. Auxiliary studies indicated that in mice with multiple NDEA-induced liver foci there is a marked reduction in basal MT levels and in response to zinc induction. Further studies showed that spontaneous transformation of a liver cell line resulted in a marked reduction in basal MT production which was associated with a marked increase in sensitivity to cadmium toxicity. Thus it appears that cadmium may be toxic to initiated cells within liver and lung because the MT gene is only poorly expressed in these particular premalignant or transformed cells. The mechanisms involved in cadmium-induced tumor suppression and its potential as a possible chemotherapeutic strategy deserve further study.

POSSIBLE ROLE OF MT-LIKE PROTEINS IN ALZHEIMER'S DISEASE:

Roles for MT in diseases other than cancer have also been discovered and create exciting new avenues for further study. This potentially includes the debilitating neurological disorder called Alzheimer's disease (AD). Uchida (1993), Tsuji et al. (1992) and Nakajima, et al. (1992) all discussed findings with regard to Growth Inhibitory Factor (GIF) which was determined to be a 68 amino acid MT-like protein with additional 1 amino acid insert and a 6 amino acid insert in the N-terminal and the C-terminal portions, respectively, compared to MT. Molecular cloning of the cDNA for human GIF revealed striking homology to MTs (Nakajima, et al. 1992). GIF gene and MT genes are both located on chromosome 16. These results indicate that GIF is a new molecule of the MT family, although distinct differences exist such as a the lack of induction response for GIF in mouse brain by metal treatments

(Nakajima, et al., 1992). GIF was shown to inhibit neurotrophic activity in neonatal rat cerebral cortical neurons in vitro. GIF protein and mRNA are located only in the central nervous system and unlike MT are not found in liver or kidney. GIF is quite abundant in normal human brain but its presence is greatly reduced in the brain from AD patients. In the AD cortex, GIF and GIF mRNA are drastically reduced, suggesting that the GIF gene is down-regulated in AD and that this factor is important in the pathogenesis of the disease (Uchida, 1993).

AREAS FOR FURTHER RESEARCH:

Although MT appears to be expressed during periods of rapid growth in several normal tissues, a comprehensive survey of MT expression in tumor cells has not been performed. The initial work in the area is quite encouraging (Cherian, 1993; Kay et al., 1992; Suzuki et al., 1992; Waalkes et al., 1993). However, a systematic data base on MT expression in tumors should be developed, particularly since the limited available evidence is ambiguous. Of the small number of studies available concerning human or rodent tumors, several indicate MT is highly concentrated while others indicate MT can be quite deficient. Based on these disparate findings, it would be imprudent to generalize in regard to all tumors about their ability to express the MT gene, or in fact the ability to express it at all stages of development and progression in any individual tumor. Since there is evidence that expression of MT can be important in tumor cell resistance to certain chemotherapeutics these data may prove valuable in cancer management. Several important questions could be answered with such a data base, including whether the stage of tumor progression or metastases modify the expression of MT, or if MT could be used as a tumor cell marker or, in serology, as a diagnostic tool. Data concerning MT expression in tumors may also help define tumor cell metal metabolism, especially for zinc and copper, and possibly lead to further points of chemotherapeutic attack.

There are several other areas in which research into the medical implications of MT shows great promise. This includes, among others, the definition of roles for MT in metabolic disorders of metal metabolism. In this regard work is proceeding with animal models of Wilson's disease (Sugawara et al., 1992; Kodama et al., 1992; Sakurai et al., 1992; Kawahara et al., 1992). Furthermore, efforts directed at manipulation of MT gene expression in transgenic mice (Palmiter et al., 1993) should help reveal the functions of MT not only in development and physiology but also in pathogenesis of diseases, including cancer. There is also some indication than MT may play a role in development of hepatic cirrhosis (Nomura et al., 1992) and its role in other degenerative diseases should be investigated. Initial work has begun on defining the degradation of MT (Klaassen et al., 1993), and this may be an important aspect in the utilization of MT-associated essential metals or in disease states arising form improper utilization or storage of such metals.

CONCLUSIONS:

Much of research on MT has concerned its role in either metal homeostasis or toxicity and, more recently, the molecular control mechanisms of its genetic expression. Another area that is now emerging in MT research is its medical implications, including its roles in disease states, such as cancer or Alzheimer's disease, and their therapy. An important aspect of this, with clear human relevance, is the role of MT induction in normal tissue of cancer patients as a potential adjunct for reducing the toxic side effects of electrophilic chemotherapeutics, including CDDP. There is also clear evidence that MT may play a role in tumor cell pathobiology including in development of tumor cell resistance to electrophilic chemotherapeutics. Thus the sensitivity of tumor cells to chemotherapy by CDDP and perhaps other metals may depend at least in part on the extent of MT expression in tumors. Some tumors appear to be quite sensitive to cadmium, which correlates with poor expression of MT, and this may prove to be of value in the future as a chemotherapeutic. A comprehensive survey of MT expression in tumors would be of great value, but is as yet unavailable. Many other exciting new avenues for research exist within the area of the medical implications of MT.

REFERENCES:

Cherian, M.G. (1993) In Suzuki, K.T., Imura, N., and Kimura, M. (Eds.), Metallothionein III, Basel, Switzerland, Birkhäuser Verlag.

Kawahara, S., Aoki, Y., Takeichi, N., Misawa, S., Suzuki, K.T. (1992) Abstracts of The 3rd International Meeting on Metallothionein, p. 139.

Kay, J., Schmid, K., Ellis, I.O., Elmes, M.E., and Jasani, B. (1992) Abstracts of The 3rd International Meeting on Metallothionein, p. 141.

Kitazawa, H., Yamabe, Y., Imura, N., Naganuma, A. (1992) Abstracts of The 3rd International Meeting on Metallothionein, p. 92.

Klaassen, C.D., Choudhuri, S., McKim, J.M., Jr., Lehman-McKeeman, L. and Kershaw, W.C. (1993) In Suzuki, K.T., Imura, N., and Kimura, M. (Eds.), Metallothionein III, Basel, Switzerland, Birkhäuser Verlag.

Kodama, H., Tunakawa, A., Abe, T., Takama, M., Takahashi, I. (1992) Abstracts of The 3rd International Meeting on Metallothionein, p. 118.

Kondo, Y., Yamagata, K., Satoh, M., Imura, N., Akitmoto, M. (1993) In Suzuki, K.T., Imura, N., and Kimura, M. (Eds.), Metallothionein III, Basel, Switzerland, Birkhäuser Verlag.

Koropatnick, J. (1992) Abstracts of The 3rd International Meeting on Metallothionein, p. 135.

Lazo, J.S., Yang, Y.-Y., Woo, E, Kuo, S.-M. and Saijo, N.(1993) In Suzuki, K.T., Imura, N., amd Kimura, M. (Eds.), Metallothionein III, Basel, Switzerland, Birkhäuser Verlag.

Nakajima, T., Watanabe, H., Kobayashi, H., Yuasa, T., Uchida, Y., Ihara, Y., Miyatake, T., and Tsuji, S. (1992) Abstracts of The 3rd International Meeting on Metallothionein, p. 33.

Naganuma, A., Satoh, M., and Imura, N. (1993) In Suzuki, K.T., Imura, N. and Kimura, M. (Eds.), Metallothionein III, Basel, Switzerland, Birkhäuser Verlag.

Nomura, N., Ohmori, K., Asaka, N., Muranaka, Y., Kojima, J., and Kimura, M. (1992) Abstracts of The 3rd International Meeting on Metallothionein, p. 125.

Onosaka, S., Tanaka, K., Doi, M., and Okahara, K. (1978). Eisei Kagaku 24:128-131.

Palmiter, R.D., Sandgren, E.P., Masters, B.A., and Brinster, R.L. (1993) In Suzuki, K.T., Imura, N., and Kimura, M. (Eds.), Metallothionein III, Basel, Switzerland, Birkhäuser Verlag.

Petering, D.H., Quesada, M., Dughish, M., Krull, S., Gan, T., Lemkuil, D., Pattanaik, A., Byrnes, R.W., Savas, M., Whelan, C.F., and Shaw, C.F. (1993) In Suzuki, K.T., Imura, N. and Kimura, M. (Eds.), Metallothionein III, Basel, Switzerland, Birkhäuser Verlag.

Sakurai, H., Kamada, H., Fukudome, A., Hirai, S., Kimura, M., Otaki, N., Nakajima, K., Kawano, K., and Hagino, T. (1992) Abstracts of The 3rd International Meeting on Metallothionein, p. 119.

Saijo, N., Miura, K. and Kasahara (1993) In Suzuki, K.T., Imura, N. and Kimura, M (Eds.), Metallothionein III, Basel, Switzerland, Birkhäuser Verlag.

Satoh, M., Kloth, D.M., Toyoda, H., Naganuma, A., Cherian, M.G., and Imura, N. (1992) Abstracts of The 3rd International Meeting on Metallothionein, p. 138.

Sugawara, N., Sugawara, C., Katakura, M., Li, D., Sato, M., Mori, M., Nishimura, H., and Nishimura, N. (1992) Abstracts of The 3rd International Meeting on Metallothionein, p. 81.

Suzuki, K., Nakajima, K., Kawaharada, U., Makita, H., Otaki, N., and M. Kimura (1992) Abstracts of The 3rd International Meeting on Metallothionein, p. 31.

Takeda, A., Sato, T., Hoshino, A., Tamano, T., Okada, S. (1992) Abstracts of The 3rd International Meeting on Metallothionein, p. 117.

Toyoda, H., Mizushima, T., Satoh, M., Chiba, H., Naganuma, A., and Imura, N. (1992) Abstracts of The 3rd International Meeting on Metallothionein, p. 140.

Tsuji, S., Kobayashi, H., Uchida, Y., Ihara, Y. and Miyatake, T. (1992) Abstracts of The 3rd International Meeting on Metallothionein, p. 64.

Uchida, Y. (1993) In Suzuki, K.T., Imura, N., and Kimura, M. (Eds.), Metallothionein III, Basel, Switzerland, Birkhäuser Verlag.

Waalkes, M.P., Diwan, B.A., Coogan, T.P., Weghorst, T.P., Ward, J.M., Rice, J.M., Cherian, M.G. and Goyer, R.A. (1993) In Suzuki, K.T., Imura, N. and Kimura, M. (Eds.), Metallothionein III, Basel, Switzerland, Birkhäuser Verlag.

Metallothionein III
K. T. Suzuki, N. Imura and M. Kimura (eds)

UTILIZATION OF METALLOTHIONEIN INDUCER IN CANCER THERAPY

Akira Naganuma, Masahiko Satoh and Nobumasa Imura

Department of Public Health and Molecular Toxicology, School of Pharmaceutical Sciences, Kitasato University, Minato-ku, Tokyo 108, Japan.

SUMMARY: Highly effective anticancer agents usually have severe side effects which prevent their effective use in cancer chemotherapy. Protective effects of preinduction of metallothionein (MT) synthesis by MT inducers against side effects of various anticancer agents have been examined in mice. Among several metal compounds examined Bi compounds showed high ability to depress the lethal and renal toxicity of cisplatin (*cis*-DDP) without affecting its antitumor activity. This specific effect of Bi on the toxicity of *cis*-DDP appears to be attributable to its tissue specific induction of MT. Bi induces MT in normal tissues, but not in tumor tissue inoculated in mice. Bi compounds also prevented the heart and bone marrow toxicity of adriamycin and bone marrow lesion caused by γ-irradiation. Moreover, the lethal toxicity of bleomycin, pepromycin or 5-fluorouracil was significanly depressed by pretreatment of mice with Bi. These results suggest that preinduction of MT synthesis by Bi compounds is useful in preventing the adverse effects of antitumor drugs and γ-irradiation in cancer therapy.

INTRODUCTION

Many anticancer drugs with various action mechanisms have recently been developed. However, severe side effects of current anticancer agents generally interfere with their effective use in cancer chemotherapy. Metallothionein (MT), is a cysteine rich protein of low molecular weight which is induced by various metals such as Zn, Cd and Hg (Webb, 1979). Although the physiological roles of MT have not been completely clarified yet, it is known that MT detoxifies some heavy metals (Leber and Miya, 1976; Yoshikawa, 1974) and acts as a free radical scavenger (Shiraishi *et al.*, 1982; Thornalley and Vasak, 1985). Binding of alkylating agents to MT has also been reported and a high concentration of MT was found in the cells resistant to alkylating agents (Endresen *et al.*, 1983). Some heavy metal complexes, alkylating agents and free radical-generating antibiotics are currently used in cancer chemotherapy as anticancer drugs. Therefore, MT may protect the patients from the toxicity of these anticancer agents.

CISPLATIN

Cisplatin (*cis*-DDP) is a potent antitumor drug containing a heavy metal, platinum in its coordination complex structure and has been widely used in cancer chemotherapy (Rosenberg *et al.*, 1969). However, application of this platinum compound has been limited by its severe nephrotoxicity (Madias and Harrington, 1978). MT has been observed to bind with platinum in the liver and kidney of *cis*-DDP treated rats (Sharma and Edwards, 1983; Zelazowski *et al.*, 1984). Bakka et al.

Table I
Effect of pretreatment with metal compounds on lethal toxicity of *cis*-DDP

Pretreatment		
Metals	Doses (μmol/kg)	Survival rate (%)
Control	–	0
Cu	40	100
Bi	40	70
Zn	100	60
Hg	5	40
Cd	10	70
Ni	50	30
Co	150	30

Mice were administered *cis*-DDP (35 μmol/kg; s.c.) 24 hr after injection of the metal compounds.

Table II
Effect of p.o. preadministration of BSN on lethal toxicity of *cis*-DDP and MT content in kidney and liver of mice

Drugs and doses		Survival rate[a]	MT concentration[b]	
BSN (mg/kg/day)	*cis*-DDP (μmol/kg)	of mice (%)	(Hg bound, nmol/g tissue) Kidney	Liver
–	0	100	26.5 ± 6.1	31.9 ± 3.3
–	35	0		
16.7 x 5	35	50	52.2 ± 9.3	33.8 ± 4.7
33.3 x 5	35	70	104.7 ± 7.7	33.6 ± 7.2
50.0 x 5	35	100	154.7 ± 30.0	35.2 ± 6.1
66.7 x 5	35	100	198.6 ± 9.2	36.5 ± 7.0

Mice were pretreated p.o. with BSN once a day for 5days. *cis*-DDP was administered s.c. 24 hr after the last administration of BSN.
[a]Determined 20 days after cis-DDP injection. [b]Determined 24 hr after the last administration of BSN.

(1981) reported that cultured cells with a high content of MT were resistant to *cis*-DDP. These facts prompted us to examine the *in vivo* effect of preinduction of MT synthesis by heavy metals on the toxicity of *cis*-DDP.

Table I shows the effect of pretreatment of mice with Cu, Bi, Zn, Hg, Cd, Ni or Co compound on the lethal toxicity of *cis*-DDP. Among these metallothionein inducers, copper, bismuth and zinc efficiently prevented the lethal toxicity of *cis*-DDP (Naganuma *et al.*, 1984; Naganuma *et al.*, 1985). We further examined the effect of bismuth pretreatment on the toxicity of *cis*-DDP, because bismuth compounds are generally less toxic than the other metal compounds examined. Although some investigators reported that amino acid composition of the Bi-induced low molecular weight protein in rats is different from MT induced by Zn (Piotrowski *et al.*, 1979), we confirmed that bismuth is an effective inducer of MT, and significantly increases MT messenger RNA level in the kidney of mice and rats (Kobayashi et al., 1989). To examine the effect of pretreatment with bismuth compounds on *cis*-DDP toxicity, bismuth subnitrate (BSN) was chosen from several bismuth compounds, because BSN is used clinically as an antidiarrheal drug. BSN was administered orally to mice once a day for 5 days, and *cis*-DDP was administered 24 hr after the last injection of BSN. As shown in Table II, the oral administration of BSN increased MT concentration especially in the kidney of the mice. The pretreatment with BSN at a dose of over 50 mg/kg/day resulted in above a 6-fold increase in the renal MT level, and completely depressed the lethal toxicity of *cis*-DDP (Naganuma *et al.*, 1987). The protective effect of BSN against the lethal toxicity of *cis*-DDP was significantly correlated with the preinduced level of MT in the kidney, but not with the MT level in the liver (Table II). This result suggests a participation of renal MT increased by BSN in the protection of the mice from the lethal toxicity of *cis*-DDP. The effective p.o. dosage of BSN (33.3 mg/kg/day) to reduce the lethal toxicity of *cis*-DDP observed in this study (Table II) is not far from that commonly used for humans (2 g/person). The preadministration of BSN (50 mg/kg/day) could also prevent *cis*-DDP renal toxicity indicated by increase in blood urea nitrogen (BUN), the bone marrow toxicity indicated by decrease in number of total leukocytes and the intestinal toxicity indicated by incidence of diarrhea (Table III). Table IV shows the effect of BSN pretreatment on antitumor activity of *cis*-DDP examined in Ehrlich tumor-bearing mice. *Cis*-DDP administration decreased the weight of solid tumor of the mice with the increase in blood urea nitrogen levels. However, pre-treatment of tumor-bearing mice with BSN efficiently depressed the lethal and renal

Table III
Effect of p.o. preadministration of BSN on renal, bone marrow and gastrointestinal toxicity of *cis*-DDP in mice

Drugs and doses			Total	Diarrhea
BSN (mg/kg/day)	*cis*-DDP (μmol/kg)	BUN (mg/100ml)	leukocytes (cells/ml)	observed (%)
0	0	19.6 ± 0.9	7025 ± 222	0
0	45	136.0 ± 17.7	5325 ± 457	100
50 x 5	45	26.1 ± 1.8	7025 ± 310	0
50 x 5	0	120.4 ± 1.3	6975 ± 195	0

Mice were pretreated p.o. with BSN once a day for 5days. *cis*-DDP was administered s.c. 24 hr after the last administration of BSN. BUN and total leukocytes were determined 4 days after *cis*-DDP injection.

Table IV
Effect of BSN pretreatment on antitumor activity and renal toxicity of *cis*-DDP in mice inoculated s.c. with Ehrlich tumor cells

Drugs and doses		BUN on day 11	Tumor weight (g)	
BSN (mg/kg/day)	*cis*-DDP (µmol/kg)	(mg/100ml)	On day 11	On day 18
0	0	18.8 ± 2.6	1.45 ± 1.01	2.93 ± 0.27
0	20	27.2 ± 4.2	0.20 ± 0.01	0.80 ± 0.20
50 x 5	20	20.4 ± 1.3	0.18 ± 0.02	0.75 ± 0.15
0	25	38.6 ± 3.4	0.09 ± 0.03	0.34 ± 0.08
50 x 5	25	21.7 ± 0.6	0.10 ± 0.02	0.32 ± 0.08
0	35	154.3 ± 24.6	0.05 ± 0.02	died
50 x 5	35	20.9 ± 1.5	0.04 ± 0.02	0.13 ± 0.01
0	45	254.4 ± 23.7	0.03 ± 0.01	died
50 x 5	45	21.9 ± 0.6	0.03 ± 0.01	0.07 ± 0.03

Mice were inoculated s.c. with Ehrlich tumor cells on day 0. BSN was administered p.o. once a day for 5days (from day 1 to 5). *cis*-DDP was administered s.c. on day 6.

toxicity of *cis*-DDP without compromising its antitumor activity and allowed the administration of a relatively high dose of *cis*-DDP. In this case, a significant increase in the renal MT concentrations was observed by BSN administration, but the tumor MT level was not changed. This may be a reason why bismuth compounds do not affect the antitumor activity of *cis*-DDP (Kondo *et al.*, 1991; Kondo *et al.*, 1992; Naganuma et al., 1987). Contrary to Bi, pretreatment of tumor-bearing mice with Zn or Cu could induce MT synthesis in the tumor, and suppressed the antitumor activity of *cis*-DDP. This property of bismuth compounds appears to be ideal for their clinical application as an adjunct in chemotherapy with *cis*-DDP.

ADRIAMYCIN

Adriamycin (ADR) is an anthracycline antibiotic and also extensively used as an anticancer drug. However, the clinical use of ADR is limited, mainly by its

Table V
Effect of p.o. preadministration of BSN on lethal toxicity of ADR in mice

Drugs and doses		Survival rate[a]
BSN (mg/kg/day)	ADR (μmol/kg)	of mice (%)
0	0	100
0	30	0
0	30	50
50 x 5	30	70
50 x 5	30	100
50 x 5	0	100

Mice were pretreated p.o. with BSN once a day for 5days. ADR was administered s.c. 24 hr after the last administration of BSN.
[a)]Determined 20 days after ADR injection.

cardiotoxicity (Lenaz and Page, 1976). One possible mechanism for the cytotoxic effect of adriamycin is the production of free radicals during its intracellular metabolism (Bachur *et al.*, 1978; Berlin and Haseltine, 1981). Recently, metallothionein was reported to possibly act as a radical scavenger (Shiraishi et al., 1982; Thornalley and Vasak, 1985). Therefore, there is a possibility that MT prevents ADR toxicity.

Table VI
Effect of pretreatment with BSN on the lipid peroxidation induced in the heart with ADR

Drugs and doses		Lipid peroxidation in heart	
BSN (mg/kg/day)	ADR (μmol/kg)	malondialdehyde (μmol/g heart)	Conjugated dienes (A_{222})
0	0	0.38 ± 0.06	0.22 ± 0.02
0	45	1.78 ± 0.24	0.91 ± 0.06
25	45	1.56 ± 0.10	0.71 ± 0.09
50	45	1.00 ± 0.24	0.51 ± 0.12
75	45	0.61 ± 0.05	0.29 ± 0.04
100	45	0.55 ± 0.16	0.27 ± 0.05

Mice were pretreated p.o. with BSN once a day for 5days. ADR was administered s.c. 24 hr after the last administration of BSN. Lipid peroxidation in the heart was determined 4 days after ADR injection.

Table V shows the effect of pretreatment of mice with BSN once a day for 5 consecutive days on the lethal toxicity of ADR. The lethal toxicity of ADR is significantly reduced by BSN preadministration (Naganuma *et al.*, 1988). This protective effect of BSN against the lethal toxicity of ADR was observed only when BSN was administered prior to the ADR injection. The levels of both malondialdehyde and conjugated dienes, determined as indicators of lipid peroxidation in the heart, increased markedly with ADR administration, but the increases were significantly prevented by the pretreatment with BSN (Table VI). The bone marrow toxicity of ADR, reflected by a decrease in the number of total leukocytes, was also depressed by the BSN-pretreatment (Naganuma et al., 1988). Table VII shows the effect of preinduction of MT by BSN on the antitumor activity of ADR in transplanted mice. The pretreatment with BSN significantly decreased the cardiotoxicity and bone marrow toxicity of ADR, but it did not affect the antitumor activity. The protective effect of BSN on cardiotoxicity and bone marrow toxicity of ADR may be due to the induction of MT synthesis by BSN in the heart and bone marrow (Fig. 1). The involvement of cardiac MT in prevention of ADR toxicity was also examined using

Table VII
Effect of pretreatment with BSN on the antitumor activity and cardiotoxicity of ADR in mice inoculated s.c. with colon 38 tumor cells

Drugs and doses		Tumor weight (g)		Lipid peroxidation
BSN (mg/kg/day)	ADR (μmol/kg)	On day 11	On day 18	(MDA, μmol/g heart)
0	0	0.35 ± 0.08	1.16 ± 0.24	0.22 ± 0.02
50	0	0.35 ± 0.10	1.12 ± 0.31	0.91 ± 0.06
0	25	0.04 ± 0.01	0.21 ± 0.09	0.71 ± 0.09
50	25	0.04 ± 0.01	0.23 ± 0.06	0.51 ± 0.12
0	35	0.02 ± 0.00	died	0.29 ± 0.04
50	35	0.01 ± 0.00	0.09 ± 0.03	0.27 ± 0.05

Mice were inoculated with colon 38 cells on day 0. BSN was administered once a day for 5days (from day 1 to day 5). ADR was administered on day 6. Lipid peroxidation in the heart was determined on day 11.

various MT-inducing metals (Zn, Cu, Co, Cd or Hg) other than Bi (Satoh *et al.*, 1988a). The ADR-induced lipid peroxidation in the heart of mice was significantly protected by preadministration of these metals. The protective effects of these metals against cardiotoxicity of ADR were clearly correlated with the MT levels in the heart induced by these metals. These results suggest that the increased MT synthesis in the hearts may be involved in the protection against the cardiotoxicity of ADR, although the increased level of cardiac MT was only about 2- to 3-fold of the control.

ADR is often used in combination with *cis*-DDP, since the modes of actions of these two antitumor drugs are different. Therefore, the effect of MT preinduction by BSN against toxic side effects of simultaneously administered ADR and *cis*-DDP was examined in mice (Satoh *et al.*, 1988b). Pretreatment of mice with BSN (p.o.; 50 mg/kg/day) once a day for 5 consecutive days significantly reduced the renal

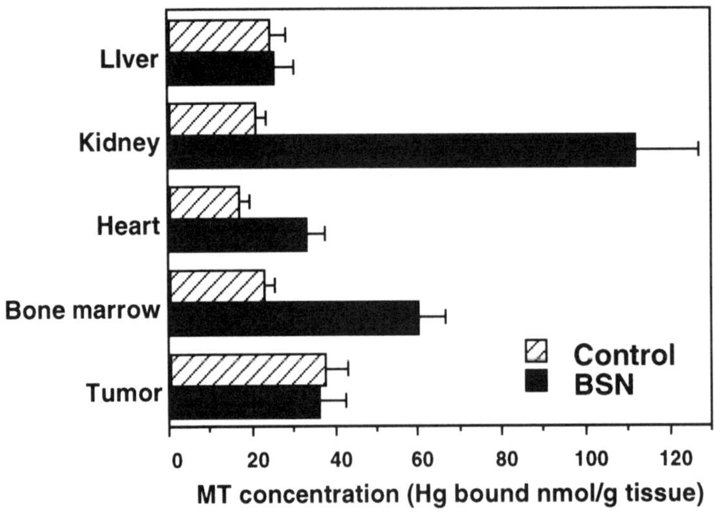

Fig. 1. MT concentrations in tissues of tumor-bearing mice after treatment with BSN

Mice were inoculated s.c. with colon 38 tumor cells. BSN was administered p.o. once a day for 5 days. MT was determined 24 hr after the last administraion of BSN.

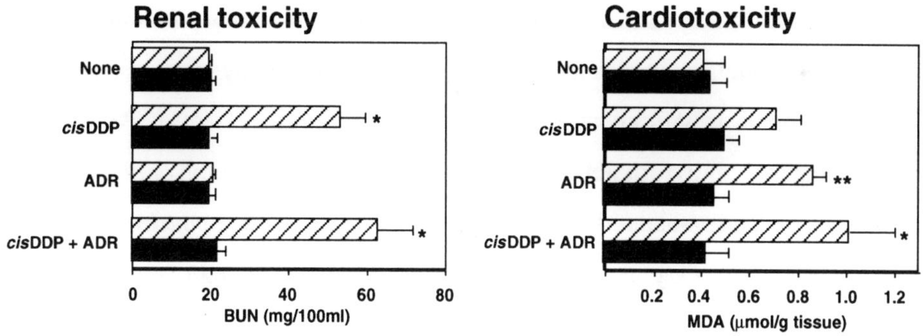

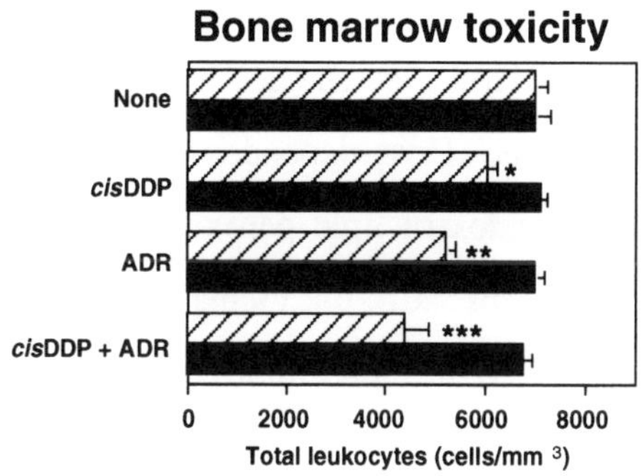

Fig. 2. Effect of preinduction of MT biosynthesis by BSN on the renal toxicity, bone marrow toxicity, and cardiotoxicity of *cis*-DDP and ADR in mice.
*,**,***:Significantly different from the control (*P<0.01, **P<0.005, *** P<0.001).
▨, BSN-untreated mice
■, BSN-pretreated mice.

toxicity of *cis*-DDP, the cardiotoxicity of ADR, and the lethal and bone marrow toxicity of both drugs observed in mice receiving ADR and *cis*-DDP simultaneously (Fig. 2). This result suggests that the preinduction of MT biosynthesis by BSN is useful for reduction of the toxic side effects of both ADR and *cis*-DDP coadministered.

γ-IRRADIATION

Cytotoxicity of irradiation is known to be due to intracellular free radial formation (Chapman and Gillespie, 1981). Bakka et al. (1982)found reduced sensitivity to radiation in Cd-resistant cells, which are rich in MT. Matsubara et al. (1986) reported that the mortality rate in irradiated mice was minimized by preadministration of MT inducing metals such as Zn and Cd. Since radiotherapy is often used clinically in combination with chemotherapy, we examined the effect of BSN on the adverse effect and antitumor activity of g-irradiation in mice (Satoh *et al.*, 1989).

BSN was preadministered to mice once a day for 5 days as in the cases of ADR and *cis*-DDP. Total body irradiation with ^{60}Co was performed 24 hr after the last administration of BSN. The number of total leukocytes (determined as an index of bone marrow toxicity) in mice irradiated by ^{60}Co without BSN pretreatment were markedly decreased in a dose-dependent manner. However, the decrease was significantly suppressed by the pretreatment with BSN. Dose-dependent increase in the level of malondialdehyde (determined as an index of lipid peroxidation) was observed in the bone marrow of mice irradiated with ^{60}Co, the increase was also significantly prevented by the pretreatment with BSN. In solid tumor-bearing mice, ^{60}Co irradiation significantly decreased the tumor weight, but it induced bone marrow toxicity indicated by decrease in number of total leukocytes. BSN-pretreatment showed no modifying effect on the tumor weight decreased by gamma-irradiation,

Table VIII
Effect of pretreatment with BSN on the antitumor activity and bone marrow toxicity of γ-ray irradiaton in mice inoculated s.c. with colon 38 tumor cells

Doses		Tumor weigt	Total leukocytes
BSN (mg/kg/day)	^{60}Co (Gy/mouse)	(g)	(cells/mm^3)
0	0	0.35 ± 0.08	0.22 ± 0.02
200 X 5	0	0.35 ± 0.10	0.91 ± 0.06
0	6	0.04 ± 0.01	0.71 ± 0.09
200 X 5	6	0.04 ± 0.01	0.51 ± 0.12

Mice were inoculated with colon 38 cells on day 0. BSN was administered orally once a day for 5 days (from day 1 to day 5), and γ-rays were irradiated on day 6. Tumor weight and total leukocytes were determined on day 16.

and significantly prevented its bone marrow toxicity (Table VIII). Thus, BSN pretreatment appears to be useful to prevent the side effects not only of chemotherapy but also of radiotherapy for neoplasms.

OTHER ANTITUMOR DRUDS

Toxicity of anticancer drugs other than ADR and *cis*-DDP were also prevented by preinduction of MT synthesis by BSN. Table IX shows the effect of pretreatment with BSN or Zn on survival rates of mice receiving various anticancer drugs (Satoh *et al.*, 1993). The pretreatment with BSN significantly protected against the lethality of cis-DDP, ADR, peplomycin (PEP) and 5-fluorouracil (5-FU). No protective effect of BSN pretreatment was observed in the mice treated with mitomycin C (MMC), vinblastine (VLB) or cyclophosphamide (CPA). When Zn was administered as an MT inducer, reduction of the lethal toxicity of cis-DDP, ADR, PEP, CPA

Table IX
Effect of pretreatment with BSN or $ZnCl_2$ on lethal toxicity of anticancer drugs in mice

Compounds	Doses	Survival rate (%) [a]		
		None	BSN	$ZnCl_2$
Control	---	100	100	100
cis-DDP	35 μmol/kg	0	100	63
Adriamycin	35 μmol/kg	0	63	57
Bleomycin	800 mg/kg	14	86	100
Peplomycin	65 μmol/kg	0	43	71
Mitomycin C	60 μmol/kg	0	14	29
Cyclophosphamide	2.5 mmol/kg	0	0	86
5-fluorouracil	4 mmol/kg	0	86	43
Vinblastine	20 μmol/kg	0	0	29

Mice were pretreated with BSN (50 mg/kg; p.o.) once a day for 5 days or $ZnCl_2$ (400 μmol/kg; s.c.) once a day for 2 days. Anticancer drugs were injected i.p. 24 hr after the last administration of BSN or $ZnCl_2$. [a)]Determined 20 days after the injection of anticancer drugs.

and 5-FU was observed. The difference between the efficacy of BSN and Zn in reducing the toxicity of the antitumor drugs examined may be ascribed to tissue specificity of these metal compounds in MT induction. Zn induces MT in the tumors and may inhibit not only the side effects of these drugs, but also their antitumor activity (Satoh *et al.*, 1993). However, these results suggest that preinduction of MT synthesis by administration of appropriate inducers may be useful for preventing the toxicities of various antitumor drugs.

The experimental results above mentioned demonstrate that the pretreatment with BSN as an MT-inducing agent may protect animals from side effects of various anticancer drugs without compromising their antitumor activity. Moreover, BSN can be used in clinical treatment with gamma-irradiation coupled with chemotherapy. The development of potent MT-inducing agents which can reduce the toxicity of antitumor drugs without influencing their antitumor activity would be desirable.

REFERENCES

Bachur, N. R., Gordon, S. L., and Gee, M. V. (1978) *Cancer Res.* **38,** 1745-1755.

Bakka, A., Endresen, L., Johnsen, A. B. S., Edminson, P. D., and Rugstad, H. E. (1981) *Toxicol. Appl. Pharmacol.* **61,** 215-226.

Bakka, A., Jonsen, A. S., Endresen, L., and Rugstad, H. E. (1982) *Experientia.* **38,** 381-383.

Berlin, V., and Haseltine, W. A. (1981) *J. Biol. Chem.* **256,** 4747-4756.

Chapman, J. D., and Gillespie, C. J. (1981) *Adv. Radiat. Biol.* **9,** 143-198.

Endresen, L., Bakka, A., and Rugstad, H. E. (1983) *Cancer Res.* **43,** 2918-2926.

Kobayashi, K., Toyoda, H., Satoh, M., Kagami, T., Naganuma, A., and Imura, N. (1989) *Toxicologist* **9**, 22.

Kondo, Y., Satoh, M., Imura, N., and Akimoto, M. (1991)*Cancer Chem. Pharmacol.* **29,** 19-23.

Kondo, Y., Satoh, M., Imura, N., and Akimoto, M. (1992) *Anticancer Res.* **12,** 2303-2308.

Leber, P. A., and Miya, T. S. (1976) *Toxicol. Appl. Pharmacol.* **37,** 403-414.

Lenaz, L., and Page, J. A. (1976) *Cancer Teat. Rev.* **3,** 111-120.

Madias, N. E., and Harrington, J. T. (1978) *Am. J. Med.* **65,** 307-314.

Matsubara, J., Shida , T., Ishioka, K., Egawa, S., Inada, T., and Machida, K. (1986) *Environ. Res.* **41,** 558-567.

Naganuma, A., Satoh, M., and Imura, N. (1984) *Res. Commun. Chem. Pathol. Pharmacol.* **46,** 265-270.

Naganuma, A., Satoh, M., and Imura, N. (1985) *Toxicol. Lett.* **24,** 203-207.

Naganuma, A., Satoh, M., and Imura, N. (1987) *Cancer Res.* **47,** 983-987.

Naganuma, A., Satoh, M., and Imura, N. (1988) *Jpn. J. Cancer Res.* **79,** 406-411.

Piotrowski, J. K., Szymanska, J. A., Mogilnicka, E. M., and Zelazowski, A. J. (1979) *Experientia, Suppl.* **34,** 363-371.

Rosenberg, R. C., Van Camp, L., Trosko, J. E., and Mansour, V. H. (1969) *Nature.* **222,** 385-386.

Satoh, M., Kloth, D. M., Kadhim, S. A., Chin, J. L., Naganuma, A., Imura, N., and Cherian, M. G. (1993) *Cancer Res.* **53,** 1829-1832.

Satoh, M., Miura, N., Naganuma, A., Matsuzaki, N., Kawamura, E., and Imura, N. (1989) *Eur. J. Cancer Clin. Oncol.* **25,** 1727-1731.

Satoh, M., Naganuma, A., and Imura, N. (1988a) *Toxicology.* **53,** 231-237.

Satoh, M., Naganuma, A., and Imura, N. (1988b) *Cancer Chem. Pharmcol.* **21,** 176-178.

Satoh, M., Tsuchiya, T., Kumada, Y., Naganuma, A., and Imura, N. (1993) *J. Trace Elem. Exp. Med.* **6,** 41-44.

Sharma, R. P., and Edwards, I. R. (1983) *Biochem. Pharmacol.* **32,** 2665-2669.

Shiraishi, N., Utsumi, K., Morimoto, S., Toja, I., Iida, S., Takeda, Y., and Aono, K. (1982) *Physiol. Chem. Phys.* **14,** 533-537.

Thornalley, P. J., and Vasak, M. (1985) *Biochim. Biophys. Acta.* **827,** 36-44.

Webb, M. (1979) In *The Chemistry, Biochemistry and Biology of Cadmium* (M. Webb, Ed.), pp. 195-266. Elsevier/North Holland, Amsterdam.

Yoshikawa, H. (1974) *Ind. Health.* **12,** 175-177.

Zelazowski, A. J., Garvey, J. S., and Hoeschele, J. D. (1984) *Arch. Biochem. Biophys.* **229,** 246-252.

Metallothionein III
K. T. Suzuki, N. Imura and M. Kimura (eds)

CLINICAL USE OF A BISMUTH COMPOUND AS AN ADJUNCT IN CEHMOTHERAPY WITH *cis*-DDP AGAINST HUMAN UROGENITAL TUMORS

Yukihiro Kondo[1], Kenji Yamagata[1], Masahiko Satoh[2], Akira Naganuma[2], Nobumasa Imura[2] and Masao Akimoto[1]

[1]Department of Urology, Nippon Medical School, 1-1-5, Sendagi, Bunkyo-ku, Tokyo, [2]Department of Public Health, School of Pharmaceutical Sciences, Kitasato University, Minato-ku, Tokyo, Japan

SUMMARY: In chemotherapy for urogenital tumors, combined treatment with multiple anticancer agents including cisplatinum (*cis*-DDP), is commonly used. Recently, we reported that oral administration of bismuth subnitrate (BSN) caused a substantial reduction of *cis*-DDP-induced nephrotoxicity in mice without compromising its antitumor activity. BSN has been used in the treatment of peptic ulcer and diarrhea for many years, and has a relatively low toxicity. A possible mechanism for its amelioration of *cis*-DDP toxicity is provided by the fact that bismuth selectively induces metallothionein (MT) synthesis in the kidneys. In many cell culture systems, MT has been shown to protect the cells against *cis*-DDP-induced cytotoxicity.

The purpose of the present study was to evaluate the effect of co-administration of sodium citrate on the intestinal absorption of BSN as well as the nephrotoxicity of *cis*-DDP and efficacy of the regimens on the renal bismuth and MT levels and nephrotoxicity of *cis*-DDP in patients bearing urogenital tumors.

Experimental groups were healthy adult male volunteers and patients bearing the urinary bladder, renal pelvis or renal cell carcinoma.

The nephrotoxicity of *cis*-DDP was evaluated by the ratios of the values of blood urea nitrogen and serum creatinine as well as 24 hr creatinine clearance measured before and after each course of chemotherapy for individual patient. It was confirmed that bismuth could hardly be taken up by the renal cell carcinoma cells, while it substantially accumulated in the normal region of human kidneys. The effects of pretreatment with bismuth subnitrate and sodium citrate on the nephrotoxicity of *cis*-DDP were examined during the multidrug chemotherapy including *cis*-DDP of more than 70 courses for 29 patients bearing urogenital tumors. The ratios in 24 hr creatinine clearance and serum creatinine values demonstrated a significant protection against *cis*-DDP induced renal dysfunction by co-administration of citrate and bismuth subnitrate which efficiently induces MT in the normal regions of human kidneys.

INTRODUCTION

cis-Diamminedichloroplatinum (*cis*-DDP) is a potent chemotherapeutic agent with activity against a wide variety of tumors, including urogenital cancers. However, the efficacy of *cis*-DDP is restricted by its severe side effects with its nephrotoxicity usually being dose-limiting. Metallothionein (MT) is an inducible cysteine-rich protein of low molecular weight , which has been reported to be protective against the toxicity of heavy metals and, therefore, may reduce the toxicity of *cis*-DDP.

It has been demonstrated in mice that oral administration of bismuth subnitrate (BSN) causes a substantial reduction in *cis*-DDP-induced nephrotoxicity without compromising its antitumor activity (Naganuma et al., 1987; Satoh et al., 1988). In addition, our subsequent experiments have revealed that the induction of MT synthesis by BSN actually protects the host not only from lesions caused by *cis*-DDP but also from those induced by Adriamycin (Naganuma et al., 1988). Furthermore, the MT concentration in renal cell carcinoma is not increased by BSN treatment (Kondo et al. , 1991).

However, the absorption rate of Bi orally administered as BSN was sometimes too low to induce satisfactory level of MT for the protection of animals against the toxicity of these anticancer agents. Further, there may be species difference in the intestinal absorption rate of Bi between human and experimental animals.

The purpose of the present study is to evaluate the effect of pretreatment with BSN and sodium citrate, a possible chelator of Bi given to enhance intestinal absorption, on *cis*-DDP-induced nephrotoxicity.

MATERIALS AND METHODS

Chemicals

cis-Diamminedichloroplatinum (*cis*-DDP) was supplied by Bristol Myers Japan Ltd. (Tokyo, Japan) and BSN was purchased from Iwaki Co., Ltd. (Tokyo, Japan).

Subjects and treatments

To examine the effect of citrate co-administration on the intestinal absorption of Bi, twelve healthy adult made volunteers aged 22 - 55 years were divided into two groups (6 each). One group was orally given BSN alone and another group was orally administered BSN and sodium citrate daily for five days and morning urine was taken after the last BSN-citrate administration. In order to examine the intestinal absorption and renal accumulation of Bi four patients bearing the renal cell carcinoma were orally administered BSN and sodium citrate daily for five days after informed consent and the kidneys with tumor were resected 24 hr after the last BSN-citrate administration. Samples from five untreated patients with renal cell carcinoma were used as the control.

For multidrug chemotherapy including *cis*-DDP 29 patients bearing various urogenital tumors were treated as indicated in Table I. During the chemotherapy with *cis*-DDP, the patients were given approximately 3000 ml of saline to maintain hydration.

Analysis

The MT and bismuth concentrations in the kidney and tumor tissue were determined at 24-h after the last administration of BSN. The MT levels were determined using the ^{203}Hg-binding assay (Piotrowski, 1973; Koyama et al., 1984).

Tissues were digested with nitric acid before analysis of the metal content. The bismuth was determined by the hydride-generation method (Brodie, 1979) using a flame atomic absorption spectrophotometer (Shimadzu model AA-640-12) equipped with a Varian VGA-76 vapor-generation accessory.

Toxicity evaluation

Blood urea nitrogen (BUN), serum creatinine levels and the 24-h creatinine clearance (24 hr-CCR) were measured as indicators of nephrotoxicity for *cis*-DDP. Nephrotoxicity was expressed as a ratio of each indicator value measured before or at 2 weeks after the course of *cis*-DDP for each individual patient.

Sephadex chromatography

The kidneys and tumors resected from the renal cell carcinoma patients were homogenized in 5 volumes of 10 mM Tris-HCl buffer (pH 7.6), and the

Table I. Patients' characteristics and numbers of courses of chemotherapy

	Standard chemotherapy	Standard chemotherapy plus BSN and citrate
Total No. of patients	14 (34)	15 (37)
Average age	59.2	54.2
Sex (M/F)	12 / 2	12/ 3
Bladder carcinoma	9 (14)	5 (10)
Renal pelvic and ureteral carcinoma	3 (12)	5 (9)
Testicular carcinoma	2 (8)	5 (18)

Five bladder carcinoma patients and five renal pelvic or ureteral carcinoma patients were given BSN (2g x 3 / day) and sodium citrate (2g x 3 / day) orally for 5 days (day 0 - day 4) before the following standard therapy. *cis*-DDP (100 mg / m^2) was given i.v. on day 5. Methotrexate(30 mg / m^2) and vinblastin (3 mg / m^2) were injected i.v. on day 4 and day 11. Five testicular carcinoma patients were given p.o. BSN with sodium citrate of the doses described above daily for 9 days (day 0 - day8). *cis*-DDP (20 mg / m^2) and etoposide (100 mg /m^2) were administered i.v. daily for 5 days (day 5 - day 9). Control groups consisted of nine bladder carcinoma, three renal pelvic or ureteral carcinoma and two testicular carcinoma patients were subjected to the standard therapy as described above without BSN-citrate treatment. Numbers in parentheses indicate the total numbers of courses of chemotherapy. These courses were repeated if necessary after 4 weeks interval. One day before the first administration of BSN in the treated groups or of the anticancer agent in the control groups in each course and two weeks after each course of the therapy serum creatinine, blood urea nitrogen and 24 hr creatinine clearance were measured as indicators for renal dysfunction.

homogenate was centrifuged at 105,000 x g for 1 h at 4 C in a Beckman ultracentrifuge. The supernatant thus obtained was applied to a Sephadex G-75 column (26.5 x 920 mm) and the column was eluted with 10 mM Tris-HCl buffer (pH 7.6). The eluates were collected in 5-ml fraction, and then the amounts of bismuth and zinc in each fraction were determined by atomic absorption spectrophotometer (Shimadzu model AA-640-12). Zn was measured by flame atomic absorption spectroscopy.

Statistics

Students' t-test was used for statistical comparison of the data between groups.

RESULTS

A pilot study carried out with two groups of healthy adult males (n = 6 each) demonstrated that the urinary excretion of bismuth in the subjects given

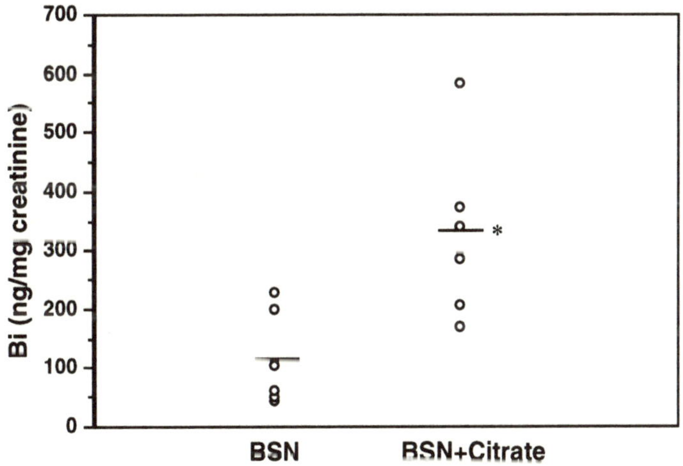

Fig. 1 Concentration of Bi in urine of adult healthy men treated with bismuth subnitrate (BSN) or BSN + citrate

BSN: p.o. treated with BSN (2g x 3 /day) for 5 days.
BSN + Citrate: p.o. treated with BSN (1g x 3 /day) + sodium citrate (2/3g x 3 /day) for 5 days.
***Significantly different from "BSN alone" (*P<0.05)**

Table II. Bismuth (Bi) concentrations in normal and malignant regions of resected human kidneys

Patient No.	Bi (ng/g tissue) Normal region	Renal cell carcinoma
Untreated		
1	17.1	2.7
2	54.1	n.d.
3	25.3	5.0
4	116.0	n.d.
5	54.1	4.8
Bi-treated		
6	9195.0	8.8
7	12043.0	33.3
8	5555.5	9.9
9	9782.5	n.d.

n.d.; not determined.

BSN plus sodium citrate was significantly higher than that in the subjects given BSN alone (Fig. 1).

The bismuth levels in normal and tumorous areas of resected kidneys of patients with renal cell carcinoma were compared. High concentrations of bismuth were found in the normal region of kidney after BSN-citrate pretreatment, while no significant increase in bismuth content was observed in the tumorous regions. Even in the untreated groups the Bi levels in the normal tissues were 5 - 10 times higher than those in the tumor tissues (Table II).

The MT level in the normal kidney tissues appeared to be increased by BSN-citrate administration, but not in the renal cell carcinoma tissues (Table III).

Distributions of bismuth and zinc in the soluble fractions of the renal tissues were examined by gel filtration. In the tissue from normal regions substantial amounts of both metals were found in the low molecular weight

Table III. Metallothionein (MT) concentrations in normal and malignant regions of resected human kidneys

Patient No.	MT (Hg bound nmol/g tissue) Normal region	Renal cell carcinoma
Untreated		
1	214.6	15.8
2	189.9	n.d.
3	203.9	74.1
4	105.8	7.5
5	165.4	9.4
Bi-treated		
6	337.6	15.8
7	678.4	19.5
8	389.4	39.2
9	415.9	10.1

n.d.; not determined.

fractions corresponding to MT fractions as well as in those of high molecular weight (Fig. 2). On the other hand, only a trace amount of bismuth was detected in MT fractions of the tissue from tumorous regions.

In order to evaluate the effect of BSN-citrate pre-administration on the nephrotoxicity of *cis*-DDP, the ratios of serum creatinine, BUN and 24 hr creatinine clearance (24 hr CCR) measured before each course to those determined 2 weeks after the course for each individual patient were calculated. As indicated in Table IV the ratio of 24 hr CCR before and after the courses is significantly higher in the BSN-citrate pretreated individuals than in the control patients. On the other hand the ratio of serum creatinine value in the BSN-citrate treated individuals is significantly lower than those in the control, although no significant alteration is observed in the ratio of BUN values.

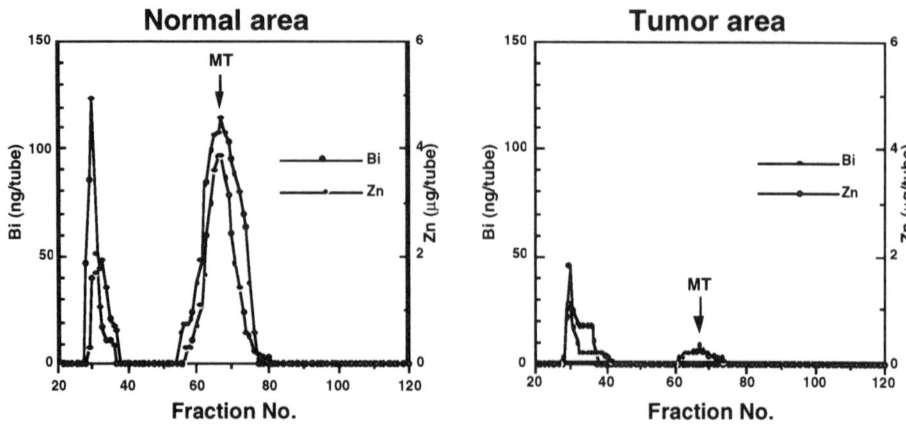

Fig. 2 Gel filtration patterns of cytosolic Bi and Zn in normal and tumorous areas of a resected kidney from a patient with renal cell carcinoma that had been administered citrate and BSN for 5 days

DISCUSSION

cis-DDP is one of the most effective anticancer agents for patients with malignant urogenital tumors. It is generally given with other anticancer agents for transitional cell carcinoma of the bladder, renal pelvis, or ureter. However, severe side effects, especially nephrotoxicity, markedly reduce its efficacy.

In chemotherapy, combinations of multiple anticancer agents are commonly used and many of these agents cause bone marrow suppression. Thus, it is important to find a method of reducing the nephrotoxicity of *cis*-DDP, since this drug has a relatively low myelotoxicity. Many efforts (Borch et al., 1980; Howell et al., 1983) have been made to reduce the nephrotoxicity of *cis*-DDP, because it limits the clinical efficacy of *cis*-DDP-based chemotherapic regimens. We have previously reported that administration of bismuth compounds prior to *cis*-DDP treatment reduce nephrotoxicity and lethality in rodents without diminishing its antitumor activity against transplanted mouse tumors (Naganuma et al., 1987).

This effect of bismuth can be explained by its ability to induce MT synthesis in the kidneys, a major target organ of *cis*-DDP, but not in the tumors. In order to develop an effective protocol for cancer chemotherapy, we examined the

Table IV. Ratios of renal toxicity indicators measured before and after each course of therapy

Indicators	Renal toxicity ratios[a)]	
	Standard chemotherapy	Standard chemotherapy plus BSN and citrate
24hr CCR	0.89 ± 0.26	1.18 ± 0.26[b)]
BUN	1.21 ± 0.45	1.02 ± 0.46
Serum creatinine	1.10 ± 0.21	0.96 ± 0.14[c)]

[a)]Values were the mean ± SD for 34 courses in the control and 37 courses in the treated patients as described in Table I.
[b),c)]Significantly different from control (P<0.001 and P<0.005, respectively).

effect of bismuth pretreatment on *cis*-DDP toxicity in patients with urogenital tumors. It was confirmed by a pilot study with healthy adult volunteers that oral co-administration of BSN with sodium citrate resulted in better intestinal absorption of bismuth compared with sole administration of BSN. Administration of BSN with citrate to the patient bearing renal cell carcinoma clearly demonstrated that the normal regions of human renal tissue took up a substantial amount of bismuth, while bismuth hardly accumulated in the tumor region. These results indicate specific protection by BSN against the renal toxicity of *cis*-DDP may occur in humans without compromising antitumor activity as so far reported by animal experiments. In trials of multidrug chemotherapy carried out for 29 patients having various urogenital carcinomas using *cis*-DDP with other anticancer agents, the ratio of three indicators for renal functions measured before and after each course of chemotherapy was calculated. The mean values of the ratios of 24 hr CCR and serum creatinine showed significant differences between the control and BSN-citrate pretreated groups. The increase in 24 hr CCR ratio in the pretreated group above 1.0 may be due to the hydration of 3,000 ml during the therapy with *cis*-DDP. In spite of the same hydration regimen as the pretreated group, the control group showed a decrease in the 24 hr CCR ratio and an increase in serum creatinine ratio, indicating a significant extent of renal dysfunction. Thus, the pretreatment with BSN and sodium citrate seems to prevent the renal damage caused by *cis*-DDP administration to patients with urogenital tumor. These findings indicate that the induction of renal MT by BSN given with sodium citrate is obviously an efficient adjunct for chemotherapy using *cis*-DDP against urogenital tumors.

REFERENCES

Borch, R.F., Katz, J.C., Lieder, P.H. and Pleasants, M.E. (1980) Proc. Natl. Acad. Sci. USA, 77, 5441-5444.
Brodie, K.G. (1979) Am. Lab. 11, 58-63.
Howell, S.B., Pfeifle, C.E., Wung, W.E. and Olshen, R.A. (1983) Cancer Res. 43, 1426-1431.
Kondo, Y., Satoh, M., Imura, N. and Akimoto, M. (1991) Cancer Chemother. Pharmacol. 29, 19-23.
Kondo, Y., Satoh, M., Imura, N. and Akimoto, M. (1992) Anticancer Res. 12, 2303--2308.
Koyama, Y., Nakatogawa, K., Naganuma, A., Imura, N., Onosaka, S. and Tanaka, K. (1984) Jpn. J. Hyg. 39, 385.
Naganuma, A., Satoh, M. and Imura, N. (1987) Cancer Res. 47, 983-987.
Naganuma, A., Satoh, M. and Imura, N. (1988) Jpn. J. Cancer Res. 79, 406-411.
Piotrowski, J.K., Bolanowska, W. and Sapata, A. (1973) Acta Biochem Pol. 20, 207.
Satoh, M., Naganuma, A. and Imura, N. (1988) Toxicology 53, 231-237.

Metallothionein III
K. T. Suzuki, N. Imura and M. Kimura (eds)

The role of metallothionein for the reduction of adverse effect of cisplatin and the induction of resistance to cisplatin

Nagahiro Saijo, M. D., Kaoru Miura, M. D., Kazuo Kasahara, M. D.

Pharmacology & Medical Oncology
National Cancer Center Research Institute and Hospital
Tsukiji 5-1-1, Chuo-ku Tokyo 104, JAPAN

Introduction

Metallothionein (MT), a metal binding protein, induced not only by heavy metals (Cherian et al., 1983; Coogan et al., 1991; Hamer 1986; Kagi et al., 1988) but also by commonly used anticancer agents has been proven to have a potential to detoxify these compounds suggesting that this substance has some role for the induction of drug resistance.(Basu et al., 1990; Bakka et al., 1981; Kelley et al., 1988; Lazo et al., 1991; Schilder et al., 1990; Teicher et al., 1987) Imura and Naganuma demonstrated that in vivo administration of bismuth nitrate (BSN) could induce MT in various tissues in mice and reduce the adverse effect of anticancer drugs. (Naganuma et al., 1987; Sato et al., 1988) In this article, the induction of MT was evaluated in cancer patients after the treatment with BSN. The protection of the disturbance of renal functions by BSN was evaluated by randomized controlled trial in patients treated with cisplatin. The role of MT is also discussed in the mechanism of cisplatin resistance.

Induction of metallothionein by the treatment with bismuth nitrate

Four patients with renal cancer and nine patients with lung cancer were accrued to this study. Informed concent was obtained from all the patients. All and four patients with renal and lung cancer, respectively received 25mg/kg of BSN twice daily for five days just before the operation. Five patients with lung cancer received no BSN. The content of MT was evaluated by Imura and Naganuma. In kidney MT content was significantly higher in normal tissue (49.19 $\pm$ 36.24 nml/gr tissue) compared with tumor tissue (6.73 $\pm$ 3.01 nmol/gr tissue). On the other hand, there was no increase of MT in patients receiving BSN not only in normal tissue but also in tumor tissue. No difference was observed in the amount of MT in the lung tumor of patients with or without BSN administration. (Table 1)

Table 1

Patient characteristics

		BSN (+)	BSN (−)
No. of patients		17	18
Age		60.4±6.4	60.6±10.6
Sex	M	11	16
	F	6	2
PS	0	1	4
	1	16	12
	2	0	2
Stage	IIIa	2	2
	IIIb	5	4
	IV	10	12
Past history of complication			
Diabetes mellitus		2	4
Hypertension		4	1
Others		3	1

This evidence suggests that the renal toxicity of anticancer agents could be reduced by BSN and that the effect of anticancer agents especially cisplatin on lung cancer would not be decreased even if BSN was administered before chemotherapy.

<u>The rabdomized trial of BSN in patients treated with cisplatin</u>

From February 1987 to March 1988 the randomized trial was conducted in National Cancer Center Hospital in order to evaluate the preventive effect of BSN on the renal function of patients treated with cisplatin. The eligibility critiria are as follows: advanced lung cancer patients without no prior cisplatin treatment, serum creatinine ≤ 1.5mg/dl and creatinine clearance ≥ 60ml/min, normal hematological function. The written informed concent was obtained from all the patients. The dose of cisplatin was 80mg/m2 and 25mg/kg of BSN was administered twice daily for five days until the day before the administration of cisplatin. Cisplatin was repeatedy administered every 3-4 weeks.

Patients were randomly assigned to with or without BSN administration. In addition to the subjective symptoms such as nausea and vomitting as well as hematological changes, renal functions such as serum creatinine and β2-microglobulin (β2-m), urinary N-acetyl-D-glucosaminidase (NAG) and β2-microglobulin were studied. The change of creatinine clearance was also determined. These examinations were repeated every course of cisplatin administration.

Patient characteristis are shown in table 2. Thirty five patients has been accrued and 17 with BSN, 18 without BSN. Majority have good performance status and there is no difference in major prognostic factors.

Table 2

Metallothionein content in lung and lung cancer

BSN (+)			BSN (−)		Ratio
Normal lung	Lung cancer	Ratio	Normal lung	Lung cancer	
3.72	3.54	1.05	5.73	8.95	0.64
3.03	8.68	0.35	6.77	9.19	0.74
6.24	2.37	2.63	10.36	14.40	0.72
7.70	9.80	0.79	4.20	4.05	1.04
			8.81	11.18	0.79
5.17±1.809	6.10±3.19		7.17±2.19	9.55±3.37	

Four patients with lung cancer were given 25 mg/kg of BSN twice daily for five days just before the operation. Five patients without BSN treatment were served as a control. Tissues of normal lung and lung cancer were obtained from each patient.
Metallothionein content : n mol/gr tissue.

All the patients received combination chemotherapy of cisplatin + vindesine with or without mitomycin C. Serum BSN concentrations ranged from 13 to 35 ng/ml and from 30 to 95ng/ml in patients received 50 and 100 mg/kg of BSN, respectively. Six and three partial responders were observed in BSN (+) and (-) groups, respectively. Majority of patients received more than two courses of combination chemotherapy.

Two patients, one per each group, were discontinued the cisplatin administration by renal toxicity. There was no difference in the frequency of acute and late emesis between two groups and in the degree of granulocytopenia and thrombocytopenia (Fig. 1, 2).

Figure 1
Changes of WBC after CDDP administration with (•) or without (○) BSN

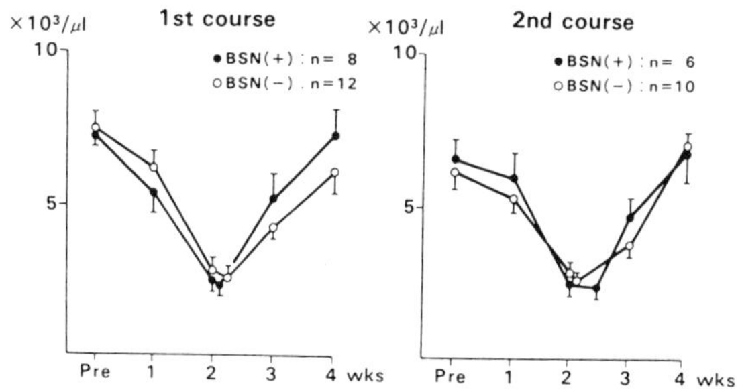

CDDP (80mg/m²) was administered and WBC was evaluated weekly.

Figure 2
Changes of thrombocyte after CDDP administration with (•) or without (o) BSN

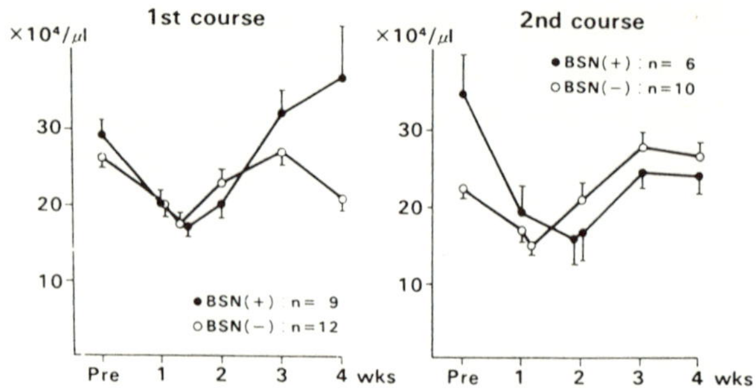

CDDP (80mg/m^2) was administered and WBC was evaluated weekly.

Serum and urine β2-m and urine NAG were evaluated serially after
cisplatin administration. Urine β2-m and NAG reached a peak level
2 days after treatment and returned to pretreatment level 7 days after (Fig. 3).

Figure 3
Changes of renal function after the administration of CDDP

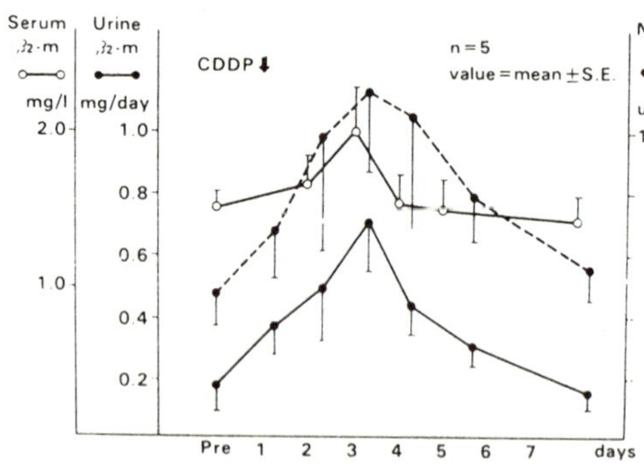

CDDP (80 mg/m^2) was administered on day 1.
Serum and urine β2microglobulin and urine NAG reached their peak levels on day 3

There was no statistical difference in serum creatinine, blood urea nitrogen and β2-m between 2 groups although all the parameters gradually increased according to the couses of chemotherapy. The values of urinary β2-m and NAG were significantly lower in BSN (+) group compared with BSN (-)group (Fig. 4).
From these data it was suggested that BSN could pevent renal dysfunction and MT induced by BSN might play an important role for this protection. In the present study majority of patients received only two courses of chemotherapy. The preventive effect of BSN on renal dysfunction should be evaluated in patients receiving many courses of cisplatin treatment.

Figure 4
Differences of β2-microglobulin and NAG in patients with (•) or without ( )BSN

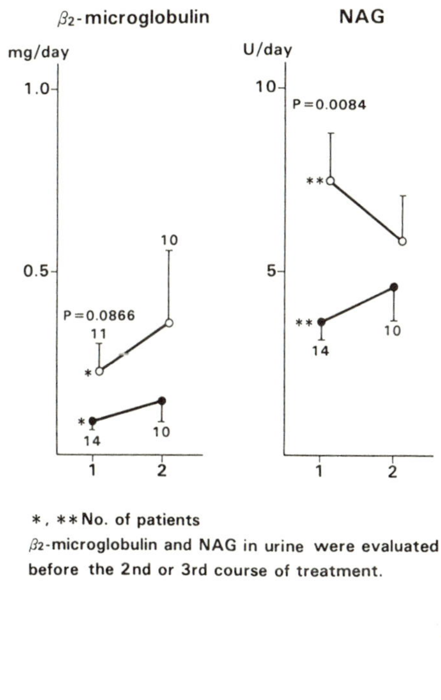

*, ** No. of patients
β2-microglobulin and NAG in urine were evaluated just before the 2nd or 3rd course of treatment.

The role of metallothionein in cisplatin resistance

The mechanisms of cisplatin resistance are multifactorial, including decreased drug accumulation, increased detoxification by thiol containing scavenger molecules such as GSH and MT, and increased repair of DNA damage (Eastman et al., 1988; Farnworth et al., 1990; Fujiwara et al., 1990; Hospers et al., 1988). Andrew et al. resported the elevation of the MT level in cadmium chloride resistant human ovarian carcinoma cells that were cross-resistant to cisplatin (Andrews et al., 1990). Mouse C127 cells transfected with the human MTII gene became resistant to cisplatin and alkylating agents. Accordingly MT may be one possible mechanism of resistance. On the other hand Schilder et al. and Farnworth resported that there was no causal relationship between MT levels and cisplatin resistance.

We have established five cisplatin resistant lung cancer cell lines such as H69/CDDP, N231/CDDP, PC-7/CDDP, PC-9/CDDP and PC-14/CDDP (Hong et al., 1988). In non small cell lung cancer cell lines such as PC-7/CDDP, PC-9/CDDP and PC-14/CDDP, the main cause of cisplatin resistance was the decreased accumulation. In PC-9/CDDP the expression of mRNA and MT protein level were decreased compared with parental cells. On the other hand there was no defect of accumulation of cisplatin in small cell lung cancer cell lines such as H69/CDDP and N231/CDDP. In two sublines of H69 such as H69/CDDP 0.2 and H69/CDDP, the formation of DNA interstrand crosslinks was significantly decreased in spite of the same amount of cisplatin uptake (Kasahara et al., 1991).

These cisplatin resistant sublines showed cross resistance to other platinum analogues and alkylating agents. The MT contents in H69/CDDP 0.2 and H69/CDDP were 1.6 and 2.6 fold higher than H69, respectively (Kasahara et al., 1991).

The amount of MT was inversely correlated with the sensitivity to cisplatin (Fig. 5).

Figure 5
Correlation of IC_{50} values with Intracellular metallothionein contents

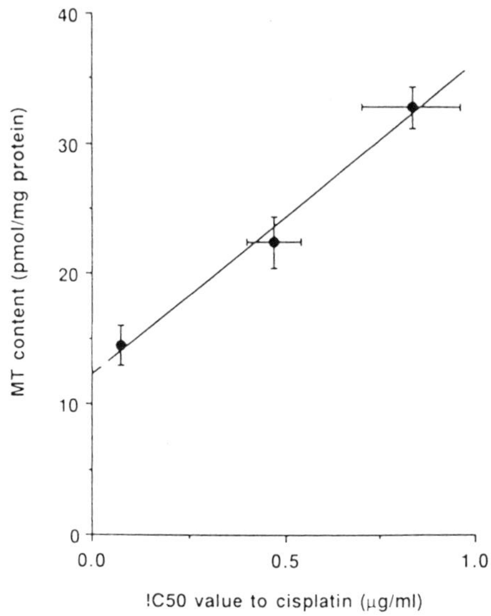

No evidence of MT gene amplification was observed in H69/CDDP 0.2 and H69/CDDP. Isoform analysis demonstrated that MTIIA mRNA was overexpressed in H69/CDDP.

Conclusion

MT could be induced in some organs treated with bismuth nitrate and the induction was considered to be organ specific. It might be possible to reduce the adverse effect of cisplatin by the administration of BSN without altering the antitumor activity. Intensive clinical trials are essential to answer this question. MT contributes to some of the mechanisms of cisplatin resistance in some cell lines. However, it is still questionable whether MT may be a primary factor controlling the toxicity and therapeutic results of anticancer drugs. Further studies are essential to examine the process of cisplatin resistance induced by MT.

Acknowledgements
This research was supported in part by a Grant-in-Aid from the Ministry of Health and Welfare for Comprehensive 10-Year Strategy for Cancer Control and from the Ministry of Education, Science and Culture. The authors also acknowledge for the funds from the Bristol-Myers Squibb Co.

References

1. Andrews, P. A. and Howell, S. B. (1990)
Cellular pharmacology of cisplatin: perspectives on mechanisms of acquired resistance.
Cancer cells, 2: 35-43.

2. Basu, A. and Lazo, J. S. (1990)
A hypotheses regarding the protective role of metallothioneins against the toxicity of DNA interactive anticancer drugs.
Toxicol. Lett ., 50: 123-135.

3. Bakka, A., Endresen, L., Johnsen, A. B. S., Edminson, P. D. and Rugustad, H. E. (1981)
Resistance against cis-dichlorodiammineplatinum in cultured cells with a high content of metallothionein.
Toxicol. Appl. Pharmacol., 61: 215-226.

4. Cherian, M. G. and Nordborg, M. (1983)
Cellular adaptation in metal toxicology and metallothionein.
Toxicology, 28: 1-15.

5. Coogan, T. P., Bare, R. and Waalkes, M. P. (1991)
Cadmium-induced DNA damage., Effect of Zinc pretreatrment.
Pro. Amer. Assoc. Cancer Res., 32: 110.

6. Eastman, A., Schulte, N., Sheibani, N., and Sorenson, C. M., (1988)
Mechanisms of resistance to platinum drugs. In, Platinium and other metal coordination compounds in cancer chemotherapy (ed. M. Nicolini), PP. 248-254. Mortinum Nijhoff, Boston.

7. Farnworth, P., Hillcoat, B. and Roos, I. (1990)
Metallothionein like proteins and cell resistance to cis-dichlorodiammenplatinum(II) in L1210 cells.
Cancer Chemother. Pharmacol., 25: 411-417.

8. Fujiwara, Y., Sugimoto, T., Kasahara, K.,Bungo, M., Yamakido, M., Tew, K. D. and Saijo, N. (1990)

8. Fujiwara, Y., Sugimoto, T., Kasahara, K.,Bungo, M., Yamakido, M., Tew, K. D. and Saijo, N. (1990)
Determinants of drug response in a cisplatin-resistant human lung cancer cell line.
Jpn . J. Cancer Res ., 81: 527-535.

9 Hamer, D. H., Metallothionein. (1986)
Ann. Rev. Biochem., 55: 913-951.

10. Hong, W. S., Saijo, N., Sasaki, Y., Minato, K., Nakano, H., Nakagawa, K., Fujiwara, Y., Nomura, M.and Twentyman, P. R. (1998) Establishment and characterization of cisplatin resistant sublines of human lung cancer cell lines.
Int. J. Cancer, 44: 462-467.

11. Hospers, G.A.P., Mulder, N. H., de Jong, B., de Ley, L., Uges, D. R. A., Fichtinger-Schepman, A. M. J., Sheper, R. J., and de Vries, E. G. E.,(1988) characterization of a human small cell lung cancer cell line with acquired resistance to cis-diammine dichloroplatinum (II) in vitro., Cancer Res., 48: 6803-6807.

12. Kagi, J. H. R. and Schaffer, A. (1988)
Biochemistry of metallothionein.
Biochemistry, 27: 8509-8515.

13. Kasahara, K., Fujiwara, Y., Nishio, K., Ohmori, T., Sugimoto, Y., Komiya, K., Matsuda, T. and Saijo, N. (1991)
Metallothionein content correlates with the sensitivity of human small cell lung cancer cell lines to cisplatin.
Cancer Res., 51: 3237-3242.

14. Kelley, S. L., Basu, A., Teicher, B. A., Hacker, M. P., Hamer, D. H. and Lazo, J. S. (1988)
Overexpression of metallothionein confers resistance to anticancer drugs.
Science, 241: 1813-1815.

15. Lazo, J. S., Reese, C., Bahnson, R. R., Teicher, B. A., Fujiwara, Y., and Saijo, N. (1991)
Metallothionein isoform expression in human lung cancer cells with acquired resistance to cisplatin.,
Proc. Amer. Assoc. Cancer Res., 32: 355.

16. Naganuma, A., Satoh, M. and Imura N. (1987)
Prevention of lethal and renal toxicity of cisdiamminedichloro platinum (II) by induction of metallothioneinsynthesis without compromising its antitumor activity in mice.
Cancer Res., 47: 983-987.

17. Satoh, M., Naganuma, A. and Imura, N. (1988)
Metallothionein induction prevents toxic side effects of cisplatin and adriamycin used in combination.
Cancer Chemother. Pharmacol., 21: 176-178.

18. Schilder, R. J., Hall, L., Monks, A., Handel, L. M., Fornace, A. J., Ozols, R. F., Fojo, A. T. and Hamilton, T. C. (1990)
Metallothionein gene expression and resistance to cisplatin in human ovarian cancer.
Int. J. Cancer, 45: 416-422.

19. Teicher, B. A., Holden, S. A., Kelly, M. J., Shea, Th. C., Cucchi, C. A., Rosowsky, A., Henner, W. D. and Frei III E. (1987)
Characterization of a human squamous carcinoma cell line resistant to cis-cliammine dichloroplatism (II).
Cancer Res., 47: 388-397.

Metallothionein III
K. T. Suzuki, N. Imura and M. Kimura (eds)

REGULATION OF METALLOTHIONEIN EXPRESSION AND CELLULAR RESISTANCE TO ELECTROPHILIC ANTINEOPLASTIC AGENTS

John S. Lazo[1], Ya-Yun Yang[1], Elizabeth Woo[1], Shiu-Ming Kuo[1] and Nagahiro Saijo[2]

[1]Department of Pharmacology, University of Pittsburgh, Pittsburgh, PA and [2]Department of Experimental Therapeutics and Medical Oncology, National Cancer Center Research Institute and Hospital, Tokyo, Japan.

SUMMARY: Metallothioneins have been implicated in reduced cellular sensitivity to electrophilic anticancer drugs as well as some mutagens and oxidants. Metallothioneins are highly inducible and transient resistance to *cis* diamminedichloroplatinum(II) (cisplatin) has been shown to occur following exposure of normal rat kidney cells to the inducer dexamethasone; this can be suppressed by coexpression of the *mos* oncogene. Some malignant human and murine cells with acquired resistance to anticancer drugs overexpress metallothioneins. In particular, we found that cisplatin resistant human head and neck squamous carcinoma cells (SCC25/CP) and small cell lung carcinoma (H69/CP) overexpressed metallothioneins. Immunohistochemical studies with antibodies to human metallothioneins have demonstrated differences in the MT subcellular localization between SCC25 and SCC25/CP cells. Multiple isoforms of human metallothionein exist but the relative importance of each in resistance to toxic electrophilic mutagens and antineoplastic drugs is not known. Thus, we developed synthetic oligonucleotides that are metallothionein isoform specific for use as probes to study the expression of individual isoforms in SCC25/CP and H69/CP cells. The most common isoform overexpressed in these cisplatin resistant cells was human metallothionein IIA. Human metallothionein IIA expression was increased in SCC25 but not SCC25/CP cells by treatment with 5-azacytidine. We used a series of restriction enzymes to assess the methylation status of the 5' regulatory region and genomic sequences of human metallothionein IIA and found no evidence for differences in methylation between SCC25 and SCC25/CP cells. One hypothesis we currently have is that changes exist in the DNA methylation status and expression of one or more cellular *trans* activating factors controlling human metallothionein IIA expression in SCC25 and SCC25/CP cells.

INTRODUCTION

DNA-interactive drugs, such as alkylating and platinating agents, represent the most efficacious and commonly used anticancer agents for solid tumors. Unfortunately, both acquired and intrinsic resistance to these drugs greatly limits their curative ability. Studies from several laboratories indicate resistance to these antineoplastic agents is the product of several distinct mechanisms including decreased drug uptake, altered DNA repair, and increased cellular thiols (Andrews and Howell, 1990; Basu and Lazo, 1990). Metallothioneins represent the major protein thiol within both normal and malignant cells. Our laboratory has focused on their role in resistance to anticancer drugs (Basu and Lazo, 1990) and the mechanisms by which cells overexpress metallothioneins.

We initially observed that cisplatin-resistant murine L1210 cells and human SW2, SCC25, SL6, and G3361 cells overexpressed metallothionein relative to their parent lines (Kelley et al., 1988). Increases in metallothionein protein levels in these cells ranged from 2 to 13 fold compared to the parental cells. Increased metallothionein levels have been seen in other cisplatin-resistant cells (Kasahara et al., 1991), although not all cisplatin resistant cells overexpress metallothionein, consistent with the multifactorial nature of cisplatin resistance (Andrews and Howell, 1990). To our knowledge, all published reports of cells, which have been made chronically resistant to cadmium and overexpress metallothionein, are cross-resistant to cisplatin; the results of Bakka et al. (1981) and Andrews et al. (1987) are good examples. In addition, Koropatnick and Pearson (1990) found both cisplatin and cadmium resistance were associated with high constitutive metallothionein expression in mouse B16 melanoma cells. Because metallothionein levels can be transiently increased in cells and tissues by a variety of pharmacological and toxicological agents (Kägi and Schäffer, 1988), we examined the possibility of transient cisplatin resistance. Normal rat kidney cells (6m2) were chosen because the kidney is a major site of cisplatin toxicity. Previous studies suggested the transforming oncogene *v-mos*, which encodes a serine/threonine kinase, suppressed dexamethasone-induced transcriptional activation of metallothionein by affecting nuclear retention of the steroid receptor complex (Basu and Lazo, 1991). Thus we used 6m2 that were infected with a modified *v-mos* oncogene construct in which expression of *v-mos* and consequently transformation was temperature sensitive. We found treatment of 6m2 cells with dexamethasone for 24 hr increased metallothionein levels by 10-fold and caused a transient 3-fold increase in resistance to cisplatin (Basu and Lazo, 1991). We believe this observation could be of great clinical importance because of the frequency of dexamethasone usage as an antiemetic with cisplatin. Interestingly, concomitant expression of the *mos* oncogene blocked cisplatin resistance. The mechanism responsible for this loss of steroid receptor responsiveness is still being examined.

Further evidence that metallothionein has a role in cellular resistance to antineoplastic agents was obtained when murine C127 cells were infected with a bovine papilloma viral construct containing the human metallothionein IIA gene. We found the transfected cells were resistant to cisplatin and alkylating agents (Kelley et al., 1988), but other investigators using the same construct did not observe similar resistance (Schilder. et al., 1990); the reasons for this difference have not been fully resolved. More recent studies suggest overexpression of metallothionein by gene transfer methods can produce cells that are more sensitive, equally sensitive or resistant to alkylating agents. For example, Lohrer and coworkers (1990), using two Chinese hamster ovary cells that were hypersensitive to methyl methane sulfonate and lacking in two independent undefined DNA repair systems, found resistance to N-methyl-N′-nitroso-N-nitroguanidine and N-nitroso-N-methylurea in one but not the second clone following transfection of the metallothionein IIA gene. More recently they have identified another clone in which overexpression of human metallothionein IIA results in increased sensitivity to these alkylating agents (Robson et al., 1992). This contextual condition for resistance coincident with or following overexpression of metallothionein could be due to differences in the subcellular location of overexpressed metallothionein or the interaction of metallothionein with other cellular constituents. The role of metallothioneins in drug resistance is further complicated by the existence and differential expression of at least 7 human isoforms of metallothionein. We have begun to examine the subcellular localization of metallothionein, the isoform composition and factors that regulate its expression in cisplatin sensitive and resistant human tumor cells.

SUBCELLULAR LOCALIZATION

Metallothionein has been identified by immunohistochemical and immunofluorescence methods to be present in the nucleus and cytoplasm of normal and malignant tissue. The subcellular distribution of metallothionein changes during development and with prolonged animal exposure to metals (Banerjee, et al., 1982; Nartey, et al., 1987). Although the functional significance of this altered subcellular distribution of metallothionein is unknown, it has been suspected that nuclear accumulation of metallothionein may occur with increased cellular metallothionein load or when cells are stimulated to undergo rapid proliferation. Studies with primary cultured adult rat hepatocyte using confocal microscopy suggest metallothionein is concentrated in the cytoplasm during G_0 and G_1 phases and in the cell nucleus in early S phase (Tsujikawa et al., 1991). We generated several rabbit antisera against human metallothionein, including one specific for human metallothionein IIA, with the goal of determining the subcellular location of metallothionein in cisplatin sensitive and resistant cells. Using immunohistochemical methods on both human squamous carcinoma cells (SCC25) and a derived cisplatin resistant line (SCC25/CP), we found more prominent staining in the SCC25/CP population, reflecting more1 metallothionein in cells, but there was significant heterogeneity in metallothionein distribution within1the population. Moreover, metallothionein subcellular distribution was markedly different between the CDDP drug sensitive and resistant populations; in SCC25/CP populations staining was mostly nuclear while in SCC25 cells staining was primarily perinuclear and cytoplasmic.

METALLOTHIONEIN ISOFORM SPECIFIC OLIGONUCLEOTIDES.

Human cells or tissues can produce 7 distinct isoforms of metallothionein, all of which exhibit extremely high amino acid homology. Because these metallothionein isoforms may not be functionally equivalent, we have begun to evaluate isoform expression in cisplatin resistant cells. We designed and synthesized oligonucleotides to the unique 5' flanking region of the 6 nonneural metallothioneins, i.e. IIA, IA, IB, IE, IF and IG. These synthetic oligodeoxyribonucleic acid probes have no significant sequence homology with any other known mammalian DNA. Metallothionein IIA mRNA was detected by Northern blot analysis in CDDP sensitive SCC25 and H69 cells and in the cisplatin resistant SCC25/CP and H69/CP cells. In H69/CP and SCC25/CP cells, which are approximately 10-fold resistant to cisplatin compared to their parental cells, the levels of metallothionein IIA mRNA were 2- to 9- fold greater than the parental cells. Results in our laboratory indicate human metallothionein IIA protein also is increased in SCC25/CP relative to SCC/25 cells. In addition we observed a 5-fold increase in steady-state mRNA levels for metallothionein IE in SCC25/CP compared to SCC25 cells; H69 and H69/CP cells did not contain constitutive mRNA levels of any other metallothionein I isoform detectable by Northern analysis, although like SCC25 and SCC25/CP, other metallothionein I isoforms could be induced by metals such as cadmium and zinc. Metallothionein IIA appears as the most prominent metallothionein isoform in the drug sensitive and resistant cells we have studied in the absence of any inducer.

TRANSCRIPTIONAL REGULATION

Because human metallothionein IIA was the most common isoform expressed in our cells, we examined further the mechanism for the increased expression of metallothionein. Our initial nuclear run-on studies suggested the increase in metallothionein IIA expression was due to elevated transcription. The majority of studies indicate DNA methylation of cytosine residues, specifically at CpG islands, suppresses gene transcription (Jones, 1986). In addition, DNA demethylating agents, such as 5-azacytidine (AZC), can stimulate the expression of metallothioneins (Foster et al., 1991). Thus we used our metallothionein IIA specific oligonucleotide to probe the effects of AZC on metallothionein IIA mRNA levels in SCC25 and SCC25/CP cells. AZC pretreatment (16μM for 72 hr) increased basal metallothionein IIA levels by approximately 3-fold in SCC25 but not SCC25/CP cells. These results were consistent with hypomethylation of either the metallothionein gene, most likely the 5′ basal regulatory region, or other genes encoding a factor that regulates metallothionein expression *in trans*. Foster et al. 1991, however, suggest AZC may alter intracellular copper in Hep G2 cells but this seems unlikely to explain our results as it would require preferential AZC effects on SCC25 but not SCC25/CP cells.

To address this question, we isolated DNA from both SCC25 and SCC25/CP cells and studied the DNA methylation status by restriction analysis with methylation sensitive enzymes. Although the analysis was limited by the number of potential methylation sites recognized by our battery of restriction enzymes, we found no evidence for differential DNA methylation in the 5′ regulatory region or the genomic sequence of metallothionein IIA in SCC25 and SCC25/CP cells. Thus, we concluded the overexpression of metallothionein IIA in SCC25/CP cells was due in part to altered methylation of a gene encoding a yet to be defined *trans*-activating factor. This is illustrated schematically in Figure 1, where the putative transcription factor is labeled "X". Current work in our laboratory is focused on defining this *trans*-activating factor.

MECHANISMS OF RESISTANCE

What is the mechanism by which metallothionein overexpression affects cellular sensitivity to electrophilic agents? Certainly metallothionein has been shown to interact with cisplatin in tissues and cells (Zelazowski et al., 1984). Incubation of cisplatin with metallothionein *in vitro* results in stoichiometric binding of platinum to sulfhydryl groups with the displacement of zinc (Pattanaik et al., 1992). Interestingly, however, both cisplatin and biologically inactive transplatin have similar metallothionein binding kinetics. Although it remains possible that the protective effect of metallothionein against electrophilic agents resides in a direct interaction between metallothionein and the electrophile, it seems more likely in light of the recent gene transfer studies with cells overexpressing metallothionein that metallothioneins interact with other cellular constituents (Koropatnick and Pearson, In Press). For example Robson et al. (1992) have suggested metallothionein may interact with elements of DNA repair systems. It was suggested more than a decade ago that metallothionein may assist in regulating intracellular zinc distribution (Udom and Brady, 1980; Li et al., 1980). Koropatnick and Pearson (1990) concluded zinc induces other factor(s) in addition to metallothionein that participate in the cisplatin resistance phenotype. These issues would now appear to be areas for future investigation.

SUMMARY

Overexpression of metallothionein has been proposed as one mechanism of cellular resistance to electrophilic anticancer drugs. Our results indicate metallothionein IIA is overexpressed in human tumor cells with acquired resistance to cisplatin. This overexpression is due to increased transcription of the metallothionein IIA gene. We speculate that differences exist in the DNA methylation status and expression of one or more cellular *trans* activating factors controlling human metallothionein IIA expression in SCC25 and SCC25/CP cells.

Acknowledgments

This work was supported in part by an American Cancer Society Grant CH486, a fellowship from the Japanese Foundation for Promotion of Cancer Research (J.S.L.), and a Grant-in-Aid from the Ministry of Health and Welfare for the Comprehensive 10-Year Strategy for Cancer Control (Japan).

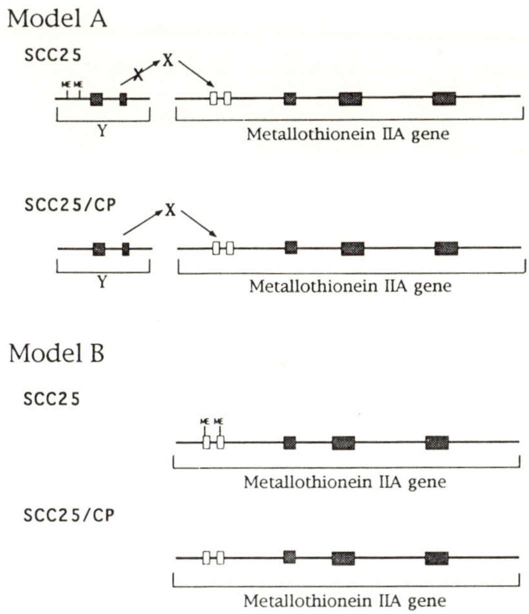

Figure 1. Two schematic models of metallothionein regulation in cisplatin sensitive and resistant cells. Model A. A putative DNA regulatory region (Y) is hypermethylated in SCC25 cells and hypomethylated in SCC25/CP cells. Its gene product "X" interacts with basal regulatory regions (open box) in the metallothionein IIA gene. Model B. The basal regulatory regions (open box) of SCC25 are methylated preventing transcription.

REFERENCES

Andrew, P.A., Murphy, M.P. and Howell, S.B. (1987) Cancer Chemother. Pharmacol. 19: 149-154.

Andrews, P.A. and Howell, S.B. (1990) Cancer Cells 2: 35-43.

Bakka,A., Endresen, L., Johnsen, A.B.S., Edminson, P.D. and Rugstad, H.E. (1981) Toxicol. Appl. Pharmacol. 61: 215-226.

Banerjee, D., Onosaka, S. and Chennan, M.G. (1982) Toxicol. Appl. Pharmacol. 24: 95-105.

Basu, A. and Lazo, J.S. (1990) Toxicol. Lett. 50: 123-135.

Basu, A. and Lazo, J.S. (1991) Cancer Res. 51: 893-896.

Foster, R., Jahroudi, N. and Gedamu, L. (1991) Biochim. Biophys. Acta. 1088: 373-379.

Kägi,J.H.R. and Schäffer, A. (1988) Biochemistry 27: 8509-8515.

Jones, P.A. (1986) Cancer Res. 46: 461-466.

Kasahara, K., Fujiwara, Y., Nishio, K., Ohmori, T., Sugimoto, Y., Matasuda, T., and Saijo, N. (1991) Cancer Res. 51: 3237-3242.

Kelley, S.L., Basu, A., Teicher, B.A., Hacker, M.P., Hamer, D.H. and Lazo, J.S. (1988) Science 241: 1813-1815.

Koropatnick, J. and Pearson, J. (1990) Somat. Cell Mol. Genet. 16: 529-537.

Koropatnick, J. and Pearson, J. (In Press) Mol. Carcinog.

Li, T.Y., Kraker, A.J., Shaw III, C.F. and Petering D.H. (1980) Proc. Natl. Acad. Sci. 77: 6334-6338.

Lohrer, H., Robson, T., Grindley, H., Foster, S. and Hall, A. (1990) Carcinogenesis 11: 1937-1941.

Nartey, N., Cherian, M.G. and Banerjee, D. (1987) Am. J. Pathol. 129: 177-182.

Pattanaik, A., Bachowski, G., Laib, J., Lemkuil,D.,Shaw III, C.F., Petering, D.H., Hitchcock, A., and Saryan, L. (1992) J. Biol. Chem. 267: 16121-16128.

Robson, T., Hall, A. and Lohrer, H. (1992) Mutation Res. 274: 177-185.

Schilder, R.J., Hall, L. Monks, A., Handel,L.M., Fornace, A.J.Jr., Ozols, R.F., Fojo, A.T. and Hamilton, T.C. (1990) Int. J. Cancer 45: 416-422.

Tsujikawa, K., Imai, T., Kakutani, M., Kayamori, Y., Mimura, T., Otaki, N., Kimura, M., Fukuyama, R., and Shimizu, N. (1991) FEBS Lett. 283: 239-242.

Zelazowski, A.J., Garvey, J.S. and Hoeschele, J.D. (1984) Biochem. and Biophysics 229: 246-252.

Udom, A.O. and Brady, F.O. (1980) Biochem. J. 187: 329-335.

Metallothionein III
K.T. Suzuki, N. Imura and M. Kimura (eds)

THE TUMOR SUPPRESSIVE ACTIVITY OF CADMIUM

M.P. Waalkes[1], B.A. Diwan[2], T.P. Coogan[1], C.M.Weghorst[1], J.M. Ward[1], J.M. Rice[1], M.G. Cherian[3], and R.A. Goyer[3]

[1]National Cancer Institute-FCRDC, Frederick, MD, USA, [2]BCDP-Program Resources Incorporated, FCRDC, Frederick, MD, USA and [3]University of Western Ontario, London, ON, Canada

SUMMARY: The mouse liver has been widely used for the evaluation of carcinogenic and/or tumor promoting efficacy of various compounds, though little is known about the actions of metallic carcinogens in this system. Thus, the ability of cadmium (Cd) to initiate or promote tumors in mouse liver was studied. In promotion studies, N-nitrosodiethylamine (NDEA) was given as an initiator followed by Cd in drinking water. NDEA caused an elevation of liver tumor incidence over control which was prevented, rather than increased, by Cd. NDEA-induced lung tumor incidence was also reduced by Cd to control levels. In initiation studies, Cd was given to mice before an established promoting regimen of sodium barbital (BB) in drinking water. Mice treated with BB showed an elevation of liver tumor multiplicity over control that was prevented by Cd in a dose-related manner. Cd injected alone also reduced spontaneous liver tumor multiplicity. In a subsequent study, the time sequence of Cd inhibition of NDEA-initiated tumor formation was investigated. NDEA-induced liver or lung tumor incidence and/or multiplicity were inhibited by Cd even when Cd was given months after NDEA. Evidence of a specific deficiency of metallothionein (MT) in tumor cells was seen immunohistologically in NDEA-induced hepatic and lung lesions. These results indicate that Cd reduces tumor formation in the mouse liver and lung regardless of route of exposure, sequence of administration, or exposure interval. Auxiliary studies showed that spontaneous transformation of a liver cell line resulted in a marked reduction in basal MT production, which was associated with a marked increase in susceptibility to Cd toxicity. The possibility exists that Cd has a specific toxicity toward initiated cells within liver and lung because the MT gene is only poorly expressed in these premalignant or transformed cells. This chemotherapeutic potential of Cd deserves further study.

Cadmium (Cd) is a very toxic transition metal, human exposure to which has generally increased over the years. Cd is a potent animal carcinogen and has been shown to induce tumors of the testes (Gunn et al., 1963, 1964; Waalkes et al., 1988a, 1989; Waalkes and Rehm 1992), lungs (Takenaka et al., 1983; Oldiges et al., 1989; Glaser et al., 1990), injection site (Heath et al., 1962; Gunn et al., 1963, 1964; Waalkes et al., 1988a, 1989), prostate (Waalkes et al., 1988a, 1989, 1992a; Waalkes and Rehm, 1992) and hematopoietic system (Waalkes et al., 1992b; Waalkes and Rehm, 1992). Cd is also a suspected human carcinogen and has been linked in some studies with respiratory and prostatic tumors in occupationally and/or environmentally exposed individuals (see Waalkes and Oberdörster, 1990 and Waalkes et al., 1992c for review).

The multistage model of chemical carcinogenesis was originally developed to describe the process of tumor development in mouse skin but has since been widely applied to a variety of tissues including liver, lung, and kidney in mice and other species. In such models the carcinogenic process consists of at least two more or less discrete stages (Boutwell, 1974; Pitot, 1983; Farber, 1984; Harris, 1991). First, an irreversible phase (initiation) occurs in which the affected cells are transformed so that, under appropriate stimulation, they have the potential to eventually form tumors. A subsequent series of reversible steps (promotion) then occurs in which the initiated cells are stimulated to grow or divide, a process of clonal expansion that can progress to tumor development. Relatively little is known about how metals interact within such systems. In the case of Cd only a few studies appear to have been carried out in established initiation/promotion model systems (Kurokawa et al., 1985; 1989). In these, Cd was used exclusively as a promoter, with generally inconclusive results.

The original goal of our studies was to assess possible initiating and promoting effects of Cd in the B6C3F1 mouse. The ability of numerous compounds to initiate or promote carcinogenesis within various tissues of these mice is well established (Ward et al., 1983, 1986; Diwan et al., 1986; Diwan and Rice 1989). In particular, the ability of the N-nitrosoalkylamines to initiate carcinogenic events and of certian barbiturates to promote the carcinogenic process is well defined in liver, kidney and thyroid. Therefore, studies were designed to determine: 1) the ability of Cd to promote carcinogenesis after initiation with N-nitrosodiethylamine (NDEA) in liver and lung and 2) the effect of Cd as an initiator and sodium barbital (5,5-diethylbarbituric acid, sodium salt; BB) as a promoter on different organ systems (Waalkes et al., 1991a).

In the first study, groups of male B6C3F1 mice (initial group size of 50) were given a single dose of NDEA (90 mg/kg bw), i.p., in tricaprylin at 4 weeks of age and $CdCl_2$ at 500 or 1000 ppm Cd in the drinking water starting 2 weeks after NDEA. Controls received tricaprylin and/or unaltered drinking water. Groups of 10 mice were killed at 16, 24, and 36 weeks and the remainder were killed

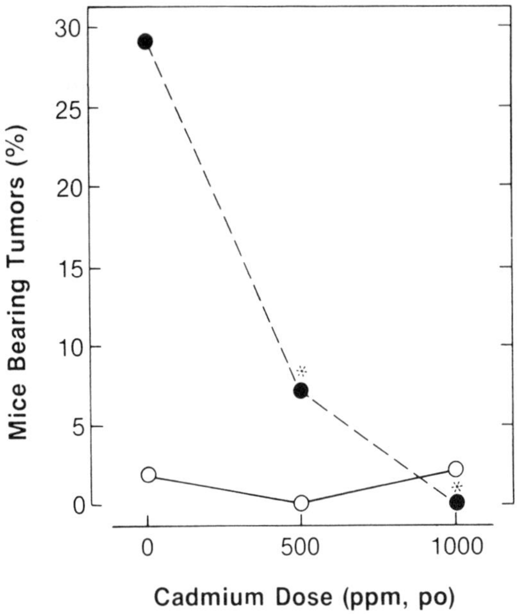

Fig. 1. Proportional incidence of male B6C3F1 mice bearing hepatic tumors with (o) or without (o) NDEA treatment and receiving Cd in the drinking water. Tumors were assessed during 52 weeks after NDEA treatment. Asterisks indicate a significant ($p \leq 0.05$) difference from appropriate control. (Modified from Waalkes et al., 1991a).

at 52 weeks after NDEA, at which time tissues were processed and analyzed for lesion numbers and types by light microscopy.

Oral Cd treatment markedly reduced the carcinogenic effects of NDEA in the livers of these mice (Figure 1). NDEA induced an elevation in liver tumors (13 tumor bearing mice/45 mice examined) over controls (1/48). This NDEA-induced increase was prevented in a dose related fashion by Cd (NDEA + 500 ppm Cd, 3/42; NDEA + 1000 ppm Cd, 0/47). Cd treatment alone did not increase or decrease liver tumor incidence compared to control. NDEA typically induced basophilic adenomas within the liver. The ability of NDEA to induce lung tumors, typically adenomas of alveolar cell origin, was also significantly reduced by Cd exposure in a dose related fashion (Waalkes et al. 1991a).

It is important to note that Cd exposure did have some effect on body weight and weight gain, particularly in mice receiving 1000 ppm Cd. At the lower dose of Cd (500 ppm) the body weight depression was much more moderate and these animals weighed on average only 18% less than controls during the first 9 months of Cd exposure. Depressed body weight or caloric restriction has been associated with reduced tumor incidence in several cases (eg. White et al. 1947; Visscher et al. 1942; Tannenbaum, 1945). However, these reductions are generally much more modest than the 73 to 100% reduction in liver tumor incidence seen with Cd treatment. Furthermore, in an analysis of a series of National Testing Program bioassays using male B6C3F1 mice, Rao et al. (1987) found that body weight depression was not significantly associated with reduced liver or lung tumors. Beyond this, several agents are effective as promoters in the B6C3F1 mouse liver despite growth reductions similar to that seen with oral Cd in the present study (Hagiwara and Ward, 1986; Ward et al. 1983). Thus, body weight depression is not the sole explanation for the antineoplastic effect of Cd.

A second study attempted to use Cd as the initiator and barbital as the promoter (Waalkes et al. 1991a). Cd was given as a single s.c. dose of either 22.5 or 20.0 μmol/kg to five week old male B6C3F1 mice. These doses were shown in preliminary studies to induce acute, but survivable, hepatocellular necrosis. Starting two weeks later, BB was then given ad libitum in the drinking water at 500 ppm. Controls received injection vehicle (saline) and/or unaltered drinking water. Mice were killed 92 weeks after Cd. BB alone markedly increased liver tumor multiplicity (7.44 tumors/liver) over control (2.24). This was prevented by Cd in a dose-related fashion (20 μmol Cd/kg + BB, 3.93; 22.5 μmol/kg + BB, 1.87). Cd alone also reduced spontaneous liver tumor multiplicity. Tumor incidence rates were also reduced by Cd for both barbital promotion of spontaneously occurring tumors and spontaneously occurring tumors themselves. Cd in this study had no effect on body weights either when given alone or in combination with barbital. This clearly indicates that the suppressive effects of Cd on tumor development in this case cannot solely be attributed to reduced growth of the host.

In summary of these initial studies, Cd had similar effects when given either as the initiator or as the promoter in the B6C3F1 mouse system. Cd, given prior to BB, a known multi-tissue tumor promoter, inhibited liver neoplastic foci formation and liver tumor incidence. This occurred both with BB promotion and with spontaneously occurring neoplasms. Likewise, Cd, given after NDEA, a known liver and lung tumor initiator, inhibited liver and lung carcinogenesis. Although these initial results indicated that Cd is an effective antineoplastic agent, since Cd was given either throughout the entire study or only at a single time point it was impossible to ascertain if any specific stages of the carcinogenic

TABLE I
Effect of Varying Cd Exposure Periods on NDEA-Induced Liver Tumor Formation in Male B6C3F1 Mice

Treatment[a]	Liver Tumor Incidence (%)[b]
Control	5/25 (20.0%)
NDEA alone	22/25 (88.0%)
NDEA + Cd (2-48 weeks)	*2/25 (8.0%)
NDEA + Cd (4-48 weeks)	*1/25 (4.0%)
NDEA + Cd (8-48 weeks)	*1/25 (4.0%)
NDEA + Cd (16-48 weeks)	*2/25 (8.0%)
NDEA + Cd (32-48 weeks)	*6/24 (25.0%)
Cd alone (2-48 weeks)	*0/25 (0.0%)

[a]A single i.p. dose of NDEA (90 mg/kg) was given to 5 week old mice and Cd was given ad libitum in the drinking water at 1000 ppm for time periods indicated.

[b]Expressed as tumor bearing mice/total mice examined (%). An asterisk shows significant difference ($p \leq 0.05$) from NDEA alone.

process are more or less affected by the Cd treatment. Therefore, the goal of our follow up study was to determine the effects of Cd exposure on tumor incidence at various times following initiation of the carcinogenic process. Specifically, in this regard the time course of Cd inhibition of NDEA carcinogenesis was determined (Waalkes et al., 1993).

Studies were designed in which Cd exposure periods were varied. NDEA was given as a single i.p. dose of 90 mg/kg to five week old male B6C3F1 mice and, starting at various times after NDEA, $CdCl_2$ was given ad libitum in the drinking water at 1000 ppm. Cd treatment was initiated either 2, 4, 8, 16, or 32 weeks after NDEA and the study was terminated at 48 weeks. Controls received injection vehicle and/or unaltered water. Several previous studies from our laboratory using this treatment protocol for NDEA have established that at 32 weeks after NDEA treatment in male B6C3F1 mice, well developed preneoplastic liver lesions and neoplastic lung lesions are seen with high frequency (Diwan et al., 1986).

Cd prevented NDEA-induced elevations in liver tumor incidence over control regardless of treatment period (Table I). In fact, Cd exposure caused a marked reduction in liver tumor incidence even when given as late as 32 weeks after NDEA, a time point at which proliferative lesions in the liver are well advanced. Cd alone also abolished the occurrence of spontaneous liver tumors.

Likewise NDEA-induced increases in lung tumor multiplicity (7.28

tumors/lung) over control (0.04) were also reduced by Cd regardless of treatment period (NDEA + Cd: 2-52 weeks, 2.40; 4-48 weeks, 2.24; 8-48 weeks, 1.44; 16-48 weeks, 2.00; 32-48 weeks, 3.67). Not only were the lung tumors fewer in number with Cd treatment but there was a marked reduction in tumor size. This is reflected as a generalized reduction of pulmonary tumor diameter at all time points for initiation of Cd treatment, with the single exception of 32 weeks.

Cd was effective in reducing NDEA-induced liver or lung tumor incidence and/or multiplicity in B6C3F1 mice even when given as late as 32 weeks after NDEA treatment. This is a point at which NDEA-induced proliferative lesions in the liver are generally well developed and lung tumors have formed (Diwan et al., 1986). Therefore, even a relatively short course of Cd given after neoplastic lesions are well developed can be an effective suppressant of tumor formation in the B6C3F1 mouse.

Overall, these results lead to the conclusion that Cd is an effective anti-tumor agent in the male B6C3F1 mouse liver and lung when given either as an initiator or a promoter and even after neoplastic lesions have substantially progressed. The exact basis of this effect is unknown, although it would appear that Cd may have a certain enhanced or specific cytotoxicity for initiated cells. Such cells could either be those occurring spontaneously or those brought about after insult with carcinogens. In this regard Cd can be a very cytotoxic agent (Stacey et al., 1980; Waalkes et al., 1988b) and the extent to which a cell is resistant or susceptible to Cd-induced cytotoxicity or cytolethality often appears to be related to the cellular production capacity for metallothionein (MT; Waalkes et al., 1985; 1988b; Liu et al. 1990). Various control mechanisms exist for the MT gene and synthetic capacity for MT varies widely among tissues (Onosaka and Cherian, 1981; Waalkes and Klaassen, 1985) and at different developmental stages (Andrews et al., 1984; Waalkes and Klaassen, 1984).

Since it is suspected that the Cd-binding protein MT plays an important role in resistance or susceptibility to Cd cytotoxicity, ancillary studies were performed to determine the content of MT and the responsiveness of the MT gene in mouse liver or lung tumors (Waalkes et al., 1993). The content of MT was determined immunohistochemically in mouse liver adenomas and carcinomas induced by NDEA. MT was localized immunohistochemically with the peroxidase-antiperoxidase method of Banerjee, et al. (1982) using an antibody developed in rabbits to rat MT (Banerjee et al., 1982;

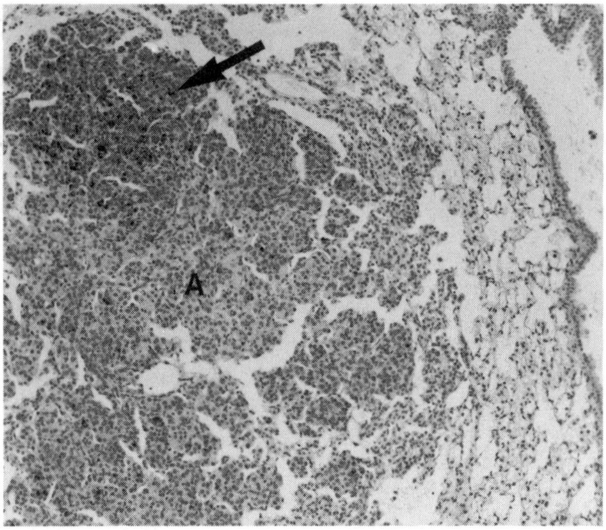

Fig. 2. Immunohistochemical localization of MT in a lung section containing a large adenoma (A). Limited focal staining does occur in few cells (eg. arrow) but the vast majority of the tumor is negative. Peroxidase-antiperoxidase immunohistochemistry with hematoxylin counter staining. X 100.

Panemangalore, et al., 1983). The liver tumors (not shown) typically stained much less consistently than the surrounding non-neoplastic tissue and were largely negative. NDEA-induced lung adenomas were also consistently negative for MT staining (Figure 2). In contrast normal liver shows a fairly uniform basal staining for MT (not shown). Thus, it appears that basal levels of MT may be relatively low in liver or lung tumors induced by NDEA in the B6C3F1 mouse.

MT was also measured biochemically by the Cd-heme method (Onosaka and Cherian, 1981). Here this method was applied to mice bearing histologically confirmed preneoplastic foci from earlier NDEA treatment (Waalkes et al., 1993). The mice were also treated with zinc (500 μmole/kg, s.c., -24 hr), a standard inducing agent for MT synthesis. In the control animal, basal levels of hepatic MT (13.0 μg/g wet weight) were markedly increased by zinc (420 μg/g). However, not only were basal levels of MT much lower in the livers of foci-bearing animals (1.4 μg/g) but zinc was clearly less effective in stimulating the production of MT synthesis in these animals (255 μg/g). Thus, it appears that the livers of animals bearing preneoplastic lesions may contain less MT and may be less able to produce MT upon stimulation.

Other experiments were directed at the study of MT gene expression during liver cell transformation (Coogan et al., 1992). TRL 1215 is a liver cell-derived line that will undergo spontaneous transformation with continued passage. This is indicated by altered morphology at high passage levels as well as growth in soft agar. After transformation, there was a marked reduction of 86% in the basal levels of MT (8.3 $\mu g/10^6$ cells) in these cells as compared to non-transformed cells (61.0 $\mu g/10^6$ cells). This was accompanied by marked enhancement in susceptibility of these transformed cells to the cytotoxic effects of Cd in vitro.

Cd is clearly very cytotoxic in various cell types including hepatocytes and tumor cell lines (Stacey et al., 1980; Waalkes et al., 1988b; Kasahara et al., 1991). The finding that tumor formation can be inhibited by Cd even when given at a relatively late stage in tumor development indicates that Cd may have a certain enhanced or specific cytotoxicity for initiated cells. In this regard, several lines of evidence indicate that the tumors induced by NDEA in mouse lung and liver do not express MT to the same extent as normal tissue even when stimulated. This suggests a reduced capacity for MT expression in these particular tumors. There is good evidence that the ability to synthesize MT is often a key aspect to the ability to resist Cd cytotoxicity (Waalkes et al., 1988b; Kasahara et al., 1991; Lui et al., 1990) and active expression of MT can markedly reduce the toxicity of cadmium (Liu et al., 1990), as well as other metals or metal complexes including *cis*-diamminedichloroplatinum (CDDP; Bakka et al., 1981; Naganuma et al., 1985). In fact, the development of tumor cell resistance to CDDP has been linked to enhanced expression of MT (Kelley et al., 1988; Kasahara et al., 1991). A common cellular sensitivity to the cytotoxicity effects of both CDDP and cadmium is often seen and is thought to be based, at least partially, on the ability of the particular cell to express the MT gene (Kasahara et al., 1991). When mice are given a carcinogenic N-nitrosoalkylamine closely related to NDEA, namely N-nitrosodimethylamine, and simultaneously receive cadmium the expression of MT is markedly inhibited soon thereafter (Onosaka and Tanaka, 1991). The chronic effects of such N-nitrosoalkylamines on MT gene expression have not been defined. However, our evidence indicates that the process of transformation may be associated with a decrease in MT expression, at least in spontaneously transformed TRL 1215 cells (Coogan et al., 1992), which in turn makes the cells much more susceptible to Cd-induced cytotoxicity. The role of MT in the tumor suppressive effects of Cd clearly deserves further study.

In apparent conflict with the tumor suppressive effects of Cd in lung and liver, numerous reports indicate that Cd is a potent carcinogen in several rodent tissues (Gunn et al., 1963, 1964; Waalkes et al., 1988a, 1989, 1992a,b,c; Waalkes and Rehm, 1992), including the lung (Takenaka et al., 1983; Oldiges et al., 1989; Glaser et al., 1990). Recent results from this laboratory also indicate the liver may be a target of Cd carcinogenicity, at least in one strain of mice (Waalkes and Rehm, 1993). However, Cd can also induce or inhibit tumors in other tissues. This includes both inhibition and induction of hematopoietic tumors, such as leukemias or lymphomas (Waalkes et al., 1991b, 1992b; Waalkes and Rehm, 1992, 1993) and, at least under certain circumstances, pancreatic tumors (Poirier et al., 1983; Waalkes et al., 1988a). With transplanted tumors, Cd inhibits growth of MSB sarcomas (Kerkvliet et al., 1979) or Dunn osteosarcomas (Noda, 1990) and can prolong the survival of mice inoculated with MOPC-1040E plasmacytoma cells (Gray et al., 1982). On the other hand, cadmium exposure can enhance development of virally induced leukemia (Blakley, 1986). Similarly, CDDP, an effective cancer chemotherapeutic agent, is also potentially carcinogenic (Leopold et al., 1979; Hennings et al., 1990). Whether Cd acts as an anticarcinogen or carcinogen is probably due to a complex set of circumstances including, at least in part, the dose, time sequence of administration, site of tumor (i.e., target cells) and route of Cd administration. In this regard, generally the anticarcinogenic doses of Cd are much higher than those required for carcinogenic action. The time course of exposure, total exposure time and the route of exposure probably also have an effect. For instance, when Cd is given very soon after or before NDEA it enhances subsequent development of renal and hepatic tumors in rats (Wade et al., 1987). Likewise, only when Cd is inhaled or instilled intratracheally is it a pulmonary carcinogen (Takenaka et al., 1983; Pott et al., 1987; Oldiges et al., 1989; Glaser et al., 1990), while suppression of lung tumor formation occurs only after systemic exposure to Cd (Waalkes et al., 1991a; 1993). In any event, the action of Cd as a carcinogen or anticarcinogen is evidently due to a complex set of circumstances but could represent a continuum of effects within key progenitor cell populations ranging from cytotoxicity and transformation to cytolethality.

Although MT appears to be expressed in developmental periods of rapid growth in various tissues, a comprehensive survey of MT expression in tumor cells has not been performed. The lung and liver tumors induced by NDEA in mice do not appear to contain substantial amounts of MT. The concentration of MT in human livers containing hepatocellular carcinomas is also markedly less than in normal livers (Onosaka et al., 1986; Onosaka and Tanaka, 1991), while in seminomas immunohistochemical staining is negative for MT (Chin et al., 1993). In contrast, several tumors clearly have high MT contents, including human thyroid adenomas and adenocarcinomas (Nartey et al., 1987), and testicular embryonal carcinomas (Kontozoglou et al., 1989). In this regard most seminomas are sensitive to CDDP while testicular embryonal carcinomas are generally resistant. Skin papillomas in mice appear to have constitutively elevated expression of MT (Hashiba et al., 1989) and equine melanoma shows elevated MT concentrations (Krauter et al., 1989). Steele et al. (1990) found that even though MT expression was below detection limits in over 50% of the human colorectal cancer cell lines tested, several lines showed abundant MT gene expression, indicating that such expression can also vary widely in cells from different tumors of the same type. Similarity, in a recent study from this laboratory (Diwan et al., unpublished), murine colorectal tumors induced by methylazoxy-methanol acetate were resistant to Cd inhibition and showed evidence of some fairly intense focal staining for MT. Based on these disparate findings, it would be imprudent to generalize in regard to all tumors about their ability to express the MT gene, or in fact the ability to express it at all stages of development and progression in any individual tumor type. Some sort of systematic data base on MT expression in tumors should be developed as such findings could be useful in predicting the tumor-specific efficacy of agents such as Cd or CDDP.

In summary, our results indicate that Cd prevents or reduces tumor formation in the B6C3F1 mouse liver and lung regardless of the exposure interval and apparently by cell specific cytotoxicity. Some evidence of a specific deficiency of MT in tumor cells was seen immunohistologically in NDEA-induced hepatic lesions and pulmonary lesions. Auxiliary studies indicated that in mice bearing multiple liver foci resulting from NDEA treatment there is a marked reduction in basal MT levels and in response to zinc induction. Beyond this it appears that the MT system may be less active and less responsive to stimuli in transformed liver cells. The possibility exists that Cd has a specific toxicity toward previously initiated cells within mouse liver and lung. The mechanisms involved in Cd-induced tumor suppression and its potential as a possible chemotherapeutic strategy deserve further study.

REFERENCES

Andrews, G.K., Adamson, E.D., and Gedamu, L. (1984) Devel. Biol. 103, 294-303.

Bakka, A., Endersen, L., Johnsen, A.B.S., Edminson, P.D. and Rugstad, H.E. (1981) Toxicol. Appl. Pharmacol. 61, 215-226.

Banerjee, D., Onosaka, S., and Cherian, M. G. (1982) Toxicology 24, 95-105.

Blakley, B.R. (1986) J. Appl. Toxicol. 6, 425-429.

Boutwell, R.K. (1974) Crit. Rev. Toxicol. 2, 419-443.

Coogan, T.P., Bare, R.M. and Waalkes, M.P. (1992) The Toxicologist 12, 144.

Chin, J.L., Banerjee, D., Kadhim, S.A., Kotazaglu, T., Chauvant, P.T., and Cherian, M.G. (1993). Submitted to Cancer.

Diwan, B.A., Rice, J.M., and Ward J.M. (1986) Carcinogenesis 7, 789-794.

Diwan, B.A., and Rice, J.M. (1989) In: The Pathobiology of Neoplasia (A.E. Sirica, Ed.), Plenum Press, New York, pp.149-165.

Farber, E. (1984) Cancer Res. 44, 4217-4223.

Glaser, U., Hochrainer, D., Otto, F.J., and Oldiges, H. (1990) Chem. Environ. Toxicol. 27, 153-162.

Gray, I., Arrieh, M., and Balter, N.J. (1982) Arch. Environ. Health 37, 342-435.

Gunn, S.A., Gould, T.C. and Anderson, W.A.D. (1963) J. Natl. Cancer Inst. 31, 745-753.

Gunn, S.A., Gould, T.C. and Anderson, W.A.D. (1964) Proc. Soc. Exp. Biol. Med. 115, 653-657.

Hagiwara, A., and Ward, J.M. (1986) Fundam. Appl. Toxicol. 7, 376-386.

Harris, C.C. (1991) Cancer Res. 51, 5023-5044.

Hashiba, H., Hosoi, J., Karasawa, M., Yamada, S., Nose, K., and Kuroki, T. (1989) Molec. Carcinogen. 2, 95-100.

Heath, J. C., Daniel, M. R., Dingle, J. T., and Webb, W. (1962) Nature 193, 592-593.

Hennings, H., Shores, R.A., Poirier, M.C., Reed, E., Tarone, R.E. and Yuspa, S.H.(1990) J. Nat. Cancer Inst. 82, 836-840.

Kasahara, K., Fujiwara, Y., Nishio, K., Ohmori, T., Sugimoto, Y., Komiya, K., Matsuda, T. and Saijo, N. (1991) Cancer Res. 51, 3237-3242.

Kelley, S. L., Basu, A., Teicher, B.A., Hacker, M. P., Hamer, D.H. and Lazo, J.S. (1988) Science 241, 1813-1815.

Kerkvliet, N.I., Koller, L.D., Baecher, L.G., and Brauner, J.A. (1979) J. Nat. Cancer Inst. 63, 479-483.

Kontozoglou, T.E., Banerjee, D., and Cherian, M.G. (1989) Virchows Arch. A: Pathol. Anat. Histopathol. 415, 545-549.

Krauter, B., Nagel, W., Hartmann, H.-J. and Weser, U. (1989) Biochim. Biophys. Acta 1013, 212-217.

Kurokawa, Y., Matsushima, M., Imazawa, T., Takamura, N., Takahashi, M., and Hayashi, Y. (1985) J. Am. Coll. Toxicol. 4, 321-330.

Kurokawa, Y., Takahashi, M., Maekawa, A. and Hayashi, Y. (1989) J. Am. Coll. Toxicol., 8, 1235-1239.

Leopold, W. R., Miller, E. C., and Miller, J. A. (1979) Cancer Res. 39, 913-918.

Liu, J., Kershaw, W. C., and Klaassen, C. D. (1990) In Vitro Cell. Dev. Biol. 26, 75-79.

Naganuma, A., Satoh, M., Koyama, Y., and Imura, N. (1985) Toxicol. Lett. 24, 203-207.

Nartley, N., Cherian, M. G. and Banerjee, D. (1987) Am. J. Pathol. 129, 177-182.

Noda, K. (1990) Nippon Seikeigeka Gakkai Zasshi 64, 70-75.

Oldiges, H., Hochrainer, D., and Glaser, U. (1989) Toxicol. Environ. Chem. 23, 35-40.

Onosaka, S. and Cherian, M. G. (1981) Toxicology 22, 91-101.

Onosaka, S. and Tanaka, K. (1991) In: Metallothionein in Biology and Medicine (Klaassen, C.D. and Suzuki, K.T. (Eds.), CRC Press, Boca Raton, FL, USA, pp. 1-11.

Onosaka, S., Min, K.-S., Fukuhara, C., Tanaka, K., Tashiro, S.-I., Shimizu, I., Furuta, M., Yasutomi, T., Kobishi, K., and Yamamoto, K.-I. (1986) Toxicology 38, 261-268.
Panemangalore, M., Banerjee, D., Onosaka, S., and Cherian, M.G. (1983) Dev. Biol. 97, 95-102.
Pitot, H.C. (1983) Cancer Surveys 2, 519-537.
Poirier, L.A., Kasprzak, K.S., Hoover, K., and Wenk, M.L. (1983) Cancer Res. 43, 4575-4581.
Pott, F., Ziem, U., Reiffer, F.J., Huth, F., Ernst, H., and Mohr, U. (1987) Exp. Pathol. 32, 129-152.
Rao, G.M., Piegorsch, W.W. and Haseman, J.K. (1987) Am. J. Clin. Nutr. 45, 252-260.
Stacey, N. H., Cantilena, L. R., and Klaassen, C. D. (1980) Toxicol. Appl. Pharmacol. 53, 470-480.
Steele, G.D., Davis, S., Yow, H., Wong, J.M., Rivers, E.N., Summerhayes, I.C., Ravikumar, T.S., and Chen, L.B. (1990) Arch. Surg. 125, 493-497.
Takenaka, S., Oldiges, H., König, H., Hochrainer, D. and Oberdörster, G. (1983) J.Nat. Cancer Inst., 70, 367-373.
Tannenbaum, A. (1945) Cancer Res. 5, 616-621.
Visscher, M.B., Ball, Z.B., and Barnes, R.H. (1942) Surgery 11, 48-59.
Waalkes, M.P. and Klaassen, C.D. (1984) Toxicol. Appl. Pharmacol. 74, 314-320.
Waalkes, M.P. and Klaassen, C.D. (1985) Fund. Appl. Toxicol. 5, 473-477.
Waalkes, M.P. and Oberdörster, G. (1990) Cadmium carcinogenesis, In: Biological Effects of Heavy Metals; Volume II, Metal Carcinogenesis, (E.C. Foulkes, ed.) Boca Raton, FL, CRC Press, pp. 129-157.
Waalkes, M.P. and Rehm, S. (1992) Fund. Appl. Toxicol. 19, 512-520.
Waalkes, M.P. and Rehm, S. (1993) The Toxicologist 13, In press.
Waalkes, M.P., Wilson, M.J. and Poirier, L.A. (1985) Toxicol. Appl. Pharmacol. 81, 250-257.
Waalkes, M.P., Rehm, S., Riggs, C.W., Bare, R.M., Devor, D.E., Poirier, L.A., Wenk, M.L., Henneman, J.R. and Balaschak, M.S. (1988a) Cancer Res. 48, 4656-4663.
Waalkes, M.P., Miller, M.S., Wilson, M.J., Bare, R.M., and McDowell, A.E. (1988b) Chem.-Biol. Interact. 66, 189-204.
Waalkes, M.P., Rehm, S., Riggs, C.W., Bare, R.M., Devor, D.E., Poirier, L.A., Wenk, M.L. and Henneman, J.R. (1989) Cancer Res. 49, 4282-4288.
Waalkes, M.P., Diwan, B.A., Bare, R.M., Ward, J.M., Weghorst, C. and Rice, J.M. (1991a) Toxicol. Appl. Pharmacol, 110, 327-335.
Waalkes, M.P., Kovatch, R. and Rehm, S. (1991b) Toxicol. Appl. Pharmacol. 108, 448-456
Waalkes, M.P., Rehm, S., Perantoni, A. and Coogan, T.P. (1992a) IARC Sci. Publ., In press.
Waalkes, M.P., Rehm, S., Sass, B., and Ward, J.M. (1992b) IARC Sci. Publ., In press.
Waalkes, M.P., Coogan, T.P. & Barter, R.A. (1992c) Crit. Rev. Toxicol., 22, 175-201.
Waalkes, M.P., Diwan, D.A., Weghorst, C.A., Ward, J.M., Rice, J.M. Cherian, G.M., and Goyer, R.A. (1993). Submitted.
Wade, G.G., Mandel, R. and Ryser, H.J.-P. (1987) Cancer Res. 47, 6606-6613.
Ward, J.M., Rice, J.M., Creasia, D., Lynch, P., and Riggs, C. (1983) Carcinogenesis 4, 1021-1029.
Ward, J.M., Diwan, B.A., Ohshima, H., Hu, H., Schuller, H.M., and Rice, J.M. (1986) Environ. Health Perspect. 65, 279-291.
Ward, J.M., Rice, J.M., Creasia, D., Lynch, P., and Riggs, C. (1983) Carcinogenesis 4, 1021-1029.
White, J, White, F.R. and Milder, G.B. (1947) J. Natl. Cancer Inst. 7, 199-204.

Metallothionein III
K. T. Suzuki, N. Imura and M. Kimura (eds)

GROWTH INHIBITORY FACTOR IN BRAIN

Y.UCHIDA

Department of Neuropathology, Tokyo Metropolitan Institute of Gerontology, 35-2 Sakaecho, Itabashiku, Tokyo 173, Japan

SUMMARY: Growth inhibitory factor (GIF) is abundant in normal human brain but greatly reduced in the Alzheimer's disease (AD) brain. GIF inhibits the neurotrophic activities in AD brain extract on neonatal rat cerebral cortical neurons in vitro. GIF is a 68 amino acid metallothionein-like protein with a 1 amino acid insert and a unique 6 amino acid insert in the N-terminal and the C-terminal portions, respectively. GIF gene is located on chromosome 16, as are the metallothionein genes. GIF and GIF mRNA distribute in the central nervous system, such as cerebral cortex, cerebellum, striatum, spinal cord, but do not in liver, kidney or muscle. In the neocortex, GIF and GIF mRNA are localized in the astrocytes in the 2 to 6 layers of the gray matter, which are closely associated with neuronal perikarya and dendrites. In the AD neocortex, GIF and GIF mRNA are drastically reduced, suggesting that GIF is down-regulated in the subset of astrocytes.

INTRODUCTION

Severe loss of cortical and subcortical neurons occurs in Alzheimer's disease (AD). The trophic factor hypothesis proposed by Appel has explained neuronal loss in AD. hippocampus and cerebral cortex might fail to supply the neurotrophic factors for basal forebrain cholinergic neurons, and the lack of neurotrophic factors might lead to impairment of basal forebrain neurons (Appel, 1981). In addition to neuronal loss, AD brain is also characterized by the accumulation of senile plaques and neurofibrillary tangles (NFT) and by abortive somatodendritic sprouting (Ihara, 1988). The trophic factor hypothesis cannot explain the abortive sprouting responses in the AD brain. We therefore investigated whether neurotrophic activities decreased or increased in AD brain. The AD cerebral cortex contained higher neurotrophic activities for neonatal rat cortical neurons in vitro than non-demented aged cerebral cortex (Uchida et al.,

1988). This apparent increase in the neurotrophic activities was due to the loss of growth inhibitory activities which would suppress neurotrophic activities in normal brain (Uchida et al., 1989). The growth inhibitory activities in normal brain but not in AD brain were retained on the membrane with a molecular weight cutoff of 10 kD, and eluted in stepwise manner with 50 mM and 200 mM NaCl from DEAE-Sephacel. A growth inhibitory factor (GIF) was purified from 200mM NaCl eluted DEAE-Sephacel fractions of normal brain extracts. GIF is a new member of metallothionein family (Uchida et la., 1991). GIF and GIF mRNA are distributed only in the central nervous system but not in the peripheral tissues. In the normal cerebral cortex, GIF and GIF mRNA are localized in the subpopulation of astrocytes, which are closely associated with neuronal perikarya and dendrites. GIF may suppress abnormal sprouting responses of neurons. In AD cerebral cortex, down-regulation of GIF in astrocytes (Uchida et al., 1991; Tsuji et al., 1992) may lead to abortive sprouting of pyramidal neurons.

HIGHER NEUROTROPHIC ACTIVITY IN AD BRAIN

We examined the neurotrophic action of AD brain extract on neonatal rat cerebral cortical neurons in vitro. Cerebral cortical cells were seeded at low density on microplates precoated with gelatin-polyornithine, and fed for 5 days in a chemically defined medium containing frontal cortical extracts from AD or non-demented aged patients. Cultured cells were labeled with a monoclonal antibody to neuron-specific microtubule associated protein (MAP) 2. Neuronal survival was expressed as the total number of MAP2 positive cells or the level of anti-MAP2 binding to the culture quantified by ELISA.
Neuronal survival in the cultures with the extracts from AD brains with numerous NFT was enhanced about 4-fold of that with normal aged brain extracts (Fig. 1, middle). Neuronal survival in the cultures with the extracts from AD brains with numerous senile plaque cores but no NFT was almost the same as that with normal brain extracts (Fig. 1, right). AD brain extract stimulated not only survival but also neurite formation of cerebral cortical neurons. Neurons cultured with the extracts from AD brains with numerous NFT had larger cellbodies and more neurites. These results

indicate that AD brain has significantly higher neurotrophic activities and the degree of neurotrophic activities in the cortex is correlated with the abundance of NFT but not with senile plaques.

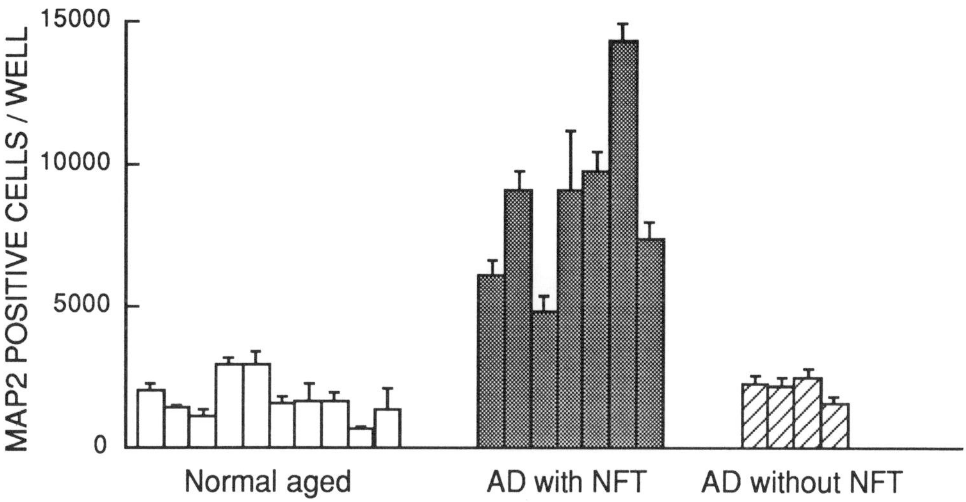

Fig. 1 Neurotrophic action of the extracts from AD brains with NFT. The brain extracts from the frontal cortex of 10 non-demented, 7 AD brains with numerous NFTs and 4 AD brains with numerous senile plaque cores but without NFT were assayed for the neurotrophic activities on neonatal rat cortical neurons in vitro.

LOSS OF GROWTH INHIBITORY ACTIVITIES IN AD BRAIN

Titration curves for neurotrophic activities in normal aged and AD brain extracts revealed the presence of both trophic and inhibitory activities in normal brain and the absence of inhibitory activities in AD brain (Fig. 2). Neurotrophic activities in normal brain extract increased with increasing protein concentration up to 31.25 µg/ml, and then declined with further increase in protein concentration. In contrast, neurotrophic activities in AD brain extract increased in a dose-dependent manner. Neurotrophic activities in AD brain extract was inhibited by the addition of normal brain extracts. These results indicate that loss of growth inhibitory activities lead to a relative increase in neurotrophic activities in AD brain.

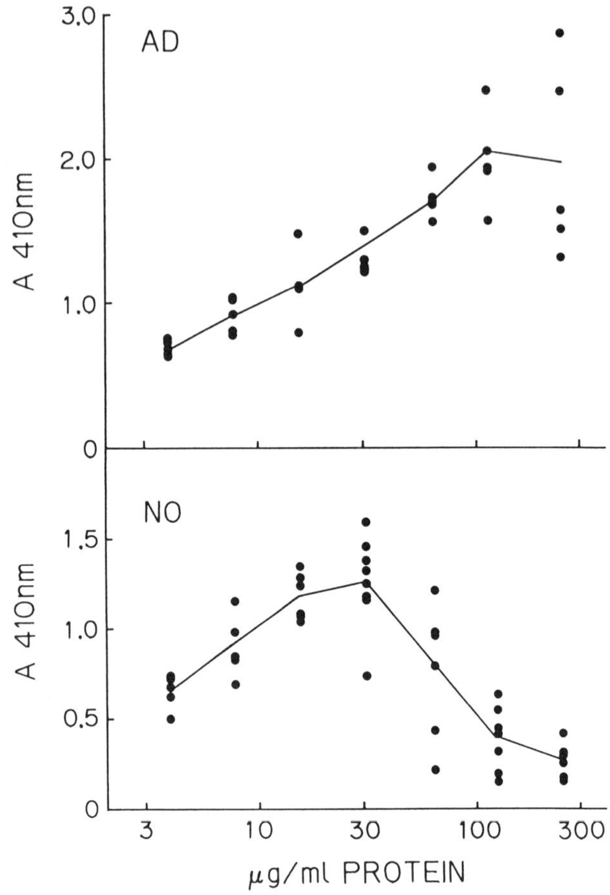

Fig. 2 Dose-response curves for neurotrophic activities in 5 AD and 8 normal brain extracts. Neurotrophic activities are expressed as the level of anti-MAP2 binding to the culture quantified by ELISA.

PURIFICATION AND CHARACTERIZATION OF A GROWTH INHIBITORY FACTOR FROM HUMAN BRAIN

PURIFICATION:

The growth inhibitory activities, which inhibited neurotrophic activities in AD brain extract, were retained on the membrane with a molecular weight cutoff of 10 kD. Then, the 10 kD retentate was applied to a DEAE-Sephacel column, and eluted in a stepwise

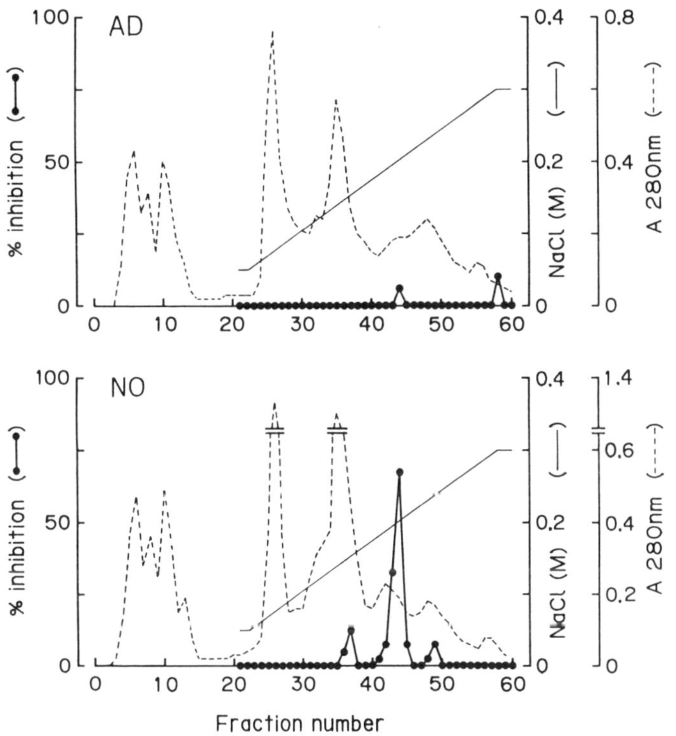

Fig. 3 Fractionation of growth inhibitory activities from AD and normal aged brains on DEAE-Sephacel column chromatography. Cerebral cortical extract from AD (upper) and normal aged (lower) patients were applied to DEAE-Sephacel columns. The columns were washed with 20 mM Tris-HCl and eluted with a linear gradient of 50 to 300 mM NaCl in 20 mM Tris-HCl (pH 7.6). The growth inhibitory activities are expressed as percent inhibition of neurotrophic activities of an AD brain extract. A strong growth inhibitory activity in normal aged brain, but very weak activity in AD brain, were eluted with 200 mM NaCl from DEAE-Sephacel columns.

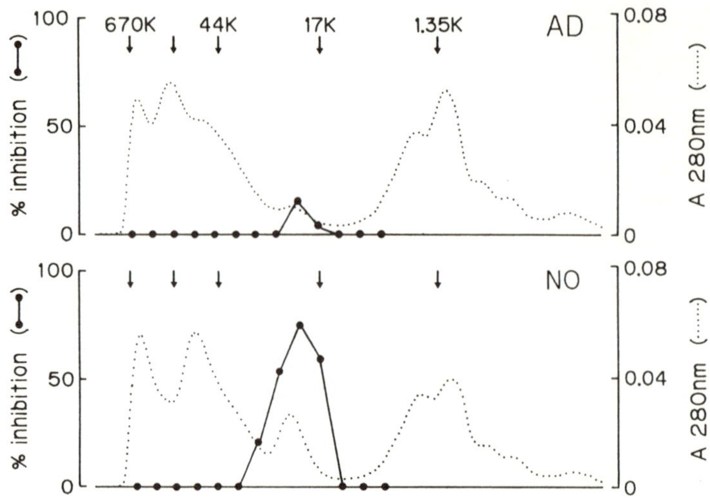

Fig. 4 Fractionation of growth inhibitory activities from DEAE-Sephacel fractions of AD (upper) and normal aged brains (lower) on gel filtration. The active fractions of normal brain extract eluted with 200 mM NaCl from DEAE-Sephacel and the fractions of AD brain extract eluted with 200 mM NaCl from DEAE-Sephacel were applied to gel filtration columns and eluted with PBS(-). A strong growth inhibitory activity in normal aged brain, but weak activity in AD brain, were eluted at 20 kD.

manner with 50, 200 and 500 mM NaCl. The inhibitory activities specific in normal brain extract were eluted with 50 and 200 mM NaCl. Based on these results, GIF was purified from normal aged brain by a combination of DEAE-Sephacel chromatography, gel filtration and reverse-phase HPLC. The active fractions eluted with 200 mM NaCl from DEAE-Sephacel (Fig. 3) were applied to a TSK G2000SW column. The inhibitory activities were eluted at 20 kD (Fig. 4). GIF was further purified from 20 kD active fractions by reverse-phase HPLC with a C18 column. Fig. 5 shows dose-response curves for inhibitory activity of GIF at each step of purification. Half-maximal concentration of purified GIF ranged from 100 to 200 ng/ml. Fig. 6 shows SDS-PAGE of GIF fractions at each step of purification. The final fraction of GIF contained a small protein with an apparent molecular weight of 6 kD under reducing conditions.

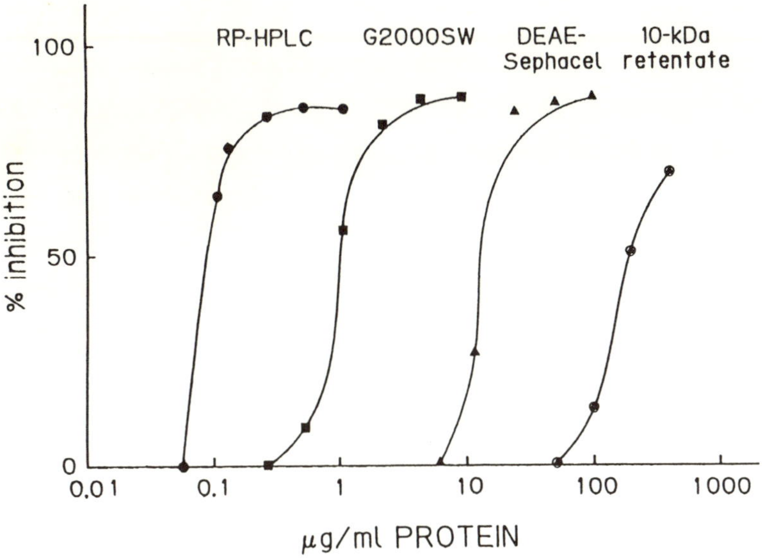

Fig. 5 Dose-response curves for growth inhibitory activity of GIF fractions.

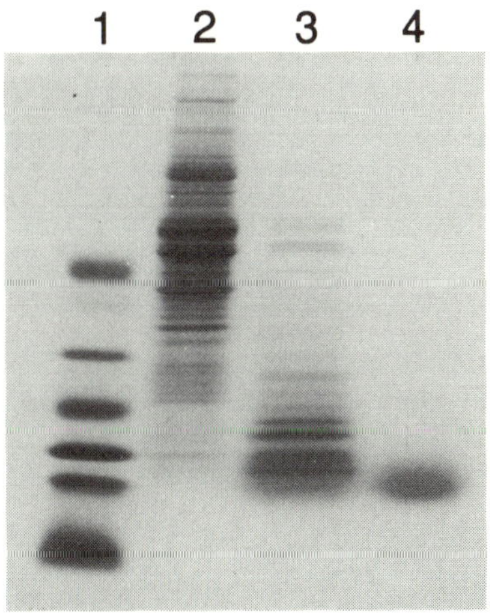

Fig. 6 SDS-PAGE of GIF fractions from each step of purification. The GIF fractions were dissolved in Laemmli's sample buffer and applied on a 7.5-20% linear gradient gel. Lane 1, molecular weight standard (46, 25.7, 18.4, 14.3, 6.2 and 3.4 + 2.3 kD); lane 2, DEAE pool; lane 3, gel filtration pool; lane 4, C18 fraction.

<u>PROTEIN SEQUENCING:</u>

Because of blocked N-terminal of GIF, peptide fragments were prepared from reduced and alkylated GIF by selective cleavage with acid at Asp-Pro or cyanogen bromide at Met-X, or by digestion with endoprotease Asp-N, V8 protease or TPCK-trypsin. The peptide fragments were separated by reverse-phase HPLC and sequenced. Fig. 7 shows complete amino acid sequence of GIF. GIF has 68 amino acids and its sequence is 70% identical to that of human metallothionein (MT) II. The most common feature is to contain 20 Cys residues, all of which are arranged in either Cys-X-X-Cys, Cys-X-Cys or Cys-Cys. GIF lacks Arg, His, Ile and Leu in addition to aromatic residues. GIF has, however, two insertion; a 1 residue insert (Thr) at residue 5 and a 6 residue insert (Glu-Ala-Ala-Glu-Ala-Glu) at residues 55-60. None of the 25 vertebrate MTs listed in the data bases contains amino acid inserts found in GIF. Moreover, the two inserts in GIF are not localized in the intron/exon boundaries in MT genes.

```
      1        10        20       30
GIF   MDPETCPCPSGGSCTCADSCKCEGCKCTSC
Ia    MDPN CSCATGGSCTCTGSCKCKECKCNSC
Ib    MDPN CSCTTGGSCACAGSCKCKECKCTSC
Ie    MDPN CSCATGGSCTCAGSCKCKECKCTSC
If    MDPN CSCAAGVSCTCAGSCKCKECKCTSC
II    MDPN CSCAAGDSCTCAGSCKCKECKCTSC
IIa   MDPN CSCAAGVSCTCASSCKCKECKCTSC
      ***  * *  * ** *  ****  *** **

              40        50        60      68
GIF   KKSCCSCCPAECEKCAKDCVCKGGEAAEAEAEKCSCCQ
Ia    KKSCCSCCPMSCAKCAQGCICKGA      SEKCSCCA
Ib    KKCCCSCCPVGCAKCAQGCVCKGS      SEKCRCCA
Ie    KKSCCSCCPVGCAKCAQGCVCKGA      SEKCSCCA
If    KKSCCSCCPVGCSKCAQGCVCKGA      SEKCSCCD
II    KKSCCSCCPVGCAKCAQGCICKGA      SDKCSCCA
IIa   KKSCCSCCPVGCAKCAQGCICKGA      SEKCSCCA
      ** ******  * ***  * ***         ** **
```

Fig. 7 Amino acid sequence of GIF and its alignment to human MTs. Bold letters indicate unique residues in GIF. Conserved residues are indicated by asterisks at the bottom.

cDNA SEQUENCING:

GIF cDNA sequencing was carried out by Tsuji et al (1992). To avoid isolation of MT cDNA instead of GIF cDNA, a unique pair of primers corresponding to GIF1-7 and GIF61-68 was designed for GIF cDNA amplification. Using these oligonucleotide primers, GIF cDNA fragments were isolated by PCR. Then to get full-length human GIF cDNAs, they screened the human brain cDNA library using the PCR product, which contained nucleotide sequences coding for GIF but not for MT. Coding sequence of GIF cDNA is 70 % identical with that of human MT 2A, but has 3 bp and 18 bp inserts which correspond to N-terminal 1 residue and C-terminal 6 residue inserts in GIF protein, respectively. Moreover, 5' and 3' non-coding regions of GIF cDNA are different from those of MT cDNAs. The chromosomal localization of human GIF gene was determined by screening a somatic cell hybrid panel with 3' non-coding segment of GIF cDNA. GIF gene is located on chromosome 16, as are the MT genes. These results confirm that GIF is a new molecule of MT family.

GROWTH INHIBITORY ACTIVITY IN GIF BUT NOT IN METALLOTHIONEIN

Native GIF purified from human brain inhibits the neurotrophic activity in AD brain extract and half-maximal concentration of native GIF is 0.2 μg/ml. Recombinant GIF purified from E. coli which was transfected with human GIF cDNA fragment and cultured with zinc had also growth inhibitory activity on neonatal rat cerebral cortical neurons, in spite of lower specific activity than that of native GIF (Half-maximal concentration of recombinant GIF was 1.5 μg/ml, Fig. 8).

The homology of GIF to MT led us to examine possible growth inhibitory activity in MT. Commercially available MTs from rabbit liver or horse kidney did not inhibit the neurotrophic activities in AD brain extract at the concentration up to 12.5 μg/ml (Fig. 8). Moreover, human brain MT fractions eluted from DEAE-Sephacel at 75 mM NaCl did not have growth inhibitory activity. Then, why GIF but not MT has growth inhibitory activity ? The active fragments from purified brain GIF contained one third of GIF polypeptide chain from N-terminal, which is quite different from that of MT. These results indicate that sequence differences in N-terminal portions of GIF and MTs lead to the functional differences between GIF and MT.

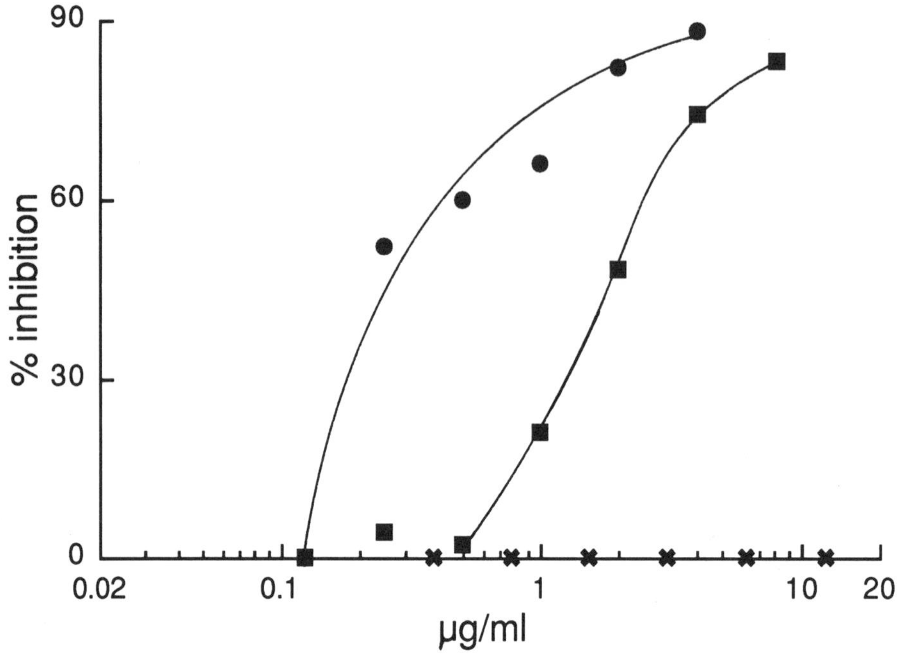

Fig. 8 The growth inhibitory activity of native GIF (●), recombinant GIF (■), and MTs (**X**) from rabbit liver and horse kidney. The growth inhibitory activity is expressed as percent inhibition of neurotrophic activities of an AD brain extract on neonatal rat cortical neurons in vitro.

DISTRIBUTION OF GIF AND GIF mRNA

We raised an antiserum against a synthetic peptide GIF 53-64 which sequence is unique to GIF. The affinity-purified antibodies reacted only with GIF but not with MT. Immunoblot with these antibodies revealed that GIF was distributed only in nervous system, such as cerebral cortex, cerebellum, striatum , but not in liver, kidney or muscle (Fig. 9). Northern blot analysis using the 3' non-coding region of GIF cDNA , which did not cross-hybridize with MT mRNAs, confirmed the GIF expression in nervous system but not in liver, kidney and testis, which are the major sites for MT mRNA expression.

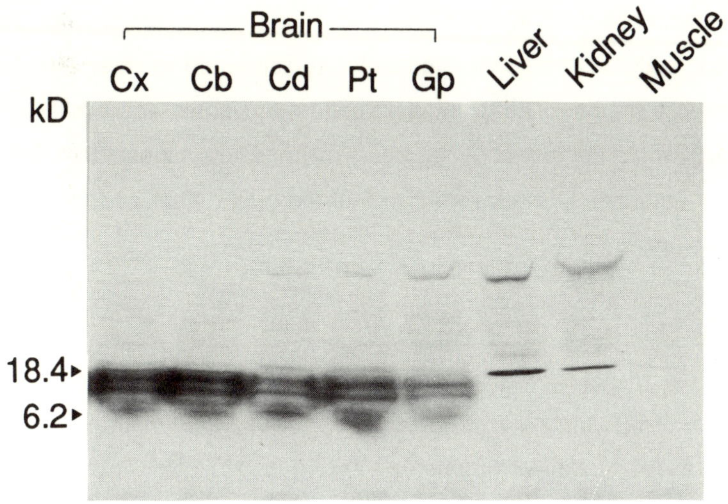

Fig. 9 Immunoblots of tissue extracts from non-demented aged patients.
Cx, cerebral cortex; Cb, cerebellum; Cd, caudate; Pt, putamen; Gp, globus pallidus.

The immunocytochemical localization of GIF was investigated in normal aged brains using the antibodies specific to GIF. The antibodies strongly stained the layer 2 to 6 of the gray matter, and weakly stained subcortical white matter, but did not stain layer 1. Higher magnification clearly showed the immunoreactivities in astrocytes but not in neurons. The immunoreactivities were distributed to astrocyte cell bodies and their fine long processes with numerous varicosities. Those stained processes formed dense meshworks in the gray matter. GIF immunoreactivities were also detected around small blood vessels or capillaries, which represented the feet of astrocytes. The distribution of GIF immunoreactivities in the gray matter is similar to that of MT immunoreactivities. Double staining for anti-GIF and anti-MT antibodies revealed that some of GIF-positive astrocytes stained with anti-MT antibodies. The distribution of GIF and MT immunoreactivities in the gray matter are reciprocal to that of glial fibrillary acidic protein (GFAP) immunoreactivities in the gray matter. GFAP antibodies stained the astrocytes in layer 1 but not layers 2 to 6 of the gray matter.

The expression site for GIF mRNA in normal aged brain was determined by in situ hybridization using RNA probe prepared by in vitro transcription of 3' non-coding region of GIF cDNA. Anti-sense signals were distributed in layer 2 to 6 of the gray matter and white matter. The distribution pattern of signals was similar to that of GIF immunoreactivity but not to that of neurons. These results suggest that GIF message is expressed in astrocyte in the layers 2 to 6 of the gray matter and white matter of the normal cerebrum.

DOWN-REGULATION OF GIF IN AD BRAIN

Immunoblot with GIF antibodies showed 6 kD intense band in normal aged brains, but only a trace in AD brains with numerous NFT (Fig. 10). Northern blot analysis also revealed down-regulation of GIF in AD cortex; The ratios of GIF mRNA to β-actin mRNA were 5.0 ± 2.1 for normal aged brains and 1.2 ± 0.4 for AD brains (Tsuji et al., 1992).

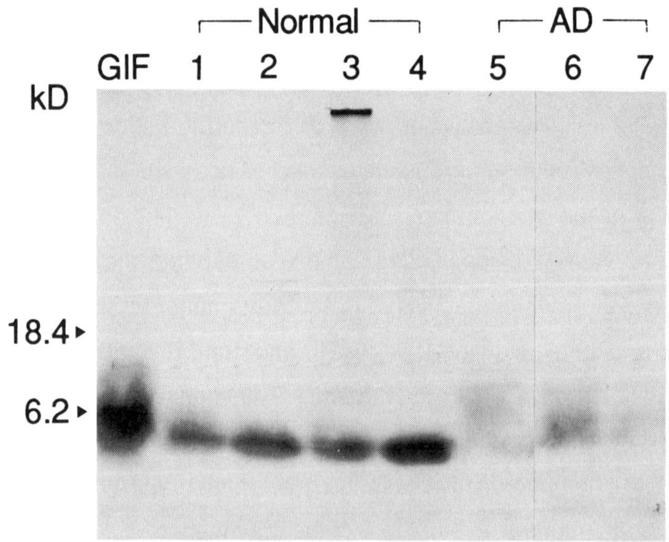

Fig. 10 Immunoblot of heat-stable extracts from normal aged and AD brains. GIF, purified GIF from human brain; lane 1-4, extracts from normal brains; lane 5-7, extracts from AD brains.

Immunohistochemical study revealed that the extents and modes of decrease in GIF in the gray matter varied among 5 AD cases: a laminar disappearance of immunoreactivities and patchy disappearance of immunoreactivities in the gray matter (Table 1). In a severely suffered case, almost of all immunoreactivities were not detected in the layers 2 to 5 of the gray matter. In three cases, immunoreactive astrocyte processes but not cell bodies disappeared in upper layers of the gray matter. In one case, patchy disappearance of immunoreactivities was noted in the gray matter. In all 5 AD cases, the moderate or strong immunoreactivities remained in lower layer (layer 5 or layers 5-6) of the gray matter. The number of GFAP-positive astrocytes dramatically increased in the region which GIF immunoreactivities disappeared. In contrast to GIF, MT immunoreactivities slightly increased in AD cerebral cortex, although there was a variation of intensity of MT immunoreactivities among the cases (Table 1). Double staining for anti-GIF and anti-MT antibodies revealed that some astrocytes were stained with both anti-GIF and anti-MT antibodies but many astrocytes were stained with only with anti-MT. The down-regulation of GIF in a subset of astrocyte which are closely associated with neuronal perikarya and dendrites may lead to the abortive sprouting of pyramidal neurons in the AD cerebral cortex.

Table 1. Immunostaining for GIF and metallothionein in temporal cortex from normal aged and AD patients.

Case	Diagnosis	GIF immunoreactivity	Metallothionein immunoreactivity
1	normal	++	±
2	normal	+	+
3	normal	++	+
4	normal	++	±
5	normal	++	±
6	AD	± (layer 2-5), + (layer 6)	±
7	AD	± (layer 3-5), + (layer 2 & 6)	±
8	AD	± (patchy), + (others)	++
9	AD	± (layer 3-6), + (layer 2)	++
10	AD	± (layer 2-4), ++ (layer 5-6)	+

++, strong; +, moderate; ±, weak

REFERENCES

Appel, S.H. (1981) Ann. Neurol. 10, 499-505.

Ihara, Y. (1988) Brain Res. 459, 138-144.

Tsuji, S., Kobayashi, H., Uchida, Y., Ihara, Y., and Miyatake, T. (1992) EMBO J. 11, 4843-4850.

Uchida, Y., Ihara, Y., and Tomonaga, M. (1988) Biochem. Biophys. Res. Commun. 150, 1263-1267.

Uchida, Y., and Tomonaga, M. (1989) Brain Res. 481, 190-193.

Uchida, Y., Takio, K., Ihara, Y., and Tomonaga, M. (1991) Neuron 7, 337-347.

Metallothionein III
K.T. Suzuki, N. Imura and M. Kimura (eds)

METALLOTHIONEIN IN TUMOR AND HOST: INTERSECTIONS OF ZINC METABOLISM, THE STRESS RESPONSE, AND TUMOR THERAPY

D.H. Petering[1], A. Quesada[2], M. Dughish[1], S. Krull[1], T. Gan[1], D. Lemkuil[1], A. Pattanaik[1], R.W. Byrnes[1], M. Savas[1], H. Whelan[3], and C.F. Shaw III[1]

[1]Department of Chemistry, University of Wisconsin-Milwaukee, Milwaukee, WI, 53201, USA, [2]Antibioticos Farma, Madrid, Spain, and [3]Medical College of Wisconsin, Milwaukee, WI, USA.

SUMMARY Substantial concentrations of zinc(Zn)- and apo- or metal unsaturated metallothionein are found in an array of different types of animal tumors. Three hypotheses to explain their presence include (1) metallothionein (MT) is an intermediate in zinc metabolism, (2) MT is a participant in gene regulation, and (3) MT is a contributor to the intracellular thiol-disulfide redox state of the cell. In support of the first hypothesis is an experiment with human muscle rhabdomyosarcoma (TE671) cells showing that as the cell cycle is slowed by incubation with thymidine and the corresponding rate of Zn transfer from the nutrient medium to Zn-metalloproteins declines, the concentration of Zn in metallothionein increases. Incubation of tumor cells with Zn to induce Zn-MT protects human glioblastoma (U373) and TE671 cells from cis-dichlorodiammine Pt(II) (cis-DDP). It also blocks some of the toxicity of H_2O_2 but not adriamycin in human leukemia HL60 cells. This mechanism of protection may have clinical relevance for brain tumors, in which dexamethasone is a palliative component of therapy. In U373 cells, dexamethasone induces Zn-MT synthesis nearly as well as Cd and provides about a 10-fold protection against cis-DDP. Induction of MT in the host during the inflammatory response to a resident tumor has been demonstrated, suggesting that elevated MT in normal tissues may affect the pharmacokinetics of antitumor drugs that react with this protein. Deliberate induction of MT in normal tissues by bismuth compounds may also block drug toxicity in the host. Pig kidney LL-CPK_1 and U373 cells respond to non-toxic concentrations of Bi-citrate by synthesizing large amounts of MT. Chemical studies support the reactivity and significant reductive capacity of the thiolate groups in MT with 5,5'-dithio-bis(2-nitrobenzoate) (DTNB) and cis-DDP. ^{111}Cd NMR studies imply that reactions of limiting amounts of DTNB with $^{111}Cd_7$-MT modify but do not destroy the clusters. In contrast, reaction with cis-DDP stoichiometrically and cooperatively destroys the clusters.

INTRODUCTION

Zinc (Zn) is a stringent nutritional requirement for dividing normal and tumor cells. After young animals are placed on a Zn deficient diet, there is both a dramatic decrease in plasma Zn within 24 hours and rapid cessation of growth (McClain et al., 1985; Petering et al., 1984). Thereafter, they can be maintained on this diet without much loss in weight or decrease in tissue Zn for several weeks (Minkel et al., 1979; Petering et al., 1984). Tumors and tumor cells in Zn deficient hosts are similarly affected (McQuitty et al., 1970; Petering et al., 1967). For example, Ehrlich ascites tumor cells can remain viable in Zn-deficient mice for more than five weeks, primed to divide rapidly as soon as nutrient Zn is reinstated (Minkel et al., 1979).

A number of human studies have suggested that plasma Zn and copper (Cu) levels change reciprocally according to the viability of resident tumors (Askari et al., 1980). An investigation of the influence of intraperitoneal Ehrlich ascites tumor cells in mice on these parameters showed that after tumor cell injection, the plasma Zn concentration rapidly decreases and Cu more slowly increases (Ujjani et al., 1986). Each response is consistent with the stimulation of an inflammatory response or acute phase stress response in the host by the tumor cells (Powanda, 1981).

Both the nutritional utilization of Zn by tumors and the redistribution of Zn during stress centrally involve metallothionein (MT). As described below, many animal tumors have substantial concentrations of MT. Furthermore, the livers of animals bearing tumors display sustained, elevated concentrations of MT, probably as a result of the generic stress response to inflammatory agents (Ujjani et al., 1986). The sections below describe the present understanding of MT in tumor-host Zn metabolism and how the efficacy of some components of clinical tumor therapy may be altered through interaction with MT in both the tumor and the host.

PRESENCE OF METALLOTHIONEIN IN TUMORS

MT is a common component of a variety of tumor cells (Table I) (Koch et al., 1980; Pattanaik et al., submitted).

Table I: Animal Tumors with Zinc- and Apometallothionein

Tumor or Normal Tissue	μg MT per g Tissue	MT-1	MT-2	Fraction ApoMT[a]
Mouse				
Melanoma B16/BL6	55	+	+	0.2
Carcinoma M25	6	+	-	0
Colon Adeno-Carcinoma 26C	74	+	+	0.9
Reticulum Cell Carcinoma M5076	16	-	+	0
Sarcoma 180	19[b]			0.9
P-388 Leukemia	5[b]			1.0
Ehrlich Mammary Adenocarcinoma	50[b]	+	+	0
Rat Kidney	9			
Mouse Liver	12[c]			

[a]Apo- or metal unsaturated MT. [b]per 10^8 cells. [c]Cd,Zn-MT present after 30 day exposure to 50 μg Cd/ml in drinking water.

An early observation of Zn-MT in Ehrlich ascites tumor cells indicated that this pool comprised about 15% of the cytosolic Zn (Koch et al., 1980; Kraker et al., 1988). Focusing on the metal content of tumor MT by comparing Sephadex G-75 profiles of metals (Zn, Cu, Cd) in untreated cell supernatants with those incubated for a few minutes with Cd^{2+} to convert all MT protein to Cd-MT, it was determined in a number of tumors that MT is unsaturated with Zn; in others the MT pool contains no Zn (Table 1, Figure 1). In another experiment, cytosol from TE671 cells was chromatographed over Sephadex G-75 (Figure 2a). No Zn or Cu was detected in the profile which was bound to MT. Nevertheless, cytosol pretreated

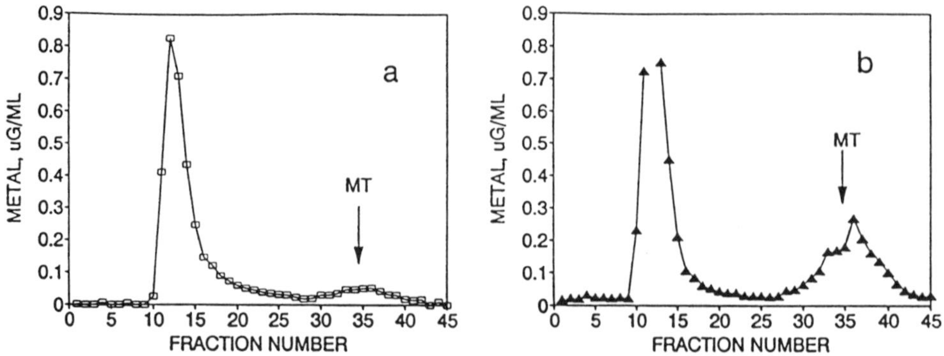

Figure 1: Sephadex G-75 Profiles of U373, Human Glioblastoma (a) 10^8 control cells grown in Dulbecco's media, supplemented with 10% fetal calf serum (FCS), (b) 10^8 control cells, cell supernatant was incubated with 6 μg/mL Cd^{2+} for 20 min prior to chromatography. No Cu was observed in the MT band of any of the samples.

with Cd^{2+} as above and then chromatographed did contain a pool of Cd-MT (data not shown). When fractions I-III from the separation shown in Figure 2a were incubated with Cd^{2+} and then rechromatographed over Sephadex G-75, a band of Cd-MT appeared only when fraction III, the MT fraction, was used (Figure 3). Some Cd in fraction I trailed into the MT band, but was not bound to MT according to further analysis. Thus, it is evident that apoMT

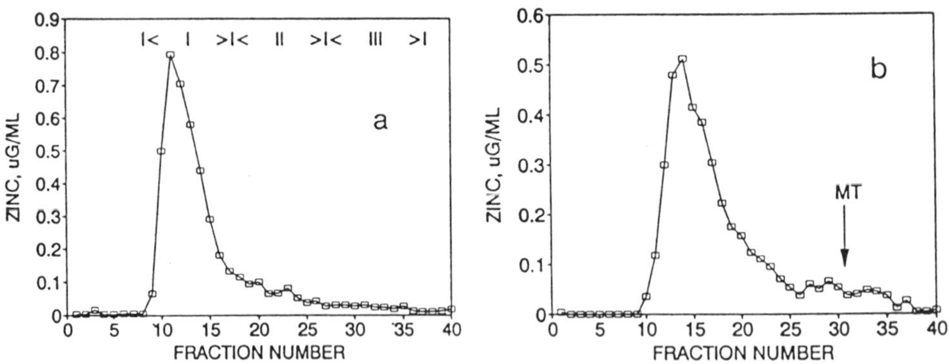

Figure 2: Variable Saturation of the MT Pool with Zn. (a) 10^8 control TE671 rhabdomyosarcoma cells grown in Dulbecco's media, supplemented with 10% fetal calf serum which double every 24 h. (b) 10^8 TE671 cells incubated 200 μM thymidine for 24 h.

migrates like Cd-MT during gel filtration and that apoMT can be observed as Cd-MT by addition of Cd^{2+} to cytosol before or after Sephadex G-75 chromatography. This conclusion was confirmed with a sample of authentic apoMT2 made from rabbit liver Zn-MT2.

These results are consistent with findings with Ehrlich cells made zinc-deficient _in vivo_ or _in vitro_, in which apoMT was also observed (Kraker et al., 1988; Krezoski et al., 1988). It has been thought previously that apoMT is immediately degraded in cells and, thus, can not be a quantitatively significant form of intracellular MT (Feldman et al., 1978). This view is based on _in vitro_ studies of the rate of proteolysis of apo- and metal-containing MTs and may not be directly relevant to cells in which the apoprotein may be present in compartments other than those such as lysosomes where biodegradation occurs. From a comparison of the values quoted for normal tissues, which either have a large basal amount of MT (rat kidney) or have been induced by Cd (mouse liver), it is also clear that substantial concentrations of MT are found in these tumors. What the functional roles of sizable concentrations of Zn-MT/apoMT in many tumors may be is an important question, particularly in light of the demonstrated need of tumors for Zn.

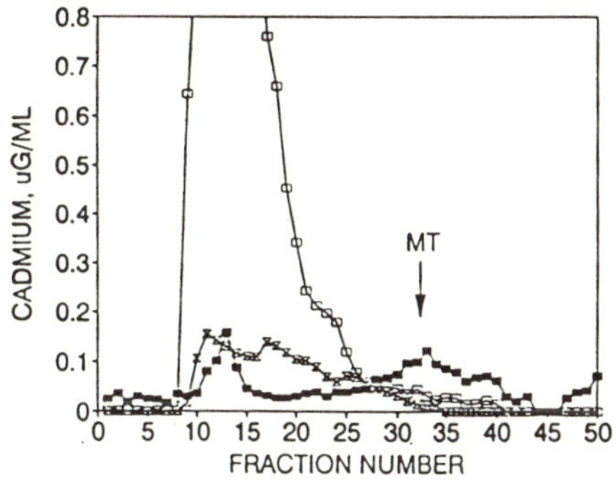

Figure 3: Sephadex-75 Chromatography of Cytosol Fractions I-III from Figure 2a. (□) fraction I, (⌶) fraction II, and (■) fraction III. Equal aliquots of Cd^{2+} were added to each fraction. The fractions were concentrated by ultrafiltration before rechromatography. Most of the Cd in fractions II and III was removed in the filtration step.

Several hypotheses deserve attention. First, Zn-MT/apoMT may be a key site linking nutrient Zn entering the cell with the final sites of binding of Zn in metalloproteins (Krezoski et al., 1988). Second, Zn-MT/apoMT may interact with Zn finger proteins or other Zn-requiring sites to control their association with Zn and, thus, their biological activity (Zeng et al., 1991a, 1991b). Third, the thiolate groups of Zn-MT/apoMT may contribute to the intracellular redox state (Vallee, 1987). Besides these potential roles in the biochemistry of tumor cells, Zn-MT/apoMT may fortuitously serve as a site of reaction of electrophilic cytotoxic agents encountered by tumors and normal tissues.

ZINC METABOLISM IN TUMOR CELLS: ROLE OF METALLOTHIONEIN

The presence of substantial concentrations of Zn-MT and apoMT in tumors raises the question of their involvement in the metabolism of the Zn in cells that have a high demand for the metal because of rapid proliferation. One hypothesis listed above is that MT is an intermediate between nutrient Zn entering the cell and its final sites of localization within the cell (Figure 4). Specifically, in this model Zn-MT donates its metal to apoproteins by direct ligand substitution.

This mechanism has been supported by studies with Ehrlich cells showing that when the extra-cellular medium is depleted of Zn, the rate of loss of Zn from the Zn-MT pool, presumably to other Zn-requiring sites, is much faster than the rate of bio-degradation of the protein (Krezoski et al., 1988). These kinetics argue that some process such as metal exchange between Zn-MT and apoproteins but not biodegradation is the rate limiting step for movement of Zn out of MT. In this experiment, the principal species undergoing degradation was apoMT. Its half-life of 5 h is not much different than that of Zn-MT (7h).

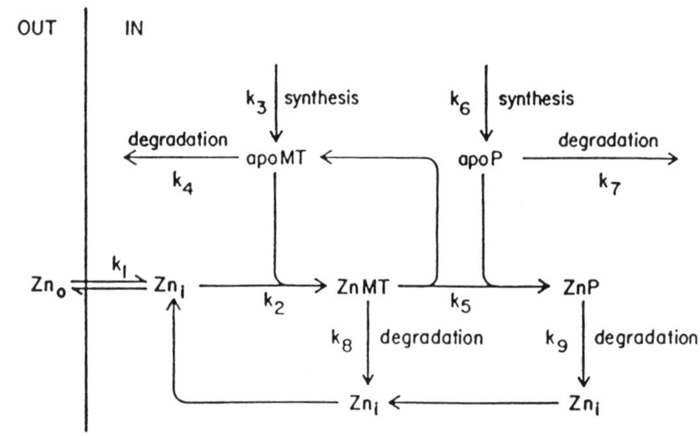

Figure 4: Model for Role of MT in Zn Metabolism.

According to Figure 4, in the steady state when nutrient zinc is available, MT would exist as Zn-MT if the rate of transfer of Zn from the extracellular medium to apoMT (k_1, k_2 processes) is faster than the rate of ligand exchange (k_5). Reciprocally, apoMT would be found if uptake of Zn into MT is rate limiting. So, one interpretation of the finding of apoMT in a tumor cell line is that the requirement to constitute Zn-proteins outstrips the capacity of the cells to accumulate Zn from the host. A slow rate of uptake of Zn into apoMT has been observed in Zn-deficient Ehrlich cells, which contain a basal concentration of apoMT, after the cells are restored to their Zn-normal growth medium (Petering et al., 1992). This is consistent with the possibility that k_1 and k_2 can become rate limiting under some circumstances.

To test the hypothesis in Figure 4, TE671 cells which contain only apoMT were utilized. It is recognized that the rate of accumulation of Zn bound to Zn-proteins (high molecular weight band in the Sephadex G-75 profile) is a reflection of the rate of the cell cycle and cell division. If Zn-protein synthesis were slowed by inhibiting the progression of the cell cycle, metal transfer from Zn-MT to apoproteins might become the overall rate controlling step and apoMT would be converted to Zn-MT. Figure 2 illustrates that when thymidine is used to slow the rate of cell division of TE671 cells by 70%, the apoMT pool (panel a) becomes completely saturated with Zn (panel b). This finding is in agreement with prediction that the saturation of the MT pool with Zn can be controlled by the proliferation rate of the cells.

METALLOTHIONEIN, INTERACTION WITH ZINC FINGER PROTEINS, AND INTRACELLULAR REDOX STATE

Kägi and coworkers have provided chemical evidence that apoMT can compete and extract Zn bound to transcription factors TFIIIA (Zn-TF) and SP-1 (reaction 1) (Zeng et al., 1991a, 1991b).

$$\text{Zn-TF} + \text{apoMT} \rightleftharpoons \text{Zn-MT} + \text{apoTF} \qquad (1)$$

From these results they hypothesized that apoMT might perturb the action of Zn-finger proteins in cells by extracting Zn and, thereby, render them unable to bind to their DNA promoter sites.

The finding of abundant apoMT in a variety of tumor cells supplies the experimental support for the existence of apoMT needed to pursue this hypothesis. Interestingly, Liu (1989) has indicated that Zn-MT donates Zn to apoglucocorticoid hormone receptor. This receptor binds Zn through 4 thiolate groups instead of with two imidazole nitrogens and two cysteine sulfhydryl residues as in the other proteins mentioned above. This difference may account for the reversal of the favored direction of reaction 1, for Zn^{2+} prefers sulfhydryl ligands and may have a larger stability constant with the glucocorticoid receptor than with TFIIIA.

The discovery of significant concentrations of MT in tumor cells including apoMT indicates that MT along with glutathione (GSH) may be a major pool of intracellular thiols. The sulfhydryl groups in both Zn-MT and apoMT are strong nucleophiles though one expects Zn-MT to be somewhat less reactive with electrophiles, because its thiolate groups are masked through binding to Zn and because of the folded structure of the holoprotein (Li et al., 1981). Furthermore, Zn-MT and presumably apoMT are good reducing agents for at least some disulfides as described below. Thus, it is possible that Zn-MT/apoMT and GSH/GSSG together set the thiol-disulfide redox state in tumor cells.

METALLOTHIONEIN AND PROTECTION AGAINST ELECTROPHILIC ANTITUMOR AGENTS AND REACTIVE SPECIES Of OXYGEN

Beginning in the early 1980s, it was shown that elevated levels of Cd-MT in Cd-resistant cells could intercept antitumor compounds with electrophilic functional groups such as chlorambucil and cis-dichlorodiammine Pt(II) (cis-DDP), and prevent their cytotoxicity (Bakka et al., 1981; Endresen et al., 1983). Interest in MT as a possible major site of reaction of electrophilic reagents in cells is linked to a similar curiosity about the importance of glutathione in these reactions (Andrews et al., 1985; Seagrave et al., 1983). Both are thiolate compounds, the latter commonly present in millimolar concentrations in cells.

The presence of large concentrations of constitutively expressed MT in tumors as indicated in Table I raises the question of the importance of its reactions with electrophilic antineoplastic agents. An early report showed that at cytotoxic concentrations of cis-DDP Pt about 30% of the drug was probably bound to MT in Ehrlich cells (Kraker et al., 1985). A recent report indicated that overexpression of MT may be a mechanism by which tumor cells become resistant to extended exposure to cis-DDP and the alkylating agents melphalan, chlorambucil, and cyclophosphamide (Kelley et al., 1988).

Experiments have been done to probe whether Zn-MT might protect against adriamycin (Adr) in a cell culture model. Using HL60 cells induced with 100 μM Zn to make enough Zn-MT so that MT-bound Zn comprised 70% of the cytosol (142 ng MT/10^8 cells), no effect on the cytotoxicity of Adr could be detected over its dose response range of 0 - 2 μM nor was there any diminution of the single strand DNA damage caused by this agent.

The basis for this study was a report that MT induction in heart muscle protected animals from the cardiotoxic effects of the drug, which have been ascribed to an oxidant stress mechanism that produces damaging hydroxyl radicals (Naganuma et al., 1988; Satoh et al., 1988). So, the influence of Zn-MT on the toxicity of H_2O_2 in HL60 cells was also examined, recognizing that it is thought to be reduced to hydroxyl radical in cells in Fenton-like reactions (Jonas et al., 1989). In contrast to the results with Adr, Zn-MT increased the concentration of hydrogen peroxide needed to inhibit cell proliferation by 75% over a 24 h period from 20-35 μM, and also decreased the amount of single strand DNA damage caused by this compound (100 μM) after 30 min exposure by 40%. The cellular concentration of Zn bound to MT was reduced by one-half during the 30 min incubation with H_2O_2, suggesting that MT had directly intercepted a reactive species to reduce the concentration of hydrogen peroxide or hydroxyl radical available to interact with DNA. This probably occurred through oxidation of the thiol groups in the protein, which released Zn bound to the clusters. A relatively large concentration of induced Zn-MT seems to be required to provide protection during this short term exposure (2.5 μmol MT thiolate groups/10^8 cells). The differential effects of elevated Zn-MT on the toxicity of Adr and H_2O_2 suggests that in HL60 cells these agents act by different mechanisms.

There are clinically relevant circumstances in which MT induction by components of the treatment regime may compromise chemotherapy. In the management of brain tumor patients, dexamethasone, a synthetic glucocorticoid, is routinely used to decrease the inflammatory response around the tumor (Davidson and Lenman, 1981). It is also commonly given to patients receiving cis-DDP before treatment to decrease vomiting. Prednisone, another glucocorticoid analog, is a component of many drug combinations which include alkylating agents (Seiden and Canellos, 1992; Spiritos and Cassileth, 1992). Dexamethasone is an inducer of MT in liver and tumor cells; prednisone also induces MT in U373 cells (Figure 5) (Karin et al., 1980). On a molar basis, dexamethasone is a better stimulus of MT synthesis in TE671 and U373 human tumor cell lines than Zn. Cells pre-treated with 1 μg/ml dexamethasone for 24 h contain 100 nmol Zn-MT/10^8 cells and require 10 times the concentration of cis-DDP to cause 50% inhibition of growth than do controls which contain 18 nmol MT/10^8 cells, largely as apoprotein.

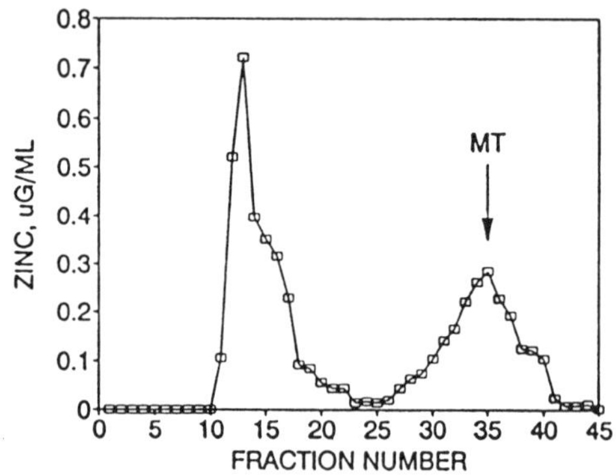

Figure 5: Induction of Zn-MT Synthesis in U373 Cells by Prednisone. Cells were exposed to 9 μg/ml prednisone for 24 h in the same medium listed in Figure 1.

Interestingly, while the combination of Zn- and apoMT protects against the toxic effects of cis-DDP, preinduction of cells with Cd, which results in a pool of Cd-MT, does not protect. In fact, these cells are 10 times more sensitive to cis-DDP than are control cells. Possibly, in the metal exchange reaction between cis-DDP and Cd-MT, the binding of Pt to the protein releases Cd^{2+} in cells, which acts as a secondary toxic agent.

INDUCTION OF METALLOTHIONEIN IN HOST TISSUES: RELATIONSHIP TO CHEMOTHERAPY

Tumors act as inflammatory agents in host organisms and induce a generalized stress response which involves an abrupt decrease in plasma Zn upon imposition of the stress. Induction of hepatic Zn-MT occurs as plasma Zn is transferred into the liver. In addition, serum Cu is elevated as ceruloplasmin is made as an acute phase reactant protein (Ujjani et al., 1986). Indications that this is probably also occurring in human cancer come from numerous studies showing changes in plasma Zn and Cu concentrations in patients with active tumors (Askari et al., 1980). To the extent that tumors stimulate this mechanism which reduces circulating Zn, they contribute to a state of metabolic wasting in which Zn is in limiting supply to sustain the needs of the organism for this nutrient (Beutler and Cerami, 1988). We have also shown previously that tumors with their large constitutive pools of apoMT compete successfully for scarce amounts of Zn in the host (Kraker et al., 1988).

According to the discussion above, the use of dexamethasone in therapeutic regimens not only might elevate the concentration of MT in tumors, it may also increase it in liver as well. Both the high background level of liver MT in organisms stressed with tumors and any additional MT induced by dexamethasone may, thus, bolster the capacity of liver to detoxify alkylating antitumor agents and thereby contribute to the observed pharmacokinetics of these drugs.

The complexity of the relationship between MT induction and chemotherapeutic effectiveness is seen in the case of cis-DDP. The dose-limiting side effect of this powerful agent is kidney toxicity directed toward the proximal tubule (Weiner and Jacobs, 1983). Platinum (II) has a strong affinity for sulfur ligands and some of the damage caused by cis-DDP in the kidney can be prevented by administration of sulfur compounds such as diethyldithiocarbamate (Dedon and Borch, 1987). Recently, it has been shown that animals pretreated with bismuth nitrate or subnitrate can be protected from the acute kidney toxicity of cis-DDP but remain sensitive to its antitumor effects (Naganuma et al., 1987). It was found that in response to bismuth, mice produce significantly elevated levels of kidney MT and it was assumed that the high density of sulfhydryl groups in this protein had reacted with the drug to inactivate it.

In initial studies to define a model to study the interactions of Bi^{3+}, MT, and cis-DDP, pig kidney cortical cells ($LL\text{-}CPK_1$ cells) were incubated with a soluble form of bismuth, Bi-citrate, and the induction of MT observed. According to Figure 6a, $LL\text{-}CPK_1$ cells make little MT until they are exposed to Bi-citrate. After 24 h of incubation, they contain a sizable pool of Zn-bound to MT, pretreatment of the cytosol with Cd followed by chromatography reveals that there is a much larger pool of MT present, probably existing as Bi,Zn-MT (Figure 6b). It is evident in these cells

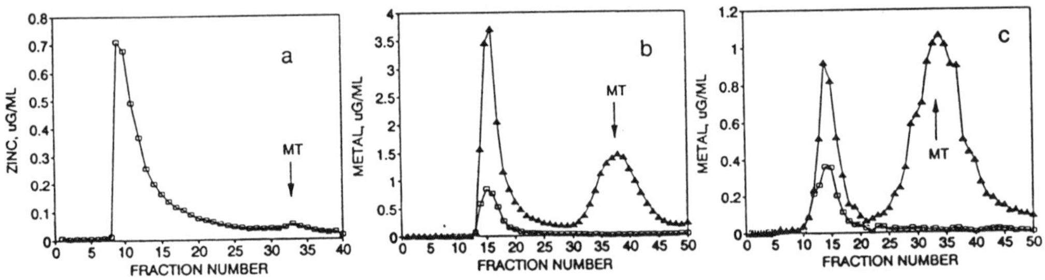

Figure 6: Induction of MT in $LL\text{-}CPK_1$ and U373 Cells by Bi-citrate. (a) $LL\text{-}CPK_1$ cells grown in medium 199 supplemented with 4% FCS. (b) $LL\text{-}CPK_1$ cells exposed to 100 μg/mL Bi as Bi-citrate for 24 h. (□) Zn and (▴) Cd profiles after 10 min incubation of supernatant with 6 μg/ml Cd^{2+}. (c) U373 cells incubated with 75 μg/mL Bi as Bi-citrate; supernatant was exposed to 6 μg/ml Cd for 20 min. (□) Zn, (▴) Cd.

that Bi is a highly effective inducer of MT. The original studies of the suppression of drug toxicity to the host by suspensions of bismuth given orally to mice indicated that normal tissues but not the tumor were protected by induction of MT (Naganuma et al., 1987; Naganuma et al., 1988; Satoh et al, 1988). Nevertheless, there is nothing special about $LL\text{-}CPK_1$ cells in this reaction, for U373 glioblastoma tumor cells also respond to Bi with the induction of large amounts of MT (Figure 6c).

REACTIONS OF THE THIOLATE GROUPS OF METALLOTHIONEIN WITH ELECTROPHILIC REAGENTS

To understand the chemical possibilities of MT as a nucleophile, mechanistic studies have been carried out which focus on the reactivity of its sulfhydryl groups with several reagents. The first detailed examination of this sort employed 5,5'-dithio-bis(2-nitrobenzoate), DTNB, as the electrophilic agent (Li et al., 1981). Over the past decade the following picture has emerged. Even though the 20 thiolate groups in the protein are masked by complexation to Zn(II), Cd(II), or Cu(I), they retain significant reactivity with this agent. In the presence of excess DTNB each of the sulfhydryl groups of the two clusters reacts in cooperative fashion by associative and dissociative pathways with the α-domain, a 4 metal, 11 thiolate cluster, being kinetically more reactive (Savas et al., 1991). Recently, it has been shown that the reaction proceeds beyond the stage of mixed disulfide formation to yield completely reduced thionitrobenzoate (TNB) and oxidized protein (reactions 2 and 3) with cysteine disulfides preferred over mixed disulfides with TNB (Savas et al., 1992).

$$(\mathrm{Cd})\mathrm{MT}{-}\mathrm{S}^- + \phi\mathrm{S{-}S}\phi \longrightarrow \mathrm{MT{-}S{-}S}\phi + \phi\mathrm{S}^- + \mathrm{Cd}^{2+} \qquad (2)$$

$$\mathrm{MT}\begin{cases}\mathrm{S{-}S}\phi \\ \mathrm{S}^-\end{cases} \longrightarrow \mathrm{MT}\begin{cases}\mathrm{S} \\ | \\ \mathrm{S}\end{cases} + \phi\mathrm{S}^- \qquad (3)$$

That is, Zn- and Cd-MT are also effective reducing agents as well as nucleophiles. Zn_7-MT reacts much slower with the physiologically interesting compound glutathione disulfide than DTNB.

An intriguing finding has been that when $^{111}Cd_7$-MT is titrated with DTNB, the ^{111}Cd signals in the protein disappear before the stoichiometric endpoint of the reaction is reached (Figure 7). This is observed despite the fact that most of the ^{111}Cd has been retained in protein after it is isolated by Sephadex G-75 chromatography (Savas et al., 1992). The monomeric and polymeric products, which contain very little TNB, react with an excess of DTNB with kinetics similar to the native protein, yielding stoichiometries of SH/Cd in the modified structure of 2.4–3.8. It seems likely that new cluster species have been produced that are in rapid chemical exchange on the NMR time scale and no longer visible by NMR spectroscopy.

Other experiments have shown that Zn-MT reacts readily with the activated double bond of N-ethylmaleimide and the nitrogen mustard, melphalan. Thus, it seems possible that MT may be a significant site of reaction in the cell of other electrophiles in the form of antitumor alkylating agents or activated carcinogenic species.

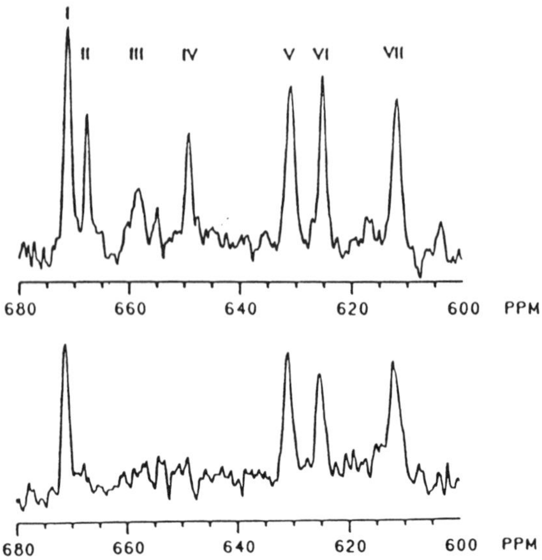

Figure 7: Reaction of $^{111}Cd_7$-MT with DTNB. (a) Native spectrum; resonances II-IV represent ^{111}Cd in the 3-metal cluster of the ß-domain. (b) Spectrum after reaction with DTNB. MT(SH): DTNB = 3:50.

Another example of reaction of the sulfhydryl groups of MT with an electrophilic center involves the interaction of cis-DDP with Zn_7- or Cd_7-MT (Pattanaik et al., 1992; Lemkuil et al., 1991). The reaction of cis-DDP with Zn_7-MT occurs with the following approximate stoichiometry:

$$10\ (NH_3)_2Pt(Cl)_2 + Zn_7\text{-MT} + 20\ H^+ \longrightarrow Pt_{10}\text{-MT} + 7\ Zn^{2+} + 20\ NH_4^+ + 20\ Cl^- \quad (4)$$

Notably, both the labile chloride ligands and the relatively inert ammine ligands are displaced as Pt binds to the thiolate groups in MT. The kinetics of reaction are biphasic, beginning with an induction phase before most of the Pt-sulfhydryl bond formation occurs. The overall rate of reaction is significantly faster than that reported for the reaction of cis-diaquo/hydroxodiammine Pt with GSH and much faster than the rate of reaction of cis-DDP with

DNA (Dedon and Borch, 1987; Johnson et al., 1980). In fact, the kinetics of reaction 4 are independent of Cl^- concentration. So, when cis-DDP enters cells it can react immediately in a bimolecular process with Zn_7-MT but must wait for slow, rate limiting chloride hydrolysis before reaction with its presumed target DNA can occur. This fact probably explains much of the protective effect of cellular MT against cis-DDP.

The preference for the sulfhydryl groups of Zn-MT may result from the following properties of the protein. First, there is a high density of thiolate groups in each domain. Some of them are accessible to the solvent (Robbins et al., 1991). Second, once reaction begins at one sulfhydryl site, it apparently occurs with rapid, cooperative reaction of the entire cluster.

The reaction of cis-DDP with Cd_7-MT follows a different kinetic path than the one involving the Zn protein (Pattanaik et al., 1992). Two kinetic steps are observed which both lead to Pt-thiolate bond formation. In addition, only about 7 Pt atoms bind to MT in the process. Although one might a priori ascribe each kinetic step to the reaction of one of the two MT clusters, experiments with the Cd_4-α domain show that even in the absence of the Cd_3-β domain it, too, reacts with cis-DDP with biphasic kinetics. This result suggests that there are alternate pathways of reaction of cis-DDP within a single cluster.

An examination of this reaction was done by NMR spectroscopy, using $^{111}Cd_7$-MT to follow the details of displacement of ^{111}Cd by Pt (Figure 7a for reference). As the protein was titrated, there was concerted reduction in resonances II-IV from the β-domains and I, and V-VII of the α-domain with some preference of reaction given to the 3-metal cluster (Lemkuil et al., 1991). The product, Pt_7-MT, seems to retain its domain structure according to experiments showing that Cd_7- and Pt_7-MT migrate similarly during Sephadex G-75 gel filtration chromatography. Nevertheless, since Cd-thiolate clusters involve tetrahedrally coordinated Cd, whereas, Pt(II) has a strong requirement for square planar coordination, the conversion of Cd_7-MT to Pt_7-MT would seem to require a much different protein geometry.

Lastly, previous studies have shown that MT can react with superoxide and hydroxyl radicals (Thornalley and Vašák, 1985). To investigate whether hydrogen peroxide can also interact with cellular Zn-MT, 200 μM H_2O_2 was incubated with 30 μM Zn-MT for 30 min at 37°C and the reaction followed by periodic assay of liberated Zn with the chromophoric ligand, 2-carboxy-2′-hydroxy-5′-sulfoformazylbenzene or zincon (Shaw et al., 1990). It was demonstrated that 22% of the Zn was released. Similar reactions were carried out in the presence of 0.5% and 1% dimethylsulfoxide (DMSO), a hydroxyl radical scavenger. Twenty-four and 20% Zn were released, respectively, showing no role for hydroxyl radicals in the release of the metal under the conditions used. This preliminary experiment shows that Zn-MT can react effectively with H_2O_2 as well as the other radical species along the pathway of reduction of dioxygen to water.

CONCLUSIONS

Large concentrations of MT have been found in a variety of animal and human tumors and cell lines. In particular, the detection for the first time of apo- or unsaturated MT in many of them deserves careful analysis in relation to the stringent requirement of growing tumors for Zn, the widespread utilization of Zn in transcription factors, and the importance of the thiol-disulfide redox state of the cell. The constitutive presence and facile induction of MT in tumors and the presence of MT at least in normal liver as a result of the host stress response to tumor suggest that the antitumor action of electrophilic drugs may be significantly affected by MT in these tissues. This concern is heightened when it is recognized that inducers of MT like dexamethasone and prednisone are frequently components of therapy. The practical considerations about the role of MT in the design of tumor chemotherapy and management are complicated. This conclusion is underscored by the possibility that some side effects of drugs such as cis-DDP may be moderated by the induction of MT in host tissues.

ACKNOWLEDGEMENTS

The authors acknowledge support from the United States National Institutes of Health through grant ES-04026.

REFERENCES

Andrews, P.A., Murphy, M.P., and Howell, S.B. (1985) Cancer Res. 45, 6250-6253.

Askari, A., Long, C.L., and Blackemore, W.S. (1980) Parenteral Enteral Nutr. 4, 561-571.

Bakka, A., Endresen, L., Johnsen, A.B.S., Edminson, P.D., and Rugstad, H.E. (1981) Toxicol. Appl. Pharmacol. 61, 215-226.

Beutler, B., and Cerami, A. (1988) Ann. Rev. Biochem. 57, 505-518.

Davidson, D.L.W., and Lenman, J.A.R. (1981) Neurological Therapeutics, Pitman Medical Limited, London, 134-140.

Dedon, P.C., and Borch, R.F. (1987) Biochem. Pharmacol. 36, 1955-1964.

Endresen, L., Bakka, A., and Rugstad, H.E. (1983) Cancer. Res. 43, 2918-2926.

Feldman, S.L., Failla, M.L., and Cousins, R.J. (1978) Biochim. Biophys. Acta 544, 638-646.

Johnson, N.P., Hoeschele, J.D., and Rahn, R.O. (1980) Chem.-Biol. Interact. 30, 151.

Jonas, S.K., Riley, P.A., and Willson, R.L. (1989) Biochem J. 264, 651-655.

Karin M. Herschman, H.R., and Weinstein, D. (1980) Biochem. Biophys. Res. Commun. 92, 1052-1059.

Kelley, S.L., Basu, A., Teicher, B.A., Hicker, M.P., Hamer, D.H., and Lazo, J.S. (1988) Science 141, 1813-1815.

Koch, J., Wielgus, S., Shankara, B., Saryan, L.A., Shaw III, C.F., and Petering, D.H. (1980) Biochem. J. 189, 95-104.

Kraker, A.J., Krakower, G., Shaw III, C.F., Petering, D.H., and Garvey, J. (1988) Cancer Res. 48, 3381-3388.

Kraker, A., Schmidt, J., Krezoski, S., and Petering, D.H. (1985) Biochem. Biophys. Res. Commun. 130, 786-792.

Krezoski, S.O., Villalobos, J., Shaw III, C.F., and Petering, D.H. (1988) Biochem. J. 255, 483-491.

Lemkuil, D., Nettesheim, D., Pattanaik, A., Shaw III, C.F., and Petering, D.H. (1991) J. Inorg. Biochem. 43(2-3), 604.

Li, T.-Y., Minkel, D.T., Shaw III, C.F., and Petering, D.H. (1981) Biochem. J. 193, 441-446.

Liu, X. (1989) Metallothionein Structure and Reactivity, Ph.D. Thesis, University of Wisconsin-Milwaukee.

McClain, C.J., Kasarskis, Jr., E.J., and Allen, J.J. (1985) Prog. Food Nutr. Sci. 9, 185-226.

McQuitty, J.T., DeWys, W.D., Monaco, L., Strain, W.H., Rob, C.G., Apgar, J., and Pories, W.J. (1970) Cancer. Res. 30, 1387-1390.

Minkel, D.T., Dolhun, P.J., Calhoun, B.L., Saryan, L.A., and Petering, D.H. (1979) Cancer Res. 39, 2451-2456.

Naganuma, A., Satoh, M., and Imura, I. (1987) Cancer Res. 47, 983-987.

Naganuma, A., Satoh, M., and Imura, I. (1988) Jpn. J. Cancer Res. 79, 406.

Pattanaik, A., Bachowski, G., Laib, J., Lemkuil, D., Shaw III, C.F., Petering, D.H., Hitchcock, A., and Saryan, L. (1992) J. Biol. Chem. 267, 16121-16128.

Pattanaik, A., Shaw III, C.F., Petering, D.H., Garvey, J., and Kraker, A. (submitted for publication).

Petering, D.H., Krezoski, S., Chen, P., Pattanaik, A., and Shaw III, C.F. (1992) Metallothioneins. Synthesis, Structure, and Properties of Metallothioneins, Phytochelatins and Metal-Thiolate Complexes, Stillman, M.J., Shaw III, C.F., and Suzuki, K.T., eds., VCH Publishers, New York, 164-185.

Petering, D.H., Loftsgaarden, J., Schneider, J., and Fowler, B.A. (1984) Environ. Health Persp. 54, 73-81.

Petering, H.G., Buskirk, H.H., and Crim, J.A. (1967) Cancer. Res. 27, 1115-1121.

Powanda, M.C. (1981) Trace Elements in the Pathogenesis and Treatment of Inflammation, Rainsford, K.D., Bune, K., and Whiteside, M.W., eds., Birkhäuser Verlag, Basel, 121-135.

Robbins, A.H., McRee, D.E., Williamson, M., Collett, S.A., Xuong, N.H., Furey, W.F., Wang, B.C., and Stout, C.D. (1991) J. Mol. Biol. 221, 1269-1293.

Satoh, M., Naganuma, A., and Imura, N. (1988) Toxicology 53, 231.

Savas, M.M., Petering, D.H., and Shaw III, C.F. (1991) Inorg. Chem. 30, 581-583.

Savas, M.M., Petering, D.H., and Shaw III, C.F. (1992) J. Inorg. Biochem., accepted.

Seagrave, J., Tobey, R.A., and Hildebrand, C.E. (1983) Biochem. Pharmacol. 32, 3017-3021.

Seiden, M.V., and Cannelos, G.P. (1992) Current Therapy in Hematology-Oncology, Brain, M.C. and Carbone, P.P. eds., B.C. Decker Publishers, Philadelphia, PA, 339-345.

Shaw III, C.F., Laib, J.E., Savas, M.M., Petering, D.H. (1990) Inorg. Chem. 29, 404-408.

Spiritos, M., and Cassileth, P. A. (1992) Current Therapy in Hematology-Oncology, Brain, M.C. and Carbone, P.P. eds., B.C. Decker Publishers, Philadelphia, PA, 346-351.

Thornalley, P.J., and Vašák, M. (1985) Biochim. Biophys. Acta 827, 36.

Ujjani, B., Krakower, G., Bachowski, G., Krezoski, S., Shaw III, C.F., and Petering, D.H. (1986) Biochem. J. 233, 99-105.

Vallee, B.L. (1987) Metallothionein II, Kägi, J.H.R. and Kojima, Y., eds., Birkhäuser Verlag, Basel, 5-16.

Weiner, M.W., and Jacobs, C. (1983) Fed. Proc. 42, 2974-2978.

Zeng, J., Heuchel, R., Schaffner, W., and Kägi, J.H.R. (1991a) FEBS Lett. 279, 310-312.

Zeng, J., Vallee, B.L., and Kägi, J.H.R. (1991b) Proc. Natl. Acad. Sci., USA 88, 9984-9988.

Metallothionein III
K. T. Suzuki, N. Imura and M. Kimura (eds)

MOLECULAR GENETICS OF METALLOTHIONEIN: AN OVERVIEW

Dean H. Hamer

Laboratory of Biochemistry, National Cancer Institute, National Institute of Health, Bethesda, MD 20892 USA

SUMMARY: Molecular genetics has contributed to three major areas of metallothionein (MT) research. 1) STRUCTURE. Because it is easier to isolate and sequence genes than proteins, molecular biology provides an important tool to study the primary structure, diversity and evolution of MT molecules. Site-directed mutagenesis provides a means to study the three-dimensional structure of MT as well. 2) REGULATION. Molecular biology was used to show that Cu regulates yeast MT gene expression by directly binding to a metal-sensitive transcription factor. Similar experiments in mammalian systems are in progress. 3) FUNCTION. The function of MT in lower organisms has been directly tested by the construction of null mutants. An important challenge for future research is to determine the still enigmatic role of MT in higher organisms by transgenic technology.

During the past decade, molecular genetics has played an increasingly important role in understanding the structure, function, and regulation of metallothionein (MT). The simple single cell eukaryote *Saccharomyces cerevisae* provides a useful example to describe some of the applications of molecular techniques to MT research.

STRUCTURE

Saccharomyces cerevisae contains a single metallothionein that is encoded by the CUP1 locus. Although yeast MT can bind in vitro to the usual range of metal ions, it is induced in vivo only by Cu, and therefore serves primarily as a Cu chelator. The structure of yeast Cu-MT was initially studied by metal chelation and reconstitution, UV and fluorescent spectroscopy, and EXAFS. The results showed that Cu-MT consists of a polynuclear Cu_8-S_{12} cluster, in which each Cu(I) ion is trigonally coordinated by branching cysteine thiolates, encased within a tightly folded polypeptide shell. This unusual structure accounts for the extraordinary ability of MT to tightly chelate and sequester Cu. Using molecular techniques, amino acid substitutions were produced at individual cysteine residues, at pairs of cysteine residues, and at the amino-terminus of the molecule. These reagents were used to examine the relationship between the structure of MT and its processing, role in Cu detoxification, and autoregulation (Wright et al., 1987; Thrower et al., 1988; Wright et al., 1988).

REGULATION

The intracellular sensor for Cu ions in *Saccharomyces cerevisae* is the MT gene transcription factor encoded by the ACE1 locus. At low Cu concentrations, ACE1 exists as an randomly folded, inactive apoprotein. At high Cu concentrations, 7-8 metal ions bind to multiple cysteine residues within the amino-terminal region of ACE1, causing it to fold into a specific DNA-binding structure that recognizes a short DNA sequence motif in the 5' promoter region of the CUP1 gene. Once ACE1 has bound to the promoter region, it stimulates transcription by interacting with the general transcription machinery (Furst et al., 1988; Culotta et al., 1989).

ACE1 acts as a highly sensitive, specific, and cooperative sensor of Cu ions. It is capable of recognizing micromolar concentrations of Cu(I), even in presence of excess divalent ions such as Zn(II) or Cd(II). Activation of the DNA-binding potential of ACE1 is highly cooperative, with an apparent Hill coefficient of 4. Thus a small change in intracellular Cu results in a large change in gene expression, resulting in increased efficiency of detoxification (Furst and Hamer, 1989).

The structure of the amino-terminal DNA and Cu-binding region of ACE1 has been analyzed by biochemical and genetic techniques. The results show that ACE1 binds to 6-8 Cu(I) ions through 11 cysteine thiolates that form a polynuclear cluster. DNA binding is dependent upon tracts of basic amino acids that are interspersed with cysteines. A remarkable and unexpected feature of this structure is its homology to that of the regulated gene product, MT (Hu et al., 1990; Dameron et al., 1991). We termed this structure the "copper fist".

FUNCTION

Yeast cells carrying CUP1 null mutations, in which the coding sequences for MT are deleted, were produced and characterized. Despite the fact that these strains produce absolutely no MT, they are perfectly viable and can accumulate Cu and incorporate it into Cu-enzymes normally. This shows that MT is not required for normal Cu metabolism in yeast cells. However, the CUP1 mutants are exquisitely sensitive to high concentrations of added Cu, demonstrating the important role of MT in Cu toxicity (Hamer et al., 1985; Hamer, 1986).

Recently we have discovered a new pathway for the metabolism of Cu by *Saccharomyces cerevisae* that involves the reduction of Cu(II) to Cu(I) at the cell surface. This reaction is catalyzed by the FRE1 cell surface oxidoreductase, a protein which is also required for the conversion of Fe(III) to Fe(II). Cells that lack FRE1 protein have increased resistance to Cu toxicity due to decreased transport of the metal into the cell.

The activity of the FRE1 complex is regulated by the CUA1 gene. This gene was discovered by a search for mutants with multiple defects in Cu metabolism and detoxification. Thus CUA1 null mutants display decreased Cu(II) reduction, decreased Cu uptake, and increased Cu resistance. In contrast, CUA1 "up" mutants have increased reduction and transport and are hypersensitive to Cu toxicity.

The CUA1 gene has been cloned and sequenced. It encodes a nuclear protein with two interesting features. First, it contains a cysteine-rich repeated motif that may serve as a Fe and/or Cu binding site. Second, the amino-terminal region is strongly homologous to ACE1. We therefore speculate that ACU1, like ACE1, is a specific DNA binding protein. We are currently searching for direct target gene of ACU1 by both biochemical and genetic techniques.

Perhaps in the future similar proteins that cooperate with MT to regulate metal metabolism in higher eukaryotes will be discovered.

Acknowledgements

I thank all of my colleagues who have collaborated on our MT research over the past decade.

REFERENCES

Culotta, V.C., Hsu, T., Hu, S., Furst, P., and Hamer, D.H. (1989). Proc. Natl. Acad. Sci. USA 86, 8377-8381.

Dameron, C.T., Winge, D.R., George, G.N., Sansone, M., Hu, S., and Hamer, D.H. (1991). Proc. Natl. Acad. Sci. USA 88, 6127-6131.

Furst, P., Hu, S., Hackett, R., and Hamer, D.H. (1988). Cell 55, 705-717.

Furst, P., and Hamer, D.H. (1989). Proc. Natl. Acad. Sci. USA 86, 5267-5271.

Hamer, D.H. (1986). Ann. Rev. Biochem. 55, 913-951.

Hamer, D.H., Thiele, D.J., and Lemontt, J.F. (1985). Science 228, 685-690.

Hu, S., Furst, P., and Hamer, D.H. (1990). The New Biologist 2, 544-555.

Thrower, A., Byrd, J., Tarbet, E., Mehra, R., Hamer, D.H., and Winge, D.R. (1988). J. Biol. Chem. 263, 7037-7042.

Winge, D.R., Nielson, K.B., Gray, W.R., and Hamer, D.H. (1985). J. Biol. Chem. 260, 14464-14470.

Write, C.F., Hamer, D.H., and McKenney, K. (1988). J. Biol. Chem. 263, 1570-1574.

Write, C., McKenney, K., Hamer, D.H., Byrd, J., and Winge, D.R. (1987). J. Biol. Chem. 262, 12912-12919.

Metallothionein III
K. T. Suzuki, N. Imura and M. Kimura (eds)

CELL-SPECIFIC EXPRESSION AND REGULATION OF THE MOUSE METALLOTHIONEIN-I and -II GENES IN THE REPRODUCTIVE TRACT AND PREIMPLANTATION EMBRYO

G.K. Andrews[1], M.T. McMaster[1], S.K. De[1] B.C. Paria[2], and S.K. Dey[2]

Department of Biochemistry and Molecular Biology[1], and Departments of Obstetrics-Gynecology and Physiology[2], University of Kansas Medical Center, Kansas City, KS. 66160-7421

SUMMARY: Studies of the mouse reproductive tract and developing embryo suggest that the metallothionein (MT) -I and -II genes are actively expressed in a cell-specific and developmentally regulated manner. These mRNAs are present at low levels in ova and preimplantation mouse embryos, and are metal-inducible only after the third cleavage. In contrast, they are constitutively abundant in male germ cells, specifically in pachytene spermatocytes and round spermatids. In the female reproductive tract, isthmus epithelial cells of the oviduct contain very low levels of these MT mRNAs on days 3 and 4 of pregnancy, and only these cells respond to exogenous metal ions by increased expression of the MT genes. In contrast, to the preimplantation period, from the time of implantation (day 5 of pregnancy) to late in gestation, the mouse embryo is surrounded by cells, interposed between the maternal and embryonic environments, that actively express the MT-I and -II genes. Expression of these genes occurs in decidual cells, placental spongiotrophoblasts, and the visceral endoderm of the yolk sac. Defining roles for MT in the reproductive tract and embryo remains a challenge. Experiments summarized within provide evidence that MT in deciduum /placenta/yolk sac acts as a barrier that restricts cadmium (Cd) from reaching the post-implantation embryo. In contrast, Cd apparently exerts direct toxic effects on the preimplantation blastocyst which is not surrounded by cells actively expressing these MT genes. Studies of the mechanisms of regulation of the MT-I and-II genes in the deciduum suggest that their expression is developmentally programmed and may be under autocrine/paracrine control. Decidual expression of a mouse MT-I promoter-human TGF-α fusion gene in transgenic mice mimics that of the endogenous MT-I gene. This finding suggests an approach for mapping decidual-specific enhancer elements in the mouse MT-I promoter.

INTRODUCTION

Cadmium (Cd) is teratogenic and embryotoxic in rodents (reviewed by Barlow and Sullivan, 1982; Vallee and Ulmer, 1972), and can cause the deciduum, placenta, and yolk sac to rapidly degenerate, or induce neurotube, limb bud and craniofacial abnormalities (Barlow and Sullivan, 1982). Importantly, Cd Transfer from mother to embryo during the teratogenic period (days 8 to 10 of gestation in the mouse) is restricted (reviewed by Dencker, et. al. 1983). Therefore, it is thought that the deciduum, placenta and yolk sac may protect the developing embryo from Cd exposure (Barlow and Sullivan, 1982; Dencker, et. al. 1983). However, the mechanism(s) of restriction of Cd entry into the embryonic environment is unknown.

The mechanisms by which essential metal homeostasis is achieved in the early embryo are also unknown. Zn deficiency can cause abnormal growth and morphogenesis (see Webb, 1987; Rogers and Hurley, 1987). Many of the physiological functions of Zn can be attributed to changes in the activity of Zn metalloenzymes.

MT may function in essential metal homeostasis and protection from metal toxicity. Thus, active expression of MT genes in the developing rodent liver (reviewed in Webb, 1987) led to the suggestion that MT is involved in fetal and neonatal development. Studies reviewed herein examined changes in MT gene expression in the mouse reproductive tract and preimplantation embryo, and the results obtained suggest that MT may play important roles in the initiation and maintenance of pregnancy.

RESULTS AND CONCLUSIONS

CD-1 mice were used, and the gestational day of pregnancy was calculated from the day on which a vaginal plug was detected (designated day 1 (D1) of pregnancy). Animals were housed in the Animal Resource Center at the University of Kansas Medical Center and used according to the institutional guidelines for the care and use of laboratory animals.

MT GENE EXPRESSION IN GERM CELLS AND PREIMPLANTATION EMBRYOS; RNA from D1 to D4 preimplantation mouse embryos was analyzed, using the reverse transcriptase-polymerase chain reaction (RT-PCR), for the presence of MT-I and MT-II mRNAs (Fig. 1) (Andrews, et. al. 1991). Both of these transcripts were detected in RNA from 1-cell fertilized eggs collected on D1 (Fig. 1) suggesting that these MT mRNAs are maternal messengers stored in the mouse egg. These transcripts were also readily detected in cleavage stage embryos (D2 & D3), and in morulae and blastocysts (Fig. 1). Thus, the zygotic MT-I and MT-II genes are coordinately expressed throughout the preimplantation period.

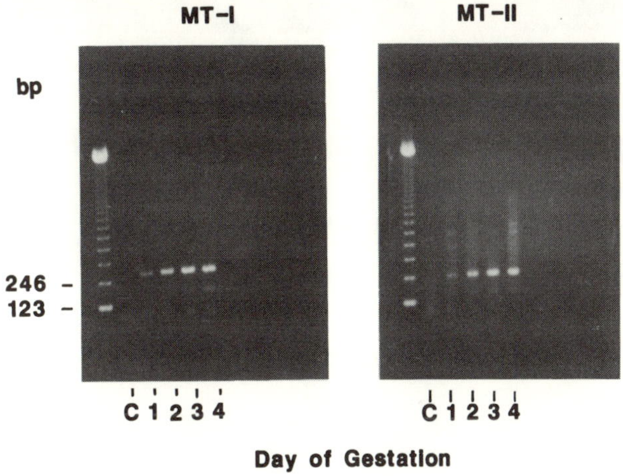

Figure 1. Detection of MT-I and MT-II mRNAs in preimplantation mouse embryos using the RT-PCR. Preimplantation mouse embryos (about 80 per group) were recovered on the indicated days of gestation, and total RNA was extracted in the presence of carrier E. coli ribosomal RNA. RNA was reverse transcribed, and subjected to 45 cycles of PCR using oligo primers specific for MT-I or MT-II mRNAs. PCR products, equivalent to that obtained from about 1 embryo equivalent of RNA, were separated by agarose-gel electrophoresis in the presence of ethidium bromide, and DNA was visualized under UV light. RT-PCR of MT-I and MT-II mRNAs in normal D1 (1-cell), D2 (2-cell), D3 (4- and 8-cell) and D4 (late morulae/blastocysts) embryos (labelled 1-4). In the control reactions (C), RNA samples were omitted from the PCR. Reverse transcriptase was omitted before PCR in parallel control reactions. Predominant reaction products were near the predicted sizes (259 & 225 base pairs (bp)) for proper amplification of the MT-I and -II cDNAs, respectively, and were much smaller that those expected from amplification of the MT genes in genomic DNA (>616 bp). Reproduced with permission from Andrews, et. al. 1991.

The temporal and spatial patterns of expression of the MT-I gene in preimplantation embryos, and the effects of exposure to Zn or Cd on levels of this mRNA were analyzed by in situ hybridization (Fig. 2). Specificity of this method has been well documented (De, et. al. 1989; Andrews, et. al. 1990; Andrews, et. al. 1991). All control and experimental samples were analyzed in parallel on the same slides.

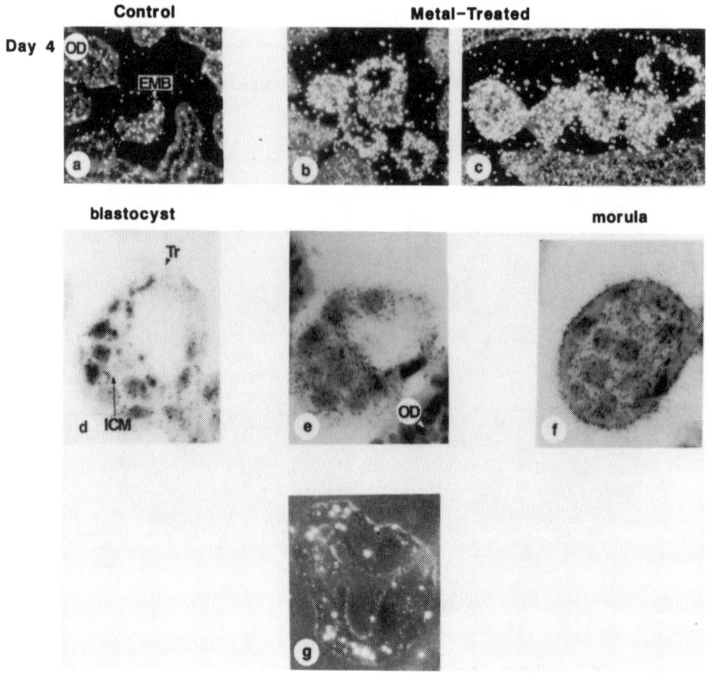

Figure 2. Detection of MT-I mRNA in preimplantation mouse embryos using in situ hybridization. Preimplantation mouse embryos on day 4 were recovered, and incubated for 5 hr in the presence or absence of Zn (50 µM) or Cd (10 µM). Embryos were transferred back into the ampulla region of the oviduct, fixed in paraformaldehyde, sectioned, and mounted onto poly-lysine coated slides. Sections within a given experiment were placed on the same slide. Slides were hybridized with a ^{35}S-labeled MT-I antisense strand RNA probe, and hybrids were detected by autoradiography for 5 days. The oviduct served as internal control to establish background signals. Day 4 embryos were at the late morula or blastocyst stages of development. Shown here are darkfield photomicrographs (200X) in which the autoradiographic grains appear as white dots (**panels a-c and g**), and brightfield photomicrographs (400X) in which the grains appear as black dots (**panels d-f**). Abbreviations are: ICM, inner cell mass; Tr, trophectoderm; EMB, embryo; OD, oviduct. Control embryos (**panel a,d**); Zn-treated embryos (**panels b,e-g**), Cd-treated embryos (**panel c**). **Panel g**, Zn-treated blastocyst hybridized with a sense strand MT probe as a control. Reproduced with permission from Andrews, et. al. 1991.

Based on autoradiographic grain counts per embryo, very low levels of MT-I mRNA were detected in preimplantation embryos on D1 to D3 (Andrews, et. al. 1991), but basal levels of MT-I mRNA increased significantly at the blastocyst stage (D4). Incubation of preimplantation embryos in medium containing Zn or Cd had little effect on MT-I mRNA levels before the 8-cell stage (D3). In contrast, essentially all blastomeres of ≥8-cell embryos were heavily labelled following metal-treatment (Fig. 2). Metal exposure resulted in a 6- to 8-fold increase in MT-I mRNA per embryo for ≥8-cell embryos. Examinations of serial sections of morulae and blastocysts indicated that metal ions induced MT-I mRNA in all cell types (Fig. 2). Both Zn and Cd were effective inducers (Fig. 2, panels b & c respectively). Thus, the MT-I gene responds dramatically to metals at and after the third cleavage (8-cell stage of development).

Low, but detectable, levels of MT-I and -II MT mRNAs were present in the mouse ovary, and metal treatment rapidly increased these levels. However, in ova at various stages of development, in situ hybridization detected little MT-I mRNA before or after metal injection. High grain densities were detected over interstitial and some granulosa cells in metal-treated ovaries. Thus, the MT genes are not constitutively expressed at high levels, nor are they detectably metal responsive in mouse ova at various stages of development (Andrews, et. al. 1991).

Testes from the sexually mature mice, contained high levels of both MT-I and -II mRNAs (De, et. al. 1991). These levels were age dependent, being low the first two weeks after birth, and reaching maximal levels in the adult (9 weeks after birth). In adult testis, in situ hybridization indicated that only cells germ cells contained high levels of MT mRNA. Pachytene spermatocytes and round spermatids isolated from adult testes, contained both MT-I and MT-II mRNAs in levels equivalent to those found in zinc-treated hepatocytes, whereas very low levels of MT mRNA were detected in isolated Sertoli cells. Isolated pachytene spermatocytes and round spermatids apparently synthesize MT-I and MT-II. Whether these cell types accumulate MT remains to be determined. MT-I mRNA levels in adult testes were not increased substantially after systemic injection of Cd, Zn or bacterial lipopolysaccharide (LPS). These studies established that the MT-I and -II genes are actively expressed in a developmentally regulated fashion in the male germ cells of the mouse. This suggests a role for MT in the process of spermatogenesis. Increased mRNA stability contributes to the accumulation of MT-I and -II transcripts in male germ cells (De, et. al. 1991), but little else is known about mechanisms regulating MT gene expression in these cell types.

MT GENE EXPRESSION IN THE REPRODUCTIVE TRACT; During the preimplantation period (D1 to D4), the embryo develops within the oviduct, and the blastocyst enters the uterine lumen on D4 morning and implantation is initiated in the evening. The uterine stroma proliferates in response to the implanting embryo and forms the deciduum which surrounds the embryo from about D6 to D10. Invasion of the trophoblasts, formation of the placenta and degeneration of the deciduum are essentially completed by D12, and the placenta and the embryo-derived visceral yolk sac surround the embryo until parturition on D21.

Low levels of MT-I mRNA were found in the oviduct (D1-D4). However, in situ hybridization established that MT-I mRNA is elevated specifically in the epithelial cells of the isthmus region of the oviduct and levels of this mRNA increase on D3 and D4. Remarkably, no other oviductal cell types contained elevated MT-I mRNA, and injection of Zn on D3 led to an increase in MT-I mRNA levels exclusively in the isthmus epithelial cells (Andrews, et. al. 1991).

Low levels of MT-I mRNA were localized on D4 in the uterine luminal epithelium. However, early on D5 MT-I mRNA was elevated in the antimesometrial portion of the uterine luminal epithelium that underlies the implanting blastocyst and in the deeper stroma, and not in uterine cells in the interimplantation regions or epithelia distal from the implanting embryo (McMaster, et. al. 1993). Later on D5, this mRNA was elevated in the primary decidual zone at the site of blastocyst implantation, while the deeper uterine stroma and myometrium were negative. From D6 to D8, MT-I mRNA was in abundance specifically in the secondary decidual zone (De, et. al. 1989).

Analysis of the relative rate of MT synthesis by native gel electrophoresis of pulse-labeled proteins (De, et. al. 1989) and of MT accumulation using the Cd-heme assay (Onosaka and Cherian, 1982) documented that MT-I and MT-II were actively synthesized and that MT accumulates to high levels in the deciduum by D8 (unpublished data). Solution hybridization showed that MT-I and MT-II mRNAs were elevated in decidua, and that they were coordinately reduced during placental development (Table I). Basal levels of MT-I mRNA in the adult liver reflect about 150 to 350 molecules per cell (Durnam and Palmiter, 1981; Durnam and Palmiter, 1983). Thus, both MT-I and MT-II mRNAs are in high abundance in the D8 deciduum.

Table I. Metallothionein-I and -II mRNA Levels in Decidua and Placentae during Gestation.

	MT Isoform[b]	
Age[a]	I	II
8	11.0	13.4
10	10.0	8.9
12	4.6	4.9
14	3.0	2.2
16	2.0	N.D

[a]Age refers to day of gestation. [b]Data represent fold-increase in MT-I and MT-II mRNA levels relative to those in control adult liver, as determined by oligodeoxyribonucleotide excess solution hybridization. Liver RNA, obtained 5 h following injection of 100 µmol/kg of $ZnCl_2$ contained 43.0- and 23.0-fold more MT-I and MT-II mRNA than control liver, respectively. Deciduum (D8), deciduum/placenta (D10) and placentae (D12 to 16) were obtained from the conceptuses from 3 to 6 pregnant females, and the results represent the average of three determinations. N.D. means not determined. Reproduced with permission from De, et. al. 1989.

Although analysis of total placental RNA showed that levels of MT-I and MT-II mRNAs declined about 6-fold by D14 (Table I), *in situ* hybridization demonstrated that spongiotrophoblasts, which are located in the outer placenta, constitutively express high levels of this mRNA (Fig. 3). In contrast, the trophoblast giant cells (outer placenta), and labyrinthine trophoblasts (inner placenta) contained little MT-I mRNA. The demarcation between the high level of MT-I mRNA in the spongiotrophoblasts, and the low level of this mRNA in the labyrinthine trophoblasts on D14 is striking (Fig. 3). These results establish that MT gene expression is elevated in a cell- and temporal-specific and manner during formation and maturation of the mouse decidua and placentae.

Earlier studies established that the endoderm cells of the visceral yolk sac also actively express the MT-I gene (Andrews, et. al. 1984). During the early stages of organogenesis (D9), MT-I and -II mRNA levels were low in the visceral yolk sac, but the abundance of these transcripts increased dramatically by D11 and remained

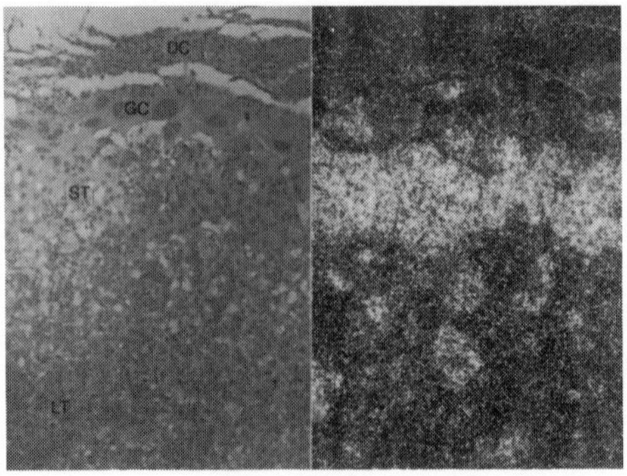

Figure 3. Localization of MT-I mRNA in the mouse placenta by <u>in situ</u> hybridization. The placenta, recovered on D14, was fixed in paraformaldehyde, sectioned, and mounted onto a poly-lysine coated slide. The slide was hybridized with a ^{35}S-labeled MT-I antisense strand RNA probe. Hybrids were detected by autoradiography for 4 days and the slide was post-stained lightly with hematoxylin. Shown here are bright field (left) and dark field (right) photomicrographs (40X magnification) of longitudinal sections near the outer edge through the mid-region of the placenta. **DC**, decidual cells; **GC**, trophoblast giant cells; **ST**, spongiotrophoblasts; **LT**, labyrinthine trophoblasts.

elevated until near the time of parturition (De, et. al. 1989). By comparison, RNA samples from the total embryo on D9 or D11 and from the uterine myometrium on D9 and D11 contained basal levels of these MT mRNAs.

These studies establish that from the time of implantation (D5) to late in gestation, the mouse embryo is surrounded by cells, interposed between the maternal and embryonic environments, that actively express the MT-I and -II genes. In distinct contrast, during the preimplantation period, the embryo is not surrounded by cells that actively express the MT genes.

<u>STUDIES OF POTENTIAL FUNCTIONS OF MT IN THE REPRODUCTIVE TRACT</u>; Stage-dependent effects of Cd exposure during the preimplantation period of pregnancy on the subsequent development and implantation of mouse embryos were examined (De, et. al. 1993). When the embryo is at the two-cell stage (D2), Cd delayed temporarily, but did not prevent implantation and was not embryolethal. In marked contrast,

the same dose of Cd administered on D4 caused pregnancy failure in all mice examined. Two-cell embryos were also found to be remarkably resistant to <u>in vitro</u> exposure to Cd. However, Cd toxicity increased with development, and morulae readily degenerated after Cd exposure. Two-cell embryos accumulated little or no ^{109}Cd (50 μM), whereas rapid accumulation and efflux of this metal occurred in blastocysts (De, et. al. 1993). Thus, pregnancy failure after Cd exposure during the preimplantation period reflects a direct embryotoxic effect of Cd, although maternal injury by Cd may also contribute. The preimplantation embryo (D4) is more susceptible to the lethal effects of Cd than is the mother (De, et. al. 1993). Therefore, sensitivity and resistance to acute metal toxicity involves mechanisms other than MT gene expression in the preimplantation mouse embryo. Furthermore, low level expression of MT in the oviduct and uterus on D4 correlates with the pronounced embryotoxic effects of Cd at this stage of pregnancy.

The relative levels of MT-I mRNA in the embryo following an acute <u>in vivo</u> or <u>in vitro</u> exposure to Cd or Zn were measured as a sensitive indicator of a rapid direct effect of these metals on the embryo (De, et. al. 1990). Animals were injected early in the morning of D10 with various doses of Zn or Cd, and embryos were collected 5 h later. Levels of MT-I mRNA were basal in D10 embryos (Fig. 4), and did not change significantly after injection of a teratogenic dosage of Cd (50 μmol/kg) (Fig. 4, left). In contrast, an injection of the equivalent amount of Zn significantly induced MT-I mRNA. To examine the possibility that the MT genes in the embryo resist induction by Cd, D10 embryos were incubated for <u>in vitro</u> for 5 h in medium with or without excess heavy metals (Cd or Zn) (Fig. 4, right). Cd was an effective inducer of embryonic MT gene expression. A comparison of the dose-response for these metals suggests that Cd is about 10-fold more potent than Zn at inducing MT mRNA in the embryo <u>in vitro</u> (Fig. 4, right). Taken together the above results suggest that little Cd rapidly reaches the D10 embryo following a teratogenic exposure to this metal. Thus, the high level of MT gene expression in the reproductive tract during the teratogenic period correlates with the apparent absence of a rapid, direct effect of Cd on the embryo.

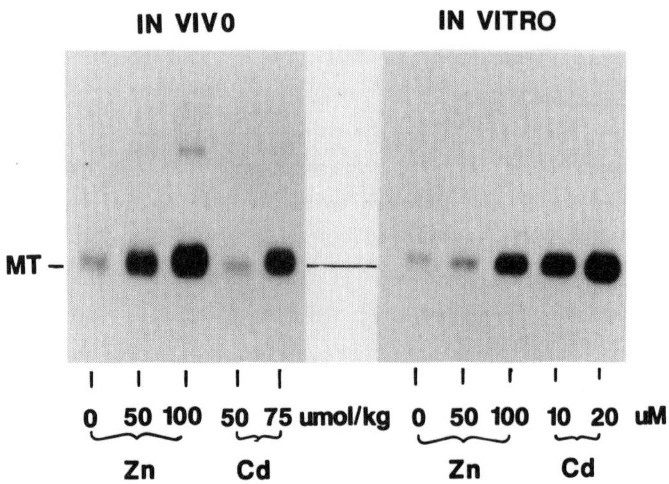

Figure 4. Northern blot analysis of MT mRNA levels in day 10 embryos exposed in vivo (left) or in vitro (right) to cadmium or zinc. Embryos (about 40 per point) were collected 5 h after a maternal sc injection of the indicated dosage of metal, or after 5 h of culture in medium containing the indicated concentrations of metal ions. Total RNA was extracted and analyzed by Northern blot hybridization using a mouse MT-I cRNA probe. Hybrids were detected by autoradiography for 5 h. Duplicate agarose gels were analyzed by acridine staining to ensure that equal amounts of RNA had been loaded. Reproduced with permission from De, et al. 1990.

STUDIES OF MECHANISMS OF REGULATION OF MT IN THE REPRODUCTIVE TRACT; Several approaches are being taken to examine the regulation of MT genes in the reproductive tract, with a current focus on the deciduum. The decidual reaction (deciduomata) can be induced on D4 in pseudopregnant mice, which are obtained by mating with a vasectomized male, by injection of 100 μl of sesame oil into the utero - tubular junction. Stromal cells throughout the uterus have divided and differentiated into decidual cells by D8. Deciduomata tissue harvested on D8 has MT-I mRNA levels equivalent to those in normal D8 decidua (De, et. al. 1989). Thus, decidual MT gene expression is not dependent on the presence of the embryo or embryo-derived factor(s). Surgical removal of the adrenal glands also had no effect on MT-I mRNA levels in deciduomata. Thus, glucocorticoid hormones are apparently not critical for decidual MT gene regulation (De, et. al. 1989). More recent studies suggest that neither metal ions (Zn) nor cytokines (IL-1β, IL-6, TNF-α) are primary mediators of decidual MT gene expression, and that decidual cells in culture continue to actively express the MT genes (unpublished data). Thus, expression of the MT genes in the decidua may reflect an autocrine/paracrine regulation or a developmentally programmed event.

To further examine this problem, a strain of transgenic mice which expresses the human TGF-α (hTGFα) cDNA under control of 750-bp of the mouse MT-I promoter (Jhappan, et. al. 1990), was examined for expression of the transgene in the deciduum and in experimentally induced deciduomata (Fig. 5). In D8 deciduum and deciduomata, levels of MT-I and hTGFα transcripts were high. In contrast, in RNA from the whole uterus at D6 these transcripts were both present at lower levels. Thus, the MT-I promoter drives apparently correct developmental expression of this reporter gene in the deciduum. Immunohistochemical detection of hTGFα further supports this conclusion (Dey, unpublished data).

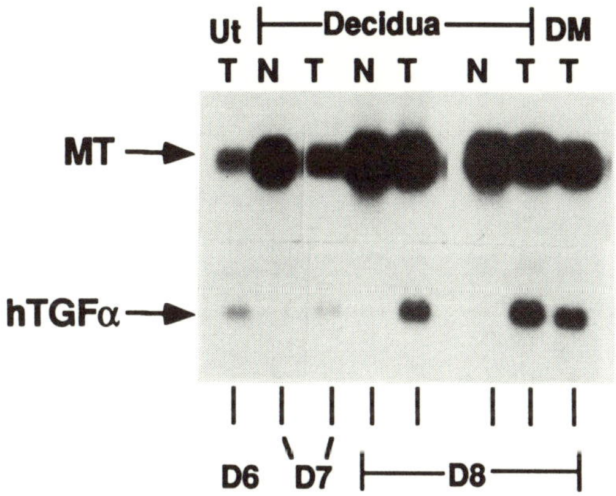

Figure 5. Developmental expression of an MT-I promoter-human TGF-α fusion gene in decidual cells of transgenic mice. A fusion gene consisting of the proximal 750-bp of the mouse MT-I promoter (Searle, et. al. 1984) driving expression of human transforming growth factor-α cDNA was used to create this transgenic strain of mice. These animals were provided by Dr. Glen Merlino (NIH) and have been described previously (Jhappan, et al. 1990). RNA was isolated from nontransgenic (N) and transgenic (T) mice and analyzed by Northern blot hybridization with cRNA probes for mouse MT-I mRNA and human TGF-α mRNA. RNA samples were as follows: day 6 (D6) total uterine RNA; day 7 (D7) decidual RNA; day 8 (D8) decidual RNA or RNA from D8 deciduomata (DM). Deciduomata was experimentally induced in uteri of D4 pseudopregnant mice, and DM was recovered by dissection 4 days after oil injection (D8).

These preliminary studies suggest that expression of the decidual MT genes must be regulated transcriptionally. Experiments in progress are exploiting transgenic technology to map promoter elements involved in regulation of expression of the MT-I and -II genes in the mouse deciduum.

ACKNOWLEDGEMENTS

This work was supported, in part, by a grant from the NIEHS (ES 04725) to GK Andrews and a grant from the NICHD (HD 12304) to SK Dey. SK De was supported by a postdoctoral fellowship from the Wesley Foundation. His current address is: Laboratory of Molecular Biology, National Institute of Mental Health, Building 36, Room 1B-08, Bethesda, MD 20892. MT McMaster was supported by a Predoctoral Graduate Research Training Fellowship from the March of Dimes Birth Defects Foundation (#18-87-05). His current address is: Department of Stomatology, HSW 604, University of California San Francisco, 3rd and Parnassus, San Francisco, CA, 94143.

REFERENCES

Andrews, G.K., Adamson, E.D. and Gedamu, L. (1984) Devel. Biol. 103: 294-303.
Andrews, G.K., Huet-Hudson, Y.M., Paria, B., McMaster, M.T., De. S.K. and Dey, S.K., (1991) Devel. Biol. 145, 13-27.
Andrews, G.K., Kage, K., Palmiter-Thomas, P. and Sarras, M.P., (1990) Pancreas, 5, 548-554.
Barlow,S.M. and Sullivan, F.M., Cadmium and its compounds, in Barlow and Sullivan, (Eds), Reproductive Hazards of Industrial Chemicals, Academic Press, London, 1982, p. 136.
De, S.K., Dey, S.K. and Andrews, G.K., (1990) Toxicology 64, 89-104.
De, S.K., Enders, G.C. and Andrews, G.K., (1991) Mol. Endo. 5, 628-636.
De, S.K., McMaster, M.T. Dey, S.K. and Andrews, G.K., (1989) Development 107, 611-621.
De, S.K., Paria, B.C., Dey, S.K. and Andrews, G.K., (1993) Toxicology, In Press.
Dencker, L., Danielsson, B., Khayat, A. and Lindgren, A., Disposition of metals in the embryo and fetus, in Clarkson, Nordberg and Sager, (Eds.), Reproductive and Developmental Toxicity of Metals, Plenum Publishers, New York, 1983, p. 607.
Durnam, D.M. and Palmiter, R.D. (1981) J. Biol. Chem. 256, 5712-5716.
Durnam, D.M and Palmiter, R.D. (1983) Anal. Biochem. 131, 385-393.
Jhappan, C., Stahle, C., Harkins, R.N., Fausto, N., Smith, G.H. and Merlino, G.T., (1990) Cell 61, 1121-1135.
McMaster, M.T., Dey, S.K. and Andrews, G.K., (1993) J. Reprod. Fert., In Press, 1993.
Onosaka, S. and Cherian, M.G. (1982) Toxicol. Appl. Pharmacol., 63, 270-274.
Rogers, J.M. and Hurley, L.S., (1987) Development 99, 231-238.
Searle, P.F., Davison, B.L., Stuart, G.W., Wilkie, T.M., Norstedt, G. and Palmiter, R.D., (1984) Mol. Cell. Biol. 4, 1221-1230.
Vallee, B.L. and Ulmer, D.D. (1972) Ann. Rev. Biochem., 41, 1-15.
Webb, M., (1987) Experi. Suppl. 52, 483-498.

Metallothionein III
K. T. Suzuki, N. Imura and M. Kimura (eds)

REGULATION OF HUMAN AND RAINBOW TROUT METALLOTHIONEIN GENES

L. Gedamu, R. Foster, N. Jahroudi, S. Samson, N. Shworak, and M. Zafarullah

Department of Biological Sciences, University of Calgary, Calgary, Alberta, Canada, T2N 1N4

SUMMARY: We have employed molecular genetics approaches to study the structure and mechanisms involved in the regulation of metallothionein (MT) genes by metal ions in two phylogenetically different organisms, human and trout. The human MTs are represented by a complex multigene family while the trout MTs are encoded by a simple gene family. Both human and trout MTs are regulated at the transcriptional level and exhibit cell-type specific and differential patterns of expression in response to metal ions. Although the trout MT regulatory region is less complex than for human MTs, the basic metal regulatory circuit involved in gene transcription is conserved between these organisms.

Metallothioneins (MTs) constitute an attractive system for biologists studying the molecular mechanisms involved in inducible gene expression. The use of molecular biology techniques has enabled researchers to isolate MT cDNA and genomic sequences from numerous evolutionarily diverse organisms for studies of transcription induction of MT mRNA as well as the genomic sequences which regulate the response to inducers. However, much of the present knowledge of MT gene structure and regulatory mechanisms is from mammals, such as mouse and human, or the lower eukaryote yeast which has a more limited metal response. Despite the importance of fish MTs in environmental toxicology (15), little is known about MT gene structure and regulation in such an important taxonomic group as teleosts. Our laboratory has studied rainbow trout (*Salmo gairdnerii*) MT gene regulation in parallel with human in order to determine the degree of conservation of MTs and metal regulatory mechanisms between two phylogenetically distant organisms. This review summarizes the progress that has been made in both human and trout MT regulation and will explain how the search for a metal regulatory transcription factor has resulted in the convergence of trout and human MT research.

CONSERVATION OF MT STRUCTURE IN HUMANS AND RAINBOW TROUT: Despite the ubiquity of MTs in the animal kingdom, there are wide differences in the number and complexity of MT genes observed in different organisms. The human (h)MTs are encoded by a multigene family of at least 12 members (27) with numerous non-functional (52) and processed psuedogenes (65) and a number of functional isoforms (the number is growing). So far, these include one MT-III isoform (41), one MT-II isoform, MT-IIA (27), and five known functional MT-I isoform genes, MT-IA (46), MT-IB (17), MT-IE (52), MT-IF (52, 66), and MT-IG (10), clustered on chromosome 16 (28, 51). The trout (t)MT system is less complex consisting of two known genes and possibly three or four genes in total (5). Our laboratory has cloned both the cDNA and genomic sequences of two trout isoforms, tMT-A and tMT-B (5, 70; S. Schieman, unpublished results).

Both the human and trout MT genes encode 61 or 62 amino acid (a.a.) polypeptides and the primary a.a. sequences of the trout MTs are comparable to human with greater than 60 % homology (5). More importantly, the functional conservation of trout and human MTs is apparent since the positions of 19 out of 20 cysteine residues are conserved in trout with the expected Cys-X-Cys or Cys-Cys arrangements which allow the co-ordination of metals (26). Such conservation strengthens a widely accepted view of common ancestral origins for teleost and mammalian MTs.

Although the introns of trout and human MT genes differ in length, the tripartite gene structure, three exons interupted by two introns at amino acids 9 and 31 (or 32 for tMT-A), is conserved among all higher eukaryotic MTs (Figure 1). Intron 2 occurs at the junction of the A- and B-domains of the polypeptides in mammals and probably trout so that the A-domain contains 11 cysteines while the B-domain has 9 (5, 70).

One important difference between the human and trout genomic sequences is the A/T richness (62 % A/T) of the tMT-B 5′ flanking regions when compared to human MT genomic sequences (75 % G/C) and other mammalian MTs (72). This may reflect a selection toward more stable G:C pairs because of the higher physiological temperature of mammals.

Regulatory Regions: Since the regulation of MT occurs mainly at the transcriptional level, most studies of MT induction have focused on observation of MT mRNA levels in response to

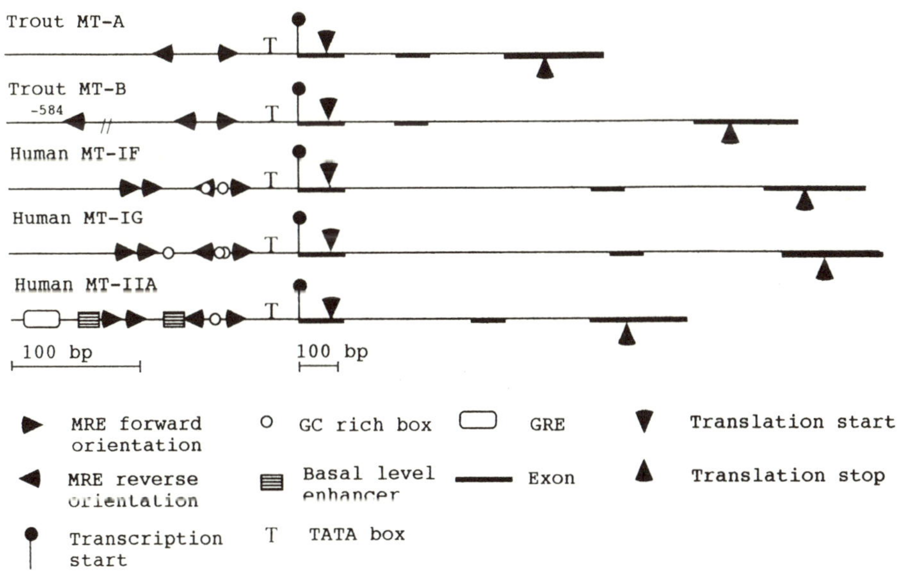

Figure 1. A schematic representation of trout MT-A and MT-B (70), and human MT-IF (66), MT-IG (10), and MT-IIA (27) genes. DNA sequence information and regulatory elements can be found in the corresponding references in the text.

various inducers and identification of the corresponding cis-acting elements. The regulatory regions of hMT-IIA, hMT-IF, hMT-IG and tMT-B are presented diagramatically in Figure 1 in the context of the putative cis-acting elements. The greater complexity of the human promoters compared to trout is obvious as they contain numerous sequences which are not found in the tMT-B 5′ regulatory region. Both hMT-IIA and hMT-I isoforms have several GC boxes (5′-GGGCGG-3′) which are possible binding sites for activating transcription factor SP1 (25). However, with the exception of the glucocorticoid responsive hMT-IE gene (52), only metals have been shown to induce the hMT-I isoforms which is reflected by the apparent simplicity of their promoter regions compared to hMT-IIA. The hMT-IIA promoter contains enhancer regions which bind factors, such as AP1 and AP2, for transcription activation by serum growth factors (22), tumor promoting phorbol esters (2, 3, 20, 22) and other activators of protein kinase A and protein kinase C pathways (20, 32, 35). As well, a glucocorticoid responsive element (GRE)(29), and a putative interferon responsive element (12) have been identified. MT-IIA has also been shown to respond to the cytokine IL-1 as well as UV light probably mediated by glucocorticoids and Ap1 respectively (30, 54, 60).

Apart from the TATA box, which is common to eukaryotes (50), the only homologous sequences of the human and trout 5′ regulatory regions are the 12 to 15 bp sequences which strongly resemble the flanking and TGCPuCNC consensus core sequences of the metal responsive elements (MREs) of mammalian MTs (62). MREs are considered to be recognized by regulatory proteins during the induction of MT transcription by heavy metal ions such as cadmium (Cd), copper (Cu), and zinc (Zn) (21, 55-59, 67).

MRE function has been studied extensively in mammals usually by transient transfection of cell cultures with MT promoter-reporter gene constructs. In our laboratory, trinucleotide mutations of the hMT-IG proximal MREa core sequence drastically lower Cd induced transcription of a reporter gene in human hepatoblastoma (Hep G2) cells (N. Shworak, unpublished observations). In order to confirm that this motif is also active in trout, the activity of the putative trout MRE sequences were tested using 5′ deletion analysis of the tMT-B 5′ flanking region. These studies revealed that the proximal 100 bp containing both MREa and MREb is sufficient for Cd and Zn induced transcription of a chloramphenicol acetyl transferase (CAT) reporter gene in trout cells (70). Deletion of the MREb results in a severe loss of activity (70) possibly indicating a co-operative interaction of MREs a and b. As is expected, deletion of MREa abolishes metal inducibility. Current studies of MRE point mutations and other tMT-B constructs should further our understanding of metal regulatory mechanisms in trout.

Although a variety of agents have been shown to induce MTs, metals are still the common inducer of all MT isoforms. Considering the possible physiological importance of trace metal metabolism, fine control of MT transcription may be required in order for MTs to best accomodate the changing metal environment or requirements of the cell. Whether this means greater induction of more effectual isoforms or a precise combination of specific MT isoforms has not been demonstrated. Even so, MT regulation has been shown to be very complex. Differential patterns of expression are observed as i) variable transcription activity in different tissues and cell types, ii) different levels of transcription among MT isoforms and iii) diverse responses to metal inducers.

<u>CELL-TYPE SPECIFIC MT EXPRESSION</u>: Tissue specific gene expression allows a specific subset of cells to acquire and maintain a specialized function needed by the organism. MT accumulation differs among tissues, being most pronounced in the liver and kidney possibly related to their physiological

role as detoxifiers (19). Also, different spatial and temporal levels of MT transcription have also been observed during development (1, 4, 39, 68) suggesting that the patterns of MT expression required by one cell type may not be appropriate for another.

Human: Endogenous expression of selected hMT isoforms is not detected in a number of human cell lines in spite of the presence of unaltered gene loci (17, 24, 53). Heguy et al. (17) demonstrated that MT-IB expression could be Cd induced in kidney carcinoma and hepatoblastoma cells (Hep G2) but only at extremely low levels in B cell lines, Hela cells, or primary cultures of fibroblasts (17). Similarily, MT-IE is Cd induced in hepatocarcinoma cells (Hep 3B2), but not mammary carcinoma or embryonic kidney cells (Hek 293) (53). In another study, metal induced endogenous MT-IF and MT-IG transcription were detected in Hep G2, Hek 293, and Hep 3B2 cells but not in lymphoblastoid derived Wi-L2 cells (10, 24, 66). These results were significant considering that hMT-IIA basal and metal induced transcription were observed in all of the cell lines tested in these studies so that MT transcription was not completely abolished. Interestingly, MT-IIA was not induced by the glucocorticoid analogue, dexamethasone (Dex) in Wi-L2 cells even though these cells have glucocorticoid surface receptors (24).

Because DNA methylation already had been implicated in the quiescence of mouse MT-I in non-expressing cell lines (8), this possibility was tested by treating the human cell lines with 5-aza-cytidine (5-azaC), a potent hypomethylating agent, prior to induction (reviewed by 6, 7). In all cases of 5-azaC treatment, metal inducible expression of the formerly silent hMT-I genes was observed, as well as the Dex induction of MT-IIA in Wi-L2 cells, implicating a role for DNA methylation in the observed cell-type specific patterns of MT expression (17, 24, 52). Direct evidence that MT regulatory regions were methylated was obtained by restriction of the genomic DNA of non-expressing cell lines with the isoschizomers, Msp I and Hpa II (17, 23). A 5′-CCGG-3′ site in the 5′ flanking DNA of hMT-IB in HeLa cells and hMT-IF in Wi-L2 cells is digestible with Msp I but not Hpa II, which is methylation sensitive, so that CpG methylation has occurred at the MT gene loci rather than a regulatory factor gene in the induction pathway (17, 23). Methylation of the GRE region of MT-IIA has also been confirmed using this method (N. Jahroudi, unpublished data) Transfection of the non-expressing cell lines with the MT 5′ regulatory regions in control of a CAT reporter gene also revealed that the MT transacting factors are present and active in the non-expressing cell lines since MT-IB, IF, and IG fusion genes exhibited metal inducible expression and Dex induction of MT-IIA was observed in Wi-L2 cells (17, 24).

Varied levels of promoter activities were also observed by comparison of transcription levels in MT expressing cell lines. For example, the hMT-IG gene has greater inducible transcription activity than hMT-IF in Hep G2 cells, as well as when transfected into MT expressing and non-expressing cell lines (10, 11, 24). Even so, hMT-IF displays greater endogenous activity than hMT-IG in Hek 293 cells, independent of methylation, since 5-azaC treatment did not increase hMT-IG expression over hMT-IF (24). DNase I digestion of Hek 293 cell genomic DNA revealed that the hMT-IF locus was more sensitive than hMT-IG to DNase I digestion (24), which may indicate that the MT-IF DNA configuration allows greater accessibility to factors which activate MT transcription in Hek 293 cells. Alterations in DNase I hypersensitivity have been studied during induction for the rat MT-I and mouse MT-I and MT-II genes (33, 63).

<u>Trout</u>: Both trout MT-A and MT-B genes are expressed in the trout cell lines, RTH-149 (hepatoma) and RTG-2 (gonadal fibroblasts). However, an embryonic epithelial cell line

from salmon (CHSE) does not express MT protein or mRNA in response to metals (36, 44-46). As with non-MT expressing human cell lines, 5-azaC treatment prior to metal exposure allows MT metal induction which displays similar kinetics to RTH-149 cells (36, 45). Also, a tMT-B promoter-CAT fusion gene displayed metal induction when transfected into CHSE cells confirming that the MT regulatory factors were present and active in this cell line (36).

DNA methylation can inhibit transcription activation by two possible mechanisms: i) causing changes in the local chromatin structure which render the genes inaccessible to activating transcription factors and ii) directly affecting the cis-acting element-regulatory factor interactions (6, 7). The actual mechanism that is occuring for MT cell-type specific regulation is not certain, but it is interesting that the chromatin effects which may decrease hMT-IG transcription in Hek 293 cells are apparently independent of methylation so that another mechanism might be in place to alter chromatin configuration.

It is tempting to propose that the mechanisms which cause MT DNA methylation and changes in chromatin structure in cell lines are also in place to allow developmental and tissue specific MT expression in vivo. First, the pattern of MT expression mimics that of animal tissues, with kidney and liver derived cell lines expressing the most isoforms and methylation was also observed in primary cultured fibroblasts so that it is not an artifact of transformation and cell line passage (17, 24, 53). Also, methylation and chromatin structure effects observed in embryonic fish (CHSE) cells and human (Hek 293) cells may reflect the importance of both for regulation of MTs during development. Finally, negative methylation effects on the hMT-IIA gene were specific to Dex induction. This may indicate that the GRE is specifically methylated or that the rest of the gene is actively demethylated by specific factors which recognize the "housekeeping" function of hMT-IIA compared with the other hMT-I isoforms but do not demethylate the more distal GRE. Heguy et al. (17) have previously suggested that controlled methylation and demethylation could be a mechanism to allow tissue specific expression for multigene families which evolved because of recent duplication events. These families might not have diverged enough to merit differential recognition of the promoter regulatory elements by tissue specific transcription factors (17). However, controlled methylation or demethylation would still require tissue or stage specific proteins to construct the appropriate methylation pattern.

DIFFERENTIAL TRANSCRIPTION REGULATION OF MT:

Human: Primer extension analysis of transcripts of all known hMT isoforms in Hep G2 cells, which do not exhibit methylation and chromatin structure effects of MT genes, has shown that only MT-IIA and MT-IE show detectable basal levels of transcription (N. Shworak, manuscript in preparation). After the addition of Cd to the cell culture medium, the different genes show the following order of transcription activity: MT-IE $\geq$ MT-IIA > MT-IG > MT-IF > MT-IA. There is a wide range of Cd induced transcription activity among the isoforms since hMT-IE transcripts were 70 fold more abundant than for MT-IA, with MT-IB transcripts not detected under these experimental conditions. Also, each gene was unique in the level of fold induction in response to Cd. A more quantitative approach by Sadhu and Gedamu (48) also has shown that, at maximum Zn induction, MT-IIA was induced to levels 6 fold greater than MT-IG and 17 fold higher than MT-IF in response to Zn.

It is interesting to note that the order of the hMT-I isoform expression in response to Cd or Zn parallels the percent divergence of the MT amino acid sequences as calculated by the method of Perler et al. (43) with highly active MT-IE exhibiting the lowest average replacement

Table I. Percent nucleotide divergences (replacement/silent) between human MT coding sequences (43).

	MT-IG	MT-IF	MT-IE	MT-IB	MT-IA
MT-IF	4.18/29.5				
MT-IE	3.32/28.4	4.18/31.4			
MT-IB	8.15/66.2	8.57/41.4	3.75/51.8		
MT-IA	5.92/35.9	8.57/21.2	4.16/19.7	8.11/42.3	
MT-IIA	2.23/33.4	4.20/33.9	2.89/39.4	5.92/59.1	4.83/28.2

divergence while MT-IA and MT-IB exhibit the greatest divergence (Table I). In fact, MT-IB has an additional cysteine residue which might affect its metal binding capacity (17). From this, it seems possible that differential transcription activities may reflect the biological function of the MT isoforms so that the polypeptides with altered metal binding capacity face less selective pressure for the maintenance of gene regulatory sequences. On the other hand, divergence of MT a.a. sequences and regulatory regions passively may have occurred together. Differential transcription activities have been reported elsewhere for MT-IA compared with MT-IIA as well as MT-IE compared with MT-IF (47, 53) and the results agree qualitatively with the above experiments.

The greater transcription activity of MT-IIA is probably related to regions of the 5′ flanking DNA which exhibit enhancer activity (16). However, MT-IF and MT-IG display comparable organization of the putative cis-acting elements on the promoters (Figure 1) so the differential activities are not as easily explained. Because MT-IF and MT-IG regulatory regions are less complex than MT-IIA, they were excellent candidates for studies of differential regulation.

To identify the regions involved in differential transcription, transfection experiments were performed using MT-IF or MT-IG gene regulatory regions fused to a CAT reporter gene. In all cell lines tested, the MT-IG promoter

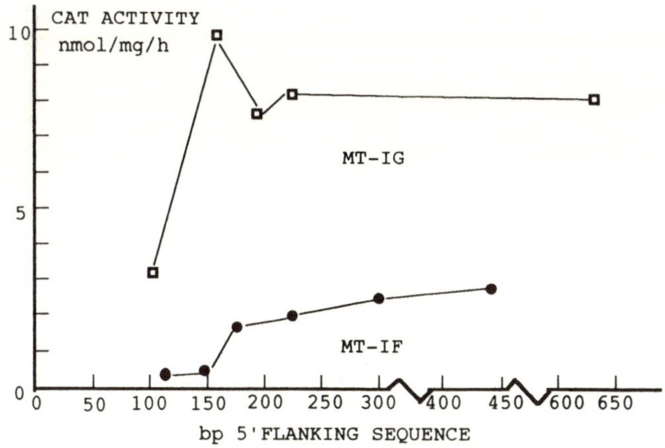

Figure 2. Comparison of the functional activity of MT-IG (upper) and MT-IF (lower) 5′ deletion fragments using transient transfection into Hep G2 cells (11)

was 3 to 5 fold more transcriptionally active than MT-IF (10, 24). Successive 5′ deletion analysis of both promoters revealed the MT-IG promoter is more compact since its proximal 105 bp can elicit a response to metals, while the comparable 111 bp fragment of MT-IF can not (Figure 2) (11). This region contains two MRE elements and two or three GC boxes in MT-IF and MT-IG respectively (10, 66). In fact, metal induced transcription of MT-IF-CAT requires 147 bp which contains two additional MREs (11). Even though differential transcription regulation of MT-IF and MT-IG can be localized to the proximal 150 bp of the promoter regions, it is interesting that it is these promoter sequences which are most conserved among hMT-I promoters (Figure 3). In fact, MT-IF and MT-IG share greater than 80 % sequence homology in this region suggesting that subtle nucleotide differences present in cis-acting elements are responsible for differential activity.

One important variation between the two promoters was that hMT-IG has a consensus TATAA box while MT-IF has a TATCA sequence (10, 52, 66). The TATA box sequence is present in most RNA polymerase II transcribed genes and is

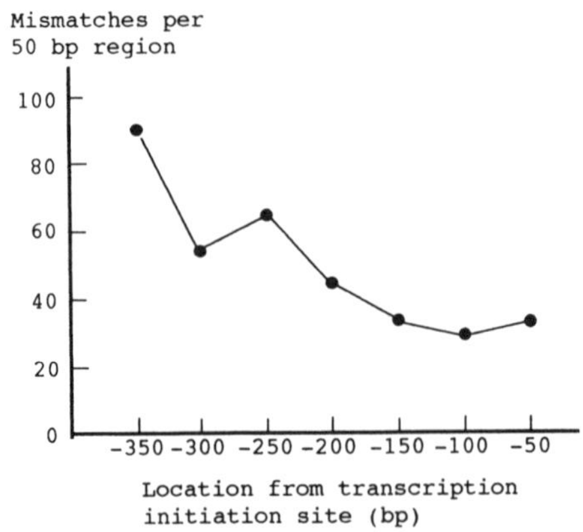

Figure 3. The proximal 5′ untranscribed regions of MT-IA(46), MT-IE (52), MT-IF (52, 66), and MT-IG (10) genes are highly conserved. The number of non-homologous nucleotides and insertions were scored as a single mismatch.

known to bind general transcription factor TFIID before assembly of the multi-protein preinitiation complex (50). The importance of the TATA box sequences for MT transcription was tested by constructing reciprocal point mutations of MT-IF (TATAA) and MT-IG (TATCA) promoters. Both transient transfection in Hep G2 cells and in vitro transcription in a rat liver nuclear extract show that MT-IF (TATAA) transcription is increased substantially, while MT-IG (TATCA) activity is nearly abolished (N. Shworak, manuscript in preparation). Similarly, Wobbe and Struhl (69) have demonstrated that the TATCA variant exhibits only 1 % of the TATAA consensus activity in vitro using either HeLa or yeast TFIID proteins. Although variation of the TATA sequence does not account completely for the differential transcription levels, the TATA box does make an important contribution. Other sequences which might be involved in differential transcription are under investigation.

<u>Trout</u>: The trout MT-A and MT-B genes also exhibit differential transcription regulation. Primer extension analysis of MT transcripts in rainbow trout hepatoma (RTH-

149) and gonadal fibroblast (RTG-2) cell lines show that MT-A is induced 2.4 fold higher than MT-B in response to maximal levels of Zn, and 1.4 fold higher in response to Cd (71-73). As with the human isoforms, it is possible that the differential transcription activities of MT-A compared to MT-B are a reflection of the biological function of each protein. Two non-conservative a.a. substitutions occur in the A-domain of tMT-A while the B-domains of MT-A and MT-B are completely conserved (5). Perhaps there is a common function for B-domains of the two tMTs with differing function of the A-domains. The metal specificity and independence of MT domains in metal binding has been demonstrated for mammalian MTs (36, 37, 61). The recent cloning of the trout MT-A gene will facilitate comparative studies of transcription regulation in order to elucidate the mechanisms of differential transcription in trout.

DIFFERENTIAL MT INDUCTION BY METALS:

Human: Differential hMT expression is also observed as varied transcription responses to different metals. This involves both the rate and the maximum level of transcript accumulation. The differential responses of MT-IIA, MT-IG, and MT-IF to Cd, Cu, and Zn in Hep G2 cells have been quantitated using probe excess titration followed by RNase protection analysis (48). MT-IF showed near equal rates of transcript accumulation for all three metals but at maximum induction, Cd and Zn induced transcripts were 2 fold higher than for Cu. This is opposite to MT-IG which responded poorly to Cd with transcripts accumulating slower and at much lower levels than for Cu or Zn. MT-IIA displayed equal responses to all three inducers. There is also evidence of metal specific post-transcriptional regulation since the level of Cu induced transcripts are constant with increasing induction time, while Zn transcripts decline because of an active process dependent on protein synthesis (49).

Trout: Varied MT transcription responses to Cd, Cu and Zn are also observed for the trout MTs. Single intraperitoneal injections of metals into whole trout resulted in MT-B induction in kidney, liver, spleen, and gills (73). However the kinetics of mRNA accumulation differed in response to each metal. Zn invoked the fastest response in liver, with an increase in MT mRNA as early as 6 hours after injection while a Cd response was observed at 12 hours. However, Zn invoked a poor response in gills compared with Cu and Cd. In the kidney and spleen, Cd induced MT mRNA levels were elvated from 12 to 48 hours while Cu and Zn induced messages declined after 26 hours (73). Differential induction by metals is also observed in RTH-149 cells where Zn produces MT levels approximately 7 to 8 fold over Cd (71-73).

The differential responses to metals are not suprising considering the varied physiological roles of the metals which bind MTs. This can range from very toxic with no known biological function, such as Cd, to metals which are involved in numerous biological processes and are nominally cytotoxic, as with Zn (14, 64). In fact, the optimal dose of Cd, Cu, or Zn that is required for maximum transcription induction in both human and trout cell lines differs for each metal (45, 66), which may be a function of differential metal uptake and efflux (9, 18, 42). Also, MTs display differential binding affinities for metals and it is possible that metal binding transcription factors might also exhibit differential affinities for the metals which in turn affect their capacity to activate transcription. Alternatively, each metal might bind different transcription factors which show differential specificity for MRE sequences. Jahroudi et al. (24) have demonstrated that the transiently transfected hMT-IG promoter is induced by Cd, Zn and Cu in Hep G2 cells but only responds to Cd and Zn in Wi-L2 cells which may be related to the levels of metal specific transcription factors. In this context, it is interesting that all studies of MRE binding factors report effects by one metal and these factors do not bind to any MRE core sequence, but only a select few (21, 56-58, 67).

CONSERVATION OF MT METAL REGULATORY MECHANISMS: The conservation of MT regulation is apparent in the preservation of the sequences of MREs among diverse organisms. This further suggests a similar conservation of the factors which are proposed to regulate MTs by interacting with MRE sequences. There are many examples in the literature which have shown that metal induced transcription is conserved among mammals (30, 34, 62). Studies in more evolutionarily diverse organisms have shown that this MT regulatory conservation can be extended across the entire spectrum of eukaryotes from *Drosophila*, to chicken, to *Xenopus* (13, 31, 40). In our laboratory, transfection studies have revealed that the tMT-B promoter is metal inducible when transfected into a human cell line (Hep G2) as well as mouse L cells (21, 70). Conversely, the hMT-IIA promoter is functional in both trout and salmon cell lines (72). This demonstrates that metal regulatory mechanisms are conserved even between fish and mammals and do not require additional mammalian transcription factors. For this reason, the trout MT-B gene is an appropriate model system for examining metal regulated transcription without complication by additional regulatory elements.

The evolutionary conservation of metal regulatory mechanisms has allowed for a cross-species approach in the search for metal regulatory transcription factors (MRFs). Recently, the trout MREa has been employed for affinity purification of an MRE binding activity in mouse L cells (21) and more recently humans (J.Imbert, personal communication). The mouse tMRE affinity fraction is able to induce transcription of the tMT-B promoter in vitro in mouse cell extracts and is dependent on Zn for this activity (21). The trout MT regulatory regions no doubt will prove useful in further searches for MRFs since the lack of GC rich regions in the trout promoter avoids co-purification of other non-MRE binding mammalian factors. DNA binding studies are currently underway using the trout MT regulatory region with mammalian nuclear proteins.

Acknowledgements: The work performed in the laboratory of L.G. was supported by a grant (to L.G.) from the Medical Research Council of Canada. We also thank the Alberta Heritage Foundation for Medical Research for studentship and fellowship awards.

REFERENCES:

1. Andrews, G., Adamson, E., and Gedamu, L. (1984) Dev. Biol. **103**: 294-303.
2. Angel, P., Poting, A., Mallick, U., Rahmsdorf, H., Schorpp, M., and Herrlich, P. (1986) Mol. Cell. Biol. **6**: 1760-1766.
3. Angel, P., Imagawa, M., Chiu, R., Stein, B., Imbra, R., Rahmsdorf, H., Jonat, C., Herrlich, P., and Karin, M. (1987) Cell **49**: 729-739.
4. Angerer, L., Kawczynski, G., Wilkenson, D. Nemer, M., and Angerer, R. (1986) Dev. Biol. **116**: 543-547.
5. Bonham, K., Zafarullah, M., Gedamu. L. (1987) DNA **6**: 519-528.
6. Cedar, H. (1988) Cell **53**: 3-4.
7. Cedar, H. and Razin, A. (1990) Biochim. Biophys. Acta **1048**: 1-8
8. Compere, S., and Palmiter, R. (1981) Cell **25**: 233-240.
9. Ettinger, M. (1987) Life Chem. Rep. **5**: 169-186.
10. Foster, R., Jahroudi, J., Varshney, U., and Gedamu, L. (1988) J. Biol. Chem. **263**: 11528-11535.
11. Foster, R., and Gedamu, L. (1991) J. Biol. Chem. **266**: 9866-9875.
12. Friedman, R., and Stark, G. (1985) Nature. **314**: 637-639.
13. Gedamu, L., Olsson, P.-E., and Zafarullah, M. (1090) *Transgenic Models in Medicine and Agriculture* 101-108.
14. Hamer, D. (1986) Ann. Rev. Biochem. **55**: 913-951.
15. Hamilton, S. and Mehrle, P. (1986) Trans. Am. Fish Soc. **115**: 596-609.
16. Haslinger, A., and Karin, M. (1985) Proc. Natl. Acad Sci. U.S.A. **82**: 8572-8596.
17. Heguy, A., West, A., Richards, R., Karin, M. (1986) Mol. Cell. Biol. **6**: 2149-2157.
18. Hempe, J., and Cousins, R. (1989) J. Nutr. **119**: 1179-1187.

19. Hunziker, P., and Kagi, J. (1985) Biochem J. **231**: 375 382.
20. Imagawa, M., Chiu, R., and Karin, M. (1987) Cell **51**: 251-260.
21. Imbert, J., Zafarullah, M., Culotta, V., Gedamu, L., and Hamer, D. (1989) Mol. Cell. Biol. **9**: 5315-5323.
22. Imbra, A., and Karin, M. (1987) Mol. Cell. Biol. **7**: 1358-1363.
23. Jahroudi, N., and Gedamu, L. (1989) Metal Ion Homeostasis 49-57.
24. Jahroudi, N., Foster, R., Price-Haughey, J., Beitel, G., Gedamu, L. (1990) J. Biol. Chem. **265**: 6506-6512.
25. Kadonaga, J., Carner, K., Masiarz, F., and Tjian, R. (1987) Cell **51**: 1079-1090.
26. Kagi, J., and Schaffer, A. (1988) Biochemistry **27**: 8509-8515.
27. Karin, M., and Richards, R. (1982) Nature **299**: 797-802.
28. Karin, M., Eddy, R., Henry, W., Shows, T., Haley, L., Byers, M. (1984) Proc. Natl. Acad. Sci. U.S.A. **81**: 5494-5498.
29. Karin, M., Haslinger, A., Holtgreve, H., Richards, R., Krauter, P., Westphal, M., Beato, M. (1984) Nature **308**: 513- 519.
30. Karin, M., Imbra, R., Heguy, A., and Wong, G. (1985) Mol. Cell. Biol. **5**: 2866-2869.
31. Kelder, B., Chen, H., and Kopchick, J. (1989) Gene **76**: 75-80.
32. Lee, W., Haslinger, A., Karin, M., and Tjian, R. (1987) Nature **325**: 368-372.
33. MacArthur, C., and Lieberman, M. (1987) J. Biol. Chem. **262**: 2161-2165.
34. Mayo, K., Warren, R., and Palmiter, R. (1982) Cell **29**: 99-108.
35. Mermod, N., Williams, T., Tjian, R. (1988) Nature **332**: 557-561.
36. Misra, S., Zafarullah, M., Price-Haughey, J., and Gedamu, L. (1989) Biochim. Biophys. Acta **1107**: 325-333.
37. Nielson, K., and Winge, D. (1984) J. Biol. Chem. **259**: 4941-4946.
38. Nielson, K., and D. Winge. (1985) J. Biol. Chem. **260**: 8698-8701.
39. Olsson, P.-E., Zafarullah, M., Foster, R., Hamor, T., and Gedamu, L. (1990) Eur. J. Biochem **193**: 229-235.
40. Otto, E., Allen, J., Young, J., Palmiter, R., Maroni, G. (1987) Mol. Cell. Biol. **7**: 1710-1715.
41. Palmiter, R., Findley, S., Whitmore, T., and Durnam, D. (1992) Proc. Natl. Acad. Sci. U.S.A. **89**: 6333-6337.
42. Pattison, S., and Cousin, R. (1986) FASEB **45**: 2805-2809.
43. Perler, F., Efstratiadis, A., Lomedico, P., Gilbert, W., Kolodner, R., Dodgson, J. (1980) Cell **20**: 555-566.
44. Price-Haughey, J., Bonham, K., and Gedamu, L. (1986) Environ. Health Perspect. **65**: 141-147.

45. Price-Haughey, J., Bonham, K., and Gedamu, L. (1987) Biochim. Biophys. Acta **908**: 158-168
46. Price-Haughey, J., and Gedamu, L. (1987) Experentia. Suppl. **52**: 465-469
47. Richards, R., Heguy, A., and Karin, M. (1984) Cell **37**:263-272.
48. Sadhu, C., and Gedamu, L. (1988) J. Biol. Chem. **263**: 2679-2684.
49. Sadhu, C., and Gedamu, L. (1989) Mol. Cell. Biol. **9**: 5738-5741.
50. Saltzman, A., and Weinman, R. (1989) FASEB J. **3**: 1723-1733.
51. Schmidt, C., Hamer, D., and McBride, O. (1984) Science **224**: 1104-1106.
52. Schmidt, C., Jubier, M., and Hamer, D. (1985) J. Biol. Chem. **260**: 7731-7737.
53. Schmidt, C., and Hamer, D. (1986) Proc. Natl. Acad. Sci. U.S.A. **83**: 3346-3350.
54. Schroeder, J., and Cousins, R. (1990) Proc. Natl. Acad. Sci. **87**: 3137-3141.
55. Searle, P., Stuart, G., and Palmiter, R. (1987) Mol. Cell. Biol. **5**: 1480-1489.
56. Searle, P. (1990) Nuc. Acids Res. **18**: 4683-
57. Seguin, C. (1991) Gene **97**: 295-300.
58. Seguin, C., and Hamer, D. (1987) Science **235**: 1383-1387.
59. Seguin, C., and Provost, J. (1988) Nuc. Acids Res. **16**: 10547-10560.
60. Stein, B., Rahmsdorf, H., Steffen, A., Litfin, M., and Herrlich, P. (1989) Mol. Cell. Biol. **9**: 5169-5181.
61. Stillman, M., and Zelazowski, A. (1988) J. Biol. Chem. **263**: 6128-6133.
62. Stuart, G., Searle, P., and Palmiter, R. (1985) Nature **317**: 828-831.
63. Taplitz, S., Calame, K., and Herschman, H. (1986) Mol. Cell. Biol. **6**: 2576-2581.
64. Thiele, D. (1992) Nucl. Acids Res. **20**: 1183-1191.
65. Varshney, U., and Gedamu, L. (1984) Gene **31**: 135-145.
66. Varshney, U., Jahroudi, N., Foster, R., and Gedamu, L. (1986) Molec. Cell. Biol. **6**: 26-37.
67. Westin, G., and Shaffner, W. (1988) EMBO J. **7**: 3763-3770.
68. Wilkenson, D., and Nemer, M. (1987) Mol. Cell. Biol. **7**: 48-58.
69. Wobbe, C., and Struhl, K. (1990) Mol. Cell. Biol. **10**: 3859-3867.
70. Zafarullah, M., Bonham, K., and Gedamu, L. (1988) Mol. Cell. Biol. **8**: 4469-4476.
71. Zafarullah, M., Olsson, P.-E., and Gedamu, L. (1990) Gene **83**: 85-93
72. Zafarullah, M., Olsson, P.-E., and Gedamu, L. (1989) Oxford Surveys on Eukaryotic Genes **6**: 110-143.
73. Zafarullah, M., Olsson, P.-E., Gedamu, L. (1990) Biochim. Biophys. Acta **1049**: 318-323.

Metallothionein III
K.T. Suzuki, N. Imura and M. Kimura (eds)

THE METALLOTHIONEIN STRUCTURAL MOTIF INVOLVED IN METALLOREGULATION

Dennis R. Winge and Charles T. Dameron

Departments of Medicine and Biochemistry, University of Utah Medical Center, Salt Lake City, Utah, USA 84132

SUMMARY: Metalloregulation in eukaryotic organisms is poorly understood. Only a limited number of physiological processes are currently known to be regulated by metal ions. The Cu-activation of gene expression in fungi is mediated by a Cu^{1+} specific sensor protein. The mechanism of Cu-activation of the sensor molecule, ACE1 (and probably AMT1), appears to be the formation of a CuS polynuclear cluster as the structural core of the proteins.

Structural similarities between ACE1, AMT1 and MT suggest that the MT motif is a good structural model to explain the metal-specific activation of the two signal transducing proteins. Metal ion specificity is achieved by the propensity of proteins with a MT motif to form multinuclear CuS centers. Coordination chemistry appears to be the driving force for Cu^{1+} metalloregulation in biology. The metalloregulation of transcriptional activation of mammalian MT genes will be intriguing as metal ion specificity is not observed. The MT motif may be observed in yet to be identified metalloregulatory proteins that regulate other processes in a Cu- or Zn-specific manner. Formation of metal:thiolate polynuclear clusters allows a significant volume of the protein structure to be altered, so metal-induced structural dynamics are possible.

REGULATION OF METALLOTHIONEIN BIOSYNTHESIS:

Metal ions are essential for many cellular processes, yet many essential metal ions are reactive and are potentially toxic (Vallee and Ulmer, 1972). Cells possess mechanisms to regulate the intracellular free metal ion concentration. These mechanisms are effective in controlling the concentration of essential as well as non-essential metal ions (Butt and Ecker, 1987; Silver and Misra, 1988). Eukaryotic organisms typically maintain free metal ion concentrations at low levels by sequestering metal ions in proteinaceous complexes such as metallothionein (MT) and ferritin (Mehra and Winge, 1991; Gadd and White, 1987). MT sequesters a variety of metal ions in animal cells and may function as a tissue storage reservoir for Zn and/or Cu (Bremner, 1987). Metal binding to MT limits the free ion concentration of several metals. MT responds to changes in metal ion flux by metal ion regulation of MT biosynthesis. The protein effectively buffers the free ion concentration of those metal ions which are capable of stimulating MT biosynthesis. Buffering efficiency is dictated by the metal binding constants and the concentration of MT. In fungal cells MT functions mainly in copper homeostasis as Cu^{1+} and Ag^{1+} are the only metal ions capable of stimulating the synthesis of MTs (Hamer, 1986).

MTs are found at low basal levels in cells and tissues, but exposure to metal ions leads to a rapid elevation in the levels of MTs (Hamer, 1986; Palmiter, 1987). Regulation of MT biosynthesis occurs predominantly at the level of transcription initiation (Palmiter, 1987). A variety of metal ions including Cd, Zn, Cu, Hg, and Ag are effective in inducing mammalian MT gene expression (Durnam and Palmiter, 1981). Unlike the metalloregulation of mammalian MT genes, the regulation of fungal MT genes is typically restricted to Cu and Ag ions (Furst et al., 1988). Copper metalloregulation in yeast is not restricted to expression of MT genes within the *CUP1* locus. Copper is also known to stimulate the expression of superoxide dismutase, catalase and glutathione peroxidase (Galiazzo et al., 1988). The Cu-regulated expression of superoxide dismutase but not glutathione peroxidase occurs transcriptionally through the same mechanism as MT metalloregulation (Carri et al., 1991; Gralla et al., 1991).

Both *cis*-acting and *trans*-acting elements are important for metalloregulation. Analyses of the DNA sequences in the 5′ upstream regions of mammalian MT genes have identified several promoter elements, designated metal-responsive elements or MREs, responsible for the metal-activated expression of MT genes (Mayo et al., 1982; Carter et al., 1984; Stuart et al., 1984; Haslinger and Karin, 1985). Metal-responsive elements have been mapped in the 5′ upstream sequence of the MT gene from *S. cerevisiae* (Imbert et al., 1990; Thiele and Hamer, 1986; Evans et al., 1990). These transcriptional elements, designated upstream-activator sequences (UAS), are often palindromic. The UAS element upstream of the MT gene consists of an inverted repeat of 16 bp (Furst et al., 1988; Hu et al., 1990). A single UAS element (UAS_c) is sufficient to confer Cu-responsiveness on reporter genes (Furst et al., 1988; Hu et al., 1990; Culotta et al., 1989; Buchman et al., 1990). The effectiveness of UAS_c to confer Cu-inducibility on downstream genes was a strong indication that this element was the binding site of the Cu-responsive *trans*-acting factor.

The metalloregulation of *cis*-acting sequences is mediated through *trans*-acting proteins. Expression of MT genes in yeasts is regulated by metal-responsive transcriptional activator proteins designated ACE1 (for <u>a</u>ctivator of <u>*C*</u>*UP*1 <u>e</u>xpression) and AMT1 (for <u>a</u>ctivator of <u>MTI</u> expression) in *S. cerevisiae* and *C. glabrata*, respectively (Thiele, 1988; Furst et al., 1988; Buchman et al., 1989; Zhou and Thiele, 1991). ACE1 is also called CUP2 (Buchman et al., 1989). Copper binding to these factors activates these proteins for binding to 5′ UAS elements (Furst et al., 1988).

In *S. cerevisiae* ACE1 protein binds specifically to the UAS_c of the MT gene in the presence of added Cu^{1+} (Furst et al., 1988; Huibregtse et al., 1989; Buchman et al., 1989). Overexpression of the ACE1 enhances the Cu-mediated expression of β-galactosidase gene fusions containing a 5′ UAS_c element (Furst et al., 1988; Szczypka and Thiele, 1989). These results clearly demonstrate that ACE1 mediates the Cu-stimulated induction of genes containing the UAS_c promoter element. Similarly, AMT1 mediates the Cu-dependent expression of *C. glabrata* MT genes and *trans*-

activates expression of reporter genes containing *C. glabrata* MT 5′ promoter sequences (Zhou and Thiele, 1991; Zhou et al., 1992).

We are interested in elucidating the molecular mechanisms by which ACE1 and AMT1 mediate the metal-dependent expression of MT genes. Other known metalloregulatory molecules (calmodulin, transferrin, Fur) exist in multiple conformers such that metal binding induces a reversible conformational switch (Anderson et al., 1990; Coy and Neilands, 1991; Meador et al., 1992; Ikura et al., 1992). The dynamics of the reversible conformational switch allows a function to be turned on or off. The tertiary fold or quaternary structure of a metalloregulatory protein such as ACE1 and AMT1 may be altered by binding of metal ions by stabilization of a particular conformer.

The sequence of ACE1 predicts a 225 residue polypeptide of M_r 24,428 (Furst et al., 1988). The polypeptide appears to be divided into two distinct domains. The N-terminal 120 residues are enriched in basic residues whereas the C-terminal segment is more acidic. There are three important features of the basic N-terminal segment of ACE1. First, the N-terminal half of ACE1 is responsible for specific DNA binding (Furst et al., 1988; Buchman et al., 1989). Second, the N-terminal segment in ACE1 contains 12 cysteinyl residues, with two CxxC and three CxC sequence motifs. Such sequence motifs are commonly observed in MTs suggesting that the motifs in ACE1 may also be important for metal binding. Specific DNA binding has been shown to be dependent on bound Cu^{1+} ions (Furst et al., 1988). Third, as with MTs there is a paucity of hydrophobic residues in the N-terminal half of ACE1.

AMT1 is homologous to ACE1 (Zhou and Thiele, 1991). There is a 50% identity in amino acid sequence in the N-terminal segment of these two proteins and total conservation in the positions of the first 11 cysteines. A comparison of ACE1 and AMT1 sequences is shown on the next page.

The abundant CxC motifs and paucity of hydrophobic residues in ACE1 and AMT1 are hallmarks of MTs. The obvious postulate is that Cu^{1+} ions bind to a MT motif within the N-terminal half of

ACE1 and AMT1 and induce a conformer that is competent for high affinity DNA binding (Furst et al., 1988). The similarities of ACE1 to MT suggest that structural features of MT itself may provide insights into the chemistry of Cu^{1+} activation of ACE1 and likely AMT1.

SEQUENCE COMPARISON OF ACE1 AND AMT1

```
     1
ACE1:MVVINGVKYACETCIRGHRAAQCTHTDGPLQMIRRKGRPSTTCGHCKELRRTKNFNPSGG
     :::::::::::..::. :.:::: : : ::  .. .::: ::: :::. :.::: ::::
AMT1:MVVINGVKYACDSCIKSHKAAQCEHNDRPLKILKPRGRPPTTCDHCKDMRKTKNVNPSGS

      61
ACE1: CMCAS-AR-RPAVGSK-EDE------TRCRCDEGEPCKCHTKRKSSRKSKGGSC
      : :    . :   :   :..         : :  ::::.:: .:: . ::
AMT1: CNCSKLEKIRQEKGITIEEDMLMSGNMDMCLCVRGEPCRCHARRKRTQKS
```

METALLOTHIONEIN STRUCTURAL MOTIF

The structures of mammalian MTs have been elucidated by NMR and x-ray diffraction techniques (Messerle et al., 1990; Robbins et al., 1991). Highlights of the 2.0 Å resolution structure of rat Cd,ZnMT (Robbins et al., 1991) that may apply to metallo-regulatory proteins will be briefly reviewed.

All proteins designated MTs exhibit repeated sequence motifs of either CxC or CxxC. Mammalian MTs typically contain 20 conserved cysteinyl residues within a polypeptide of 61-66 residues (Kagi and Schaffer, 1988; Palmiter et al., 1992). All cysteinyl sulfurs (S) participate in binding the seven divalent metal ions within the two polynuclear clusters (Robbins et al., 1991). In Cd,ZnMT, the N-terminal β domain enfolds a M_3S_9 polynuclear cluster, the C-terminal α domain enfolds a M_4S_{11} cluster. Eight of the twenty Cys residues in Cd,ZnMT have bridging sulfurs atoms; the other twelve are terminal sulfur ligands. The $Cx_{1-2}C$ motifs are observed in a variety of metal-binding proteins. Protein structures containing the CxC sequence motifs are folded in a reverse turn configuration stabilized by a hydrogen bond from the amide hydrogen of the second Cys to the sulfur of the first Cys (Robbins et al., 1991; Adman et al., 1975). The same reverse turn exists in CxxC motifs. The reverse turn brings the two sulfurs atoms in van der Waals contact and therefore in juxtaposition for metal ion ligation. In mammalian MT the two cysteinyl sulfurs in a CxC motif can coordinate the same or two different metal ions.

A second key feature of the MT structural motif is the paucity of hydrophobic residues. The structural core of mammalian MT domains is dominated by the metal-sulfur polynuclear cluster. The usual hydrophobic core seen in typical soluble, globular proteins is absent. Most soluble proteins consist of between 15 to 35% hydrophobic residues. In contrast, mammalian MTs typically contain between 0-3% hydrophobic residues; only a few MT proteins contain up to 5% hydrophobic residues. The lack of a hydrophobic core implies that the structural stabilization is not dominated by the usual solvent entropy term. Furthermore, the folding pathway must not be driven by the typical initial collapse of structures to remove hydrophobic residues from the unfavorable solvent interactions. The initial formation of a hydrophobic core cluster occurs early in the folding pathway of typical globular proteins. The lack of a hydrophobic core in MT suggests that its folding pathway is metal-dominated. Entropy still may be a significant factor in the metal-displacement of bound water molecules and may be analogous to the entropy gain from release of bound water molecules of the sugar-induced conformational change in Venus fly-trap sugar-binding proteins (Quiocho, 1986).

Solvent accessible cysteinyl sulfurs are a third salient feature of MT which may be important in metalloregulatory molecules. In the static crystal structure of MT the metal ions are shielded from solvent, yet the sulfur atoms of several of the twenty Cys are solvent accessible (Robbins et al., 1991). Clefts are present in both the N-terminal β and C-terminal α domains that reveal solvent-accessible sulfurs. The accessibility of these sulfurs may contribute to the reactivity of MT with electrophilic agents and the facile metal exchange observed (Otvos et al., 1989). MT-bound Zn^{2+} ions are rapidly displaced by free Cd^{2+} ions. The metal ions in both clusters are in rapid exchange with half-times of seconds in the β domain and minutes in the α domain (Otvos et al., 1989). Exchange is observed between both intra- and intermolecular sites (Otvos et al., 1989; Nettesheim et al., 1985). The facile metal ion exchange observed in MT is not typically seen in metalloproteins (Coleman et al., 1983).

A fourth feature of MT is the existence of multiple conformers. This is unusual in proteins. The lack of a significant hydrophobic core cluster in MT removes conformational constraints to metal-induced structural rearrangements. Detailed structural information is available only for the mammalian Cd,ZnMT conformer (Robbins et al., 1991). A NMR structure is emerging for the AgMT from *S. cerevisiae* (Narula et al., 1991). Multiple lines of evidence suggest that the conformer of a given CuMT is distinct from that of CdMT or ZnMT. First, CuMTs exhibit a higher metal binding stoichiometry with trigonal binding geometry (Nielson et al., 1985; Narula et al., 1991). Seven Zn^{2+} or Cd^{2+} ions bind to mammalian MT with tetrahedral coordination geometry (Kagi and Kojima, 1987). In contrast, twelve Cu^{1+} ions can bind to the MT with a distribution of six ions per cluster (Nielson et al., 1985). The comparison in yeast MT is Cu_7MT verses Cd_4MT (Winge et al., 1985). Bridging cysteinyl sulfurs are predicted to be common in CuMTs (George et al., 1986; George et al., 1988).

THE MT MOTIF IN METALLOREGULATION

The four mentioned features of MT are relevant to the structure of ACE1 and AMT1. Multiple CxC or CxxC sequence motifs in a metalloregulatory protein create the potential for metal ion binding. A limited hydrophobic core cluster may permit a degree of structural dynamics important for metal-induced conformers. This feature is not essential for all metalloregulatory molecules as dynamics may also involve localized domain rearrangements as in lactoferrin (Anderson et al., 1990) or changes in the oligomerization of a protein. Metal activation in ACE1 may be achieved by metal binding to a stable pre-existing apo-protein or through metal exchange as in the MT motif. There is no information on whether the inactive state of ACE1 exists as apo-ACE1 or ZnACE1. Rapid Zn^{2+} displacement in ZnACE1 by free Cu^{1+} ions occurs as in MT (Nielson et al., 1985) and this may trigger the structural reorganization necessary for activation.

A paucity of information is available on Cu^{1+} clusters in MTs. The metal binding stoichiometries and coordination geometries differ for Cu^{1+} and Cd^{2+} ions in MT, so the available structures do not provide a clear picture of the structure of CuMT. The current knowledge of CuMTs is based largely on spectroscopic studies. Spectroscopic analyses have been carried out on CuMTs with the following cluster stoichiometries: Cu_6S_9 from the β domain of rat liver, Cu_6S_7 from *Neurospora crassa*, and Cu_7S_{10} from *S. cerevisiae* (formerly reported as a Cu_8 cage) (George et al., 1986; Smith et al., 1986; George et al., 1988; Narula et al., 1991). Chemical properties of CuS clusters associated with these diverse MTs are similar. The copper ions are exclusively present as Cu^{1+}. Absorption bands in the ultraviolet are consistent with S→Cu charge transfer transitions. The presence of Cu^{1+} ions with thiolate ligands results in luminescent complexes. Photoexcitation of proteins containing Cu^{1+}-thiolate clusters with ultraviolet radiation gives rise to luminescence (Beltramini et al., 1989). The emission is Stokes shifted to energies between 580-620 nm and appears to arise largely from a $3d^9 4S^1$ excited triplet state centered on Cu(I) as reported for synthetic CuS clusters (Vogler and Kunkely, 1986; Henary and Zink, 1991; Ryu et al., 1991). There is no feature in the absorption spectrum corresponding to the excitation band. This is consistent with the expected low extinction coefficient of a spin-forbidden $3d^9 4s^1$

triplet → $3d^{10}$ singlet transition (Vogler and Kunkely, 1986). Therefore, metal centered transitions are likely to account for the observed emission in proteins containing CuS clusters, whereas a S→Cu ligand-metal charge transfer transition is responsible for the intense absorption in the ultraviolet. Unlike synthetic CuS clusters the emission in CuS-containing proteins is observed at room temperature in aqueous buffers. This is consistent with the Cu^{1+} thiolate cluster enfolded within a compact protein structure. At present, the emission properties of CuMT have a limited information content on the structure of the CuS cluster.

Heteronuclear (^{1}H-^{109}Ag) multiple quantum coherence (HMQC) NMR studies performed on AgMT from *S. cerevisiae* established the connectivities of ten Cys residues with seven bound $^{109}Ag^{1+}$ ions (Narula et al., 1991). As Ag^{1+} is isoelectronic to Cu^{1+}, AgMT is expected to be isostructural to CuMT. This study from Dr. Armitage's laboratory suggests that the original proposal of the CuS cluster stoichiometry as Cu_8S_{12} may be wrong and furthermore that mixed coordination numbers may exist in CuMT. Based on the NMR results, a new model of M_7S_{10} emerged having two of the seven metal ions in digonal coordination with the remainder trigonally ligated. Two digonal Ag^{1+} ions are consistent with the large chemical shift dispersion for Ag^{1+} nuclei in AgMT (Narula et al., 1991).

Extended X-ray absorption fine structure (EXAFS) spectroscopy is sensitive to the local atomic environment of the absorber (in this case, Cu). In contrast to x-ray edge specta, EXAFS is readily analyzed in a quantitative manner and can provide information on the type, number and mean bond distance of ligands in the first few coordination shells. Bond distances are determined to an accuracy of 0.02 Å, whereas coordination numbers are determined to about ± 25%. EXAFS of polymetallic clusters provides an average picture of the coordination environment of the various metal ions within a cage cluster. EXAFS analysis for three copper thiolate polymetallic clusters revealed a well-defined correlation between the mean Cu-S bond distances and the fraction of digonal Cu^{1+} ions within the cluster.

We previously reported copper K edge and sulfur K edge EXAFS analyses of yeast CuMT yielding a mean Cu-S distance of 2.230 Å (George et al., 1988). Copper K-edge EXAFS of a truncated mutant of CuMT (CuT48) gave a very similar bond distance of 2.236 Å (George et al., 1988). Re-analysis of the original EXAFS data using new phase and amplitude functions gave a mean Cu-S bond distance of 2.242 Å and a mean coordination number of 2.6. EXAFS data on a second yeast CuMT isolate gave a Cu-S distance of 2.250 Å. Comparison of the mean Cu-S distances in *S. cerevisiae* CuMT with those of the synthetic Cu-S clusters suggested the presence of digonal Cu^{1+} ions (Winge et al., 1993). We estimate that CuMT contains a fraction of digonal Cu(I) between 0.3 and 0.4. The fraction of digonal Cu^{1+} by HMQC NMR measurements of AgMT is likely to be 0.3. The EXAFS data set on yeast CuMT is consistent with NMR data. The importance of this observation is that the correlation of mean Cu-S bond distances with fraction of digonal copper can be extrapolated to other molecules, e.g. CuMTs and CuACE1, for which no high resolution NMR data are available. The use of EXAFS to predict the presence of digonal Cu^{1+} ions based on Cu-S distances assumes that the protein backbone does not constrain the CuS cluster to alter Cu-S bond distances. This assumption may be valid as the MT protein structure seems to be dictated by the CuS cluster.

EXAFS analyses have been carried out on the Cuβ domain of rat MT, Cu,ZnMT from pig liver, lysosomal CuMT from dog liver, and CuMT from *N. crassa* (George et al., 1986; Smith et al., 1986; Freedman et al., 1986; Abrahams et al., 1986). A mean Cu-S distances of 2.27 Å was reported for the CuMT β domain (George et al., 1986). The stoichiometry of this latter cluster is Cu_6S_9. The Cu-S distance was based on a best fit of EXAFS data occurring with a coordination number between 2-3. The Cu_6S_7 cluster in CuMT from *N. crassa* was predicted from EXAFS to consist of 3-4 sulfur atoms at 2.20 Å and perhaps two sulfurs in an outer shell at 2.72

Å (Smith et al., 1986). The best fit of the EXAFS data of lysosomal CuMT was reported to be four sulfur atoms at 2.27 Å (Freedman et al., 1986). It is not clear whether lower coordination numbers were tried in the curve fitting of the lysosomal CuMT. The mean Cu-S distance and XANES spectrum are both more consistent with predominantly trigonal coordination. Based on the correlation of Cu-S mean distances and fraction of digonal Cu^{1+}, *N. crassa* CuMT may be expected to contain a significant portion of digonal Cu^{1+}. In contrast, the Cu_6S_9 cluster in the β domain of rat MT is likely to contain predominantly trigonal coordination geometry.

Cu-Cu interactions were seen in each CuMT. The mean Cu-Cu distances are near 2.7 Å for each protein. This value is comparable with the mean Cu-Cu distances observed in synthetic tetracopper clusters (2.74 ± 0.06 Å) (Dance, 1986). The presence of a detectable Cu-Cu interaction is the best indicator that each CuMT contains a polymetallic Cu^{1+}-thiolate cluster.

The CuS polymetallic cage clusters in MT appear to be similar to CuS synthetic clusters that have been described (Dance, 1986). Structural work on synthetic polycopper thiolate clusters revealed a number of salient features that may describe the biological CuS polymetallic clusters in both MT and ACE1 (Dance, 1986). First, the clusters are held together by the bridging thiolate ligands, which are commonly doubly-bridging. Secondly, the Cu^{1+} coordination numbers can be two (digonal) or three (trigonal), and there is no evidence of preference between them. Both can occur in one cluster. Thirdly, the $Cu_x(SR)_y$ clusters are not rigid and can be subject to substantial distortion. Fourthly, Cu-S bond distances are dependent on the coordination numbers. With doubly-bridging thiolate ligands the mean Cu^{trig}-S bond distance is about 2.27 Å and the mean Cu^{dig}-S distance is about 2.16 Å. Cu-Cu bonding is only a minor energetic factor in these clusters, and probably has little influence on overall geometry.

CuS CLUSTERS IN ACE1:

Eleven Cys residues are critical for the functioning of CuACE1 in *S. cerevisiae* (Hu et al., 1990). Physical studies have been carried out with the N-terminal half of the protein expressed in *E. coli* (Dameron et al., 1991; Nakagawa et al., 1991). This segment of the ACE1 polypeptide, containing the eleven essential Cys residues, is responsible for specific DNA binding (Furst et al., 1988). DNA binding requires the presence of bound Cu^{1+} or Ag^{1+} ions (Furst et al., 1988). Cu^{1+} binding appears maximal between 6-7 mol equiv. Cu^{1+} (Dameron et al., 1991). The Cu-protein exhibits S-Cu charge transfer bands in the ultraviolet and is luminescent with emission occurring between 580-620 nm (Dameron et al., 1991; Casas-Finet et al., 1991). These optical properties are similar to those of CuMT and imply that the Cu^{1+} ions must be bound within the core of each protein such that the Cu^{1+} ions are shielded from solvent interactions.

X-ray absorption spectroscopy has been performed on *E. coli* isolates of CuACE1 as well as CuACE1 prepared by in vitro reconstitution protocols. EXAFS of CuACE1 revealed a major scatterer in the first shell that fit well with sulfur atoms. A similar mean Cu-S bond distance of 2.26 was observed by two groups independently (Dameron et al., 1991; Nakagawa et al., 1991). The data set of Nakagawa et al.(1991) was best fit with a coordination number of 2.34; the best fit of the CuACE1 samples reported by Dameron et al. (1991) had N values of 2.9 and 3.0. The mean Cu-S distance of 2.26 Å is clearly consistent with predominant trigonal Cu^{1+} coordination. From the synthetic CuS cluster data one may conclude that CuACE1 contains 5-6 Cu^{trig} and 0-1 Cu^{dig}.

Cluster formation in CuACE1 was indicated by the observed Cu-Cu interactions apparent in the EXAFS. These Cu-Cu interactions occurring at 2.72 Å are equivalent to the short Cu-Cu distances observed in CuMTs and synthetic CuS cages. The mean number of scatterers per absorbing Cu atom was 2 for both ACE1 reconstituted with Cu^{1+} in the presence of 1 mM glutathione and native yeast CuMT. A lower number of scatterers was observed in CuACE1 prepared by reconstitution procedures in the absence of glutathione (Dameron et al., 1991).

In summary, ACE1 forms a polynuclear CuS cluster that mimics CuS cages in MT and certain synthetic model compounds. There is a strong tendency for cluster formation with thiol ligands, so the similarities in cluster chemistry between CuACE1, CuMT and CuS model compounds are not surprising. An intriguing list of similarities exist between CuACE1 and CuMT leading to the postulate that MT is a good structural model of the core structure of CuACE1. The shared features are as follows. The common observation of a short Cu-Cu distance (2.7 Å) in each class of molecules implies compact polynuclear cages. The polypeptide fold in both CuMT and CuACE1 may be dictated in part by the CuS cluster. This prediction is supported by the lack of a substantial hydrophobic core in each protein. The tertiary folds in CuACE1 and CuMT contain limited periodic secondary structure elements indicating that protein stabilization is conferred by the CuS cluster. Cu(I) binding to ACE1 protects the polypeptide against proteolysis using non-specific proteases. ACE1 and MT bind Cu^{1+} and Cd^{2+} ions in different conformers. CdACE1 can readily be formed *in vitro* but the CdACE1 does not form a high affinity complex with DNA containing the specific UAS. Maximal Cd^{2+} binding occurs between 3-4 mol eq. EXAFS analysis of CdACE1 revealed a Cd-S bond distance of 2.518 Å and the curve fitting predicted tetrahedral coordination as is usually seen in CdS clusters. The mean Cd-S bond distance in MT is 2.514 Å (Robbins et al., 1991). The tetrahedral coordination of Cd^{2+} and Zn^{2+} ions in ACE1 and MT necessitates a distinct protein fold from the Cu^{1+} conformers.

The ability of ACE1 to form distinct conformers with Cu^{1+} and Cd^{2+} (or Zn^{2+}) may form the basis for the exquisite metal ion specificity in the function of ACE1. The only metal ions that yield an active conformer (Cu^{1+} and Ag^{1+}) are ions that yield homologous metal:thiolate cage clusters. The use of a multinuclear cluster as the structural core of the molecule ensures a unique conformer with Cu^{1+} and Ag^{1+}. Distinct conformers may also contribute to the activation of ACE1 through Cu^{1+}/Zn^{2+} exchange.

MODEL OF CuACE1 AS A SIGNAL TRANSDUCER

In vitro studies have demonstrated that ACE1 binds to the DNA UAS element only in the presence of Cu^{1+}. No information is available on how Cu^{1+} is presented to ACE1 *in vivo*. It is clear that ACE1 is effective in sensing the intracellular copper concentration and mediating MT biosynthesis in proportion to the magnitude of the copper pool. Glutathione is likely involved in the Cu^{1+}-activation of ACE1 by virtue of its high intracellular concentration and tendency to form CuS complexes. The Cu^{1+}:GSH complex is kinetically labile and will donate Cu^{1+} ions to either MT or ACE1. Since cells do not appear to contain any appreciable quantities of apoMT, the likely initial recipient of Cu^{1+} ions associated with GSH is ACE1. ACE1 is constitutively expressed, it is likely to be either present in the steady state as ZnACE1 or as the metal-free molecule. CuGSH can activate rapidly ACE1 *in vitro* either by heterologous metal exchange or direct binding to the metal-free protein.

CuACE1 can bind to either half of the palindromic UAS promoter sequence. The upstream half of the UAS_c palindrome appears to be the ACE1 binding site of highest affinity (Hu et al., 1990; Buchman et al., 1990). The binding of CuACE1 to the UAS element brings the C-terminal transcriptional activation domain into position for assembly of the transcriptional machinery. Transcription from the CUP1 promoter *in vitro* requires only CuACE1, general transcription factors which can be supplied as a mouse nuclear extract and RNA polymerase (Culotta et al., 1989). In the absence of Cu(I) or ACE1 only limited levels of transcription are observed.

Acknowledgement

Support from the Public Health Service grant ES03817 from the National Institutes of Health is acknowledged.

REFERENCES

Abrahams, I.L., Bremner, I., Diakun, G.P., Garner, C.D., Hasnain, S.S., Ross, I., and Vasak, M. (1986) Biochem. J. 236, 585.

Adman, E., Waterpaugh, K.D., and Jensen, L.H. (1975) Proc. Natl. Acad. Sci. USA 72, 4854.

Anderson, B.F., Baker, H.M., Norris, G.E., Rumball, G.V., and Baker, E.N. (1990) Nature 344, 784.

Beltramini, M., Giacometti, G.M., Salvato, B., Giacometti, G., Munger, K., and Lerch, K. (1989) Biochem. J. 260, 189.

Bremner, I. (1987) Experientia Suppl. 52, 81.

Buchman, C., Skroch, P., Welch, J., Fogel, S., and Karin, M. (1989) Mol. Cell. Biol. 9, 4091.

Buchman, C., Skroch, P., Dixon, W., Tullius, T.D., and Karin, M. (1990) Mol. Cell. Biol. 10, 4778.

Butt, T.R., and Ecker, D.J. (1987) Microbiol. Rev. 51, 351.

Carri, M.T., Galiazzo, F., Ciriolo, M.R., and Rotilio, G. (1991) FEBS Lett. 278, 263.

Carter, A.D., Felber, B.K., Walling, M.J., Jubier, M.F., Schmidt, C.J. and Hamer, D.H. (1984) Proc. Natl. Acad. Sci. USA 81, 7392.

Casas-Finet, J.R., Hu, S., Hamer, D., and Karpel, R.L. (1991) FEBS Lett. 281, 205.

Coleman, J.E., Nakamura, K.I., and Chlebowski, J.F. (1983) J.Biol. Chem. 258, 386.

Coy, M., and Neilands, J.B. (1991) Biochem. 30, 8201.

Culotta, V.C., Hsu, T., Hu, S., Furst, P., and Hamer, D. (1989) Proc. Natl. Acad. Sci. USA 86, 8377.

Dance, I.G. (1986) Polyhedron 5, 1037.

Dameron, C.T., Winge, D.R., George, G.N., Sansone, M., Hu, S., and Hamer, D. (1991) Proc. Natl. Acad. Sci. USA 88, 6127.

Durnam, D.M., and Palmiter, R.D. (1981) J. Biol. Chem. 256, 5712.

Evans, C.F., Engelke, D.R., and Thiele, D.J. (1990) Mol. Cell. Biol. 10, 426.

Freedman, J.H., Powers, L., and Peisach, J. (1986) Biochemistry 25, 2342.

Furst, P., Hu, S., Hackett, R., and Hamer, D.H. (1988 Cell 55, 705.

Gadd, G.M., and White, C. in Metal-Microbe Interactions, Special Publications of the Society for General Microbiology, Vol. 26, R.K. Poole and G.M. Gadd, Eds.; IRL Press: Oxford, UK 1989, p.19.

Galiazzo, F., Schiesser, A., and Rotilio, G. (1988) Biochim. Biophys. Acta 965, 46.

George, G.N., Winge, D., Stout, C.D., and Cramer, S.P. (1986) J. Inorg. Biochem. 27, 213.
George, G.N., Byrd, J., and Winge, D.R. (1988) J. Biol. Chem. 263, 8199.
Gralla, E.B., Thiele, D.J., Silar, P., and Valentine, J.S. (1991) Proc. Natl. Acad. Sci. USA 88, 8558.
Haslinger, A., and Karin, M. (1985) Proc. Natl. Acad. Sci. USA 82, 8572.
Hamer, D.H. (1986) Ann. Rev. Biochem. 55, 913.
Henary, M., and Zink, J.I. (1991) Inorg. Chem. 30, 3111.
Hu, S., Furst, P., and Hamer, D. (1990) The New Biol. 2, 544.
Huibregtse, J.M., Engelke, D.R., and Thiele, D.J. (1989) Proc. Natl. Acad. Sci. USA 86, 65.
Ikura, M., Clore, G.M., Gronenborn, A.M., Zhu, G., Klee, C.B., and Bax, A. (1992) Science 256, 632.
Imbert, J., Culotta, V., Furst, P., Gedamu, L., and Hamer, D. (1990) Advances Inorg. Biochem. 8, 140.
Kagi, J.H.R., and Kojima, Y. (1987) Experientia Suppl. 52, 25.
Kagi, J.H.R., and Schaffer, A. (1988) Biochemistry 27, 8509.
Mayo, K.E., Warren, R., and Palmiter, R.D. (1980) Cell 29, 99.
Meador, W.E., Means, A.R., and Quiocho, F.A. (1992) Science 257, 1251.
Mehra, R.K., and Winge, D.R. (1991) J. Cell. Biochem. 45, 30.
Mehra, R.K., Tarbet, E.B., Gray, W.R., and Winge, D.R. (1988) Proc. Natl. Acad. Sci. USA 85, 8815.
Messerle, B.A., Schaffer, A., Vasak, M., Kagi, J.H.R., and Wuthrich, K. (1990) J.Mol. Biol. 214, 765.
Nakagawa, K.H., Inouye, C., Hedman, B., Karin, M., Tullius, T.D., and Hodgson, K.O. (1991) J. Am. Chem. Soc. 113, 3621.
Narula, S.S., Mehra, R.K., Winge, D.R., and Armitage, I.M. (1991) J. Amer. Chem. Soc. 113, 9354.
Nettesheim, D.G., Engeseth, H.R., and Otvos, J.D. (1985) Biochemistry 24, 6744.
Nielson, K.B., Atkin, C.L., and Winge, D.R. (1985) J. Biol. Chem. 260, 5342.
Otvos, J.D., Petering, D.H., and Shaw, C.F. (1989) Comments Inorg. Chem. 9, 1.
Palmiter, R.D. (1987) Experientia Suppl. 52, 63.
Palmiter, R.D., Findley, S.D., Whitmore, T.E., and Durnam, D.M. (1992) Proc. Natl. Acad. Sci. USA 89, 6333.
Quiocho, F.A. (1986) Ann. Rev. Biochem. 55, 287.
Robbins, A.H., McRee, D.E., Williamson, M., Collett, S.A., Xuong, N.H. Furey, W., Wang, B.C., and Stout, D.C. (1991) J. Mol. Biol. 221, 1269.
Ryu, C.K., Kyle, K.R., and Ford, P.C. (1991) Inorg. Chem. 30, 3982.
Silver, S., and Misra, T.K. (1988) Ann. Rev. Microbiol. 42, 717.
Smith, T.A., Lerch, K., and Hodgson, K.O. Inorg. Chem. 25, 4677.
Stuart, G.W., Searle, P.F., Chen, H.Y., Brinster, R.L., and Palmiter, R.L. (1984) Proc. Natl. Acad. Sci. USA 81, 7318.
Szczypka, M.S., and Thiele, D.J. (1989) Mol. Cell. Biol. 9, 421.
Thiele, D.J., and Hamer, D.H. (1986) Mol. Cell. Biol. 6, 1158.
Thiele, D.J.(1988) Mol. Cell. Biol. 8, 2745.

Vallee, B.L., and Ulmer, D.D. (1972) Ann. Rev. Biochem. 41, 91.
Vogler, A., and Kunkely, H. (1986) J. Am. Chem. Soc. 108, 7211.
Winge, D.R., Dameron, C.T., George, G.N., Pickering, I.J., and Dance, I.G. in Bioinorganic Chemistry of Copper, K.D. Karlin and Z. Tyeklar, Eds.; Chapman & Hall, New York, 1992 in press.
Winge, D.R., Nielson, K.B., Gray, W.R., and Hamer, D.H. (1985) J Biol. Chem. 260, 14464.
Zhou, P., and Thiele, D.J. (1991) Proc. Natl. Acad. Sci. USA 88, 6112.
Zhou, P., Szczypka, M.S., Sosinowski, T., and Thiele, D.J. (1992) Mol. Cell. Biol. 12, 3766.

Metallothionein III
K. T. Suzuki, N. Imura and M. Kimura (eds)

METALLOTHIONEIN GENES AND THEIR REGULATION IN TRANSGENIC MICE

Richard D. Palmiter*, Eric P. Sandgren+, David M. Koeller*, Seth D. Findley* and Ralph L. Brinster+

*Howard Hughes Medical Institute, University of Washington, Seattle WA 98195
+School of Veterinary Medicine, University of Pennsylvania, Philadelphia PA 19103.

SUMMARY: Four metallothionein genes are closely linked on mouse chromosome 8. The MT-III gene is located about 20 kb 5' of the mouse MT-II gene and a recently discovered MT-IV gene lies about 18 kb 5' of the MT-III gene. We are exploring the elements that establish the MT locus and confer appropriate expression to these various MT genes. We have identified sequences flanking the MT-I and MT-II genes that confer copy-number dependent and position-independent expression of a marked MT-I gene in transgenic mice. These MT-I transgenes are expressed at a level approaching that of endogenous MT genes and they are regulated by natural and exogenous stimuli in an appropriate manner, suggesting that all the elements required for normal expression are contained within these constructs. Using this approach we have generated lines of transgenic mice that can produce up to 10 times the normal level of MT. We suspect that we have isolated the equivalent of a locus control region. Deletion analysis suggests that the organization of regulatory elements with the MT flanking sequences is complex. Mice with deficiencies in the expression of all of these MT genes are being created by homologous recombination in embryonic stem cells. Having mice that over- or under-express these MTs should ultimately help reveal the function of these unique proteins in mammalian development and physiology.

The murine metallothionein locus

An EcoRI restriction map of the murine MT locus is depicted in Fig. 1. There are 4 MT genes, all transcribed in the same direction, located within a 45 kb region. The murine MT-I gene was the first to be cloned (Durnam et al., 1980) and it was mapped to mouse chromosome 8 (Cox and Palmiter, 1983). Screening mouse genomic lambda libraries at reduced stringency did not reveal other members of the MT gene family, primarily because there is a low abundance repeat sequence (mini-satellite) that is detected with the MT-I cDNA at reduced stringency. The mouse MT-II gene was ultimately found by walking 5' of the MT-I gene and screening duplicate Southern blots of the lambda clones with ^{32}P-labelled RNA made from nuclei isolated from tissue culture cells with amplified copies of the MT locus.

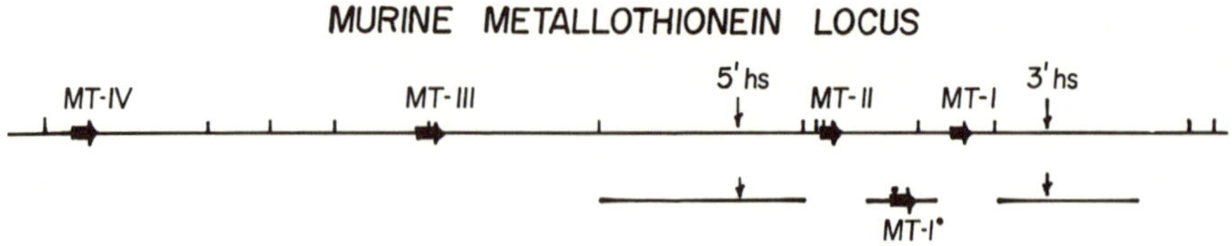

Figure 1. Organization of the murine metallothionein locus. *The four MT genes are all transcribed in the same direction. The distance between MT-IV and MT-I is about 45 kb. The vertical ticks indicate EcoRI sites. The location of the 5' and 3' DNase I hypersensitive sites (hs) mapped by MacArthur and Lieberman (1987) are indicated by vertical arrows. The location of the 10 kb and 7 kb DNA fragments flanking the MT-II and MT-I genes that have LCR function as discussed in text are indicated in the lower panel. These sequences were tested with a marked MT-I* gene.*

The duplicate blots were made from cells that either were or were not induced with zinc. By this approach we found zinc-inducible nuclear RNAs that hybridized to restriction fragments containing the MT-I gene as well as to restriction fragments that clearly did not include the MT-I gene. The latter restriction fragments were subsequently cloned and sequenced which revealed that they encoded the MT-II gene (Searle et al., 1984). The mouse MT-III gene was cloned from mouse brain cDNA by PCR using degenerate oligonucleotides based on the protein sequence of human GIF (Uchida et al., 1991) and then that cDNA was used to screen a mouse genomic lambda library (Palmiter et al., 1992). Probes from the 3' end of the MT-III lambda clone and the 5' end of the MT-II lambda clone hybridize to the same restriction fragments on genomic Southern blots. From these data we deduced that the MT-III gene is about 20 kb 5' of the MT-II gene. The next step was to prepare and screen a cosmid library made from DNA derived from cadmium-resistant hepa1A cells which have amplified their MT genes about 20 fold (Durnam and Palmiter, 1984). This library was screened with unique probes from the 5' and 3' extremes of the known MT locus with the aim of walking outwards in both directions. These cosmids were then screened at low stringency with labeled MT-I, MT-II and MT-III cDNAs to detect other linked family members that might be homologous to the known MT genes. In this way we discovered a fourth gene, MT-IV that is located about 18 kb 5' of the MT-III gene. We are currently extending this walk to see if there are yet more MT genes at this locus. In retrospect, this last approach would have worked to discover all four MT genes.

Comparisons among the murine MT genes

All the MT genes have the same basic structure with 3 exons that lie within 1 to 2 kb of genomic sequence. The intron/exon boundaries are conserved (relative to the cysteine codons) in all members of the MT family. The amino acid sequences, in one letter code, are shown in Fig. 2A. This alignment reveals that MT-III and MT-IV have amino acid insertions relative to the other MTs. The insertion site after amino acid 4 is common to both MT-III and MT-IV. Thirty-three amino acids are identical in all four MTs and another twelve are found in three of the four proteins. The number of amino acid differences between various members of the MT family are shown in Fig. 2B. This analysis reveals that the MT-I and MT-II genes are most similar and that MT-II, MT-III and MT-IV proteins differ from each other by a similar extent. While there is obvious homology in the coding regions, there is virtually no conservation of nucleotide sequence in the non-coding regions of any of these genes suggesting that these genes duplicated and diverged from each other a long time ago.

A.

```
mMT-I        MDPN CSCSTGGSCTCTSSCACKNCKCTSCKK...
mMT-II       MDPN CSCASDGSCSCAGACKCKQCKCTSCKK...
mMT-III      MDPETCPCPTGGSCTCSDKCKCKGCKCTNCKK...
mMT-IV       MDPGECTCMSGGICICGDNCKCTTCSCKTCRK...
Conserved    MDP  C C  gGsC C   CkCk CkCt CkK

...SCCSCCPVGCSKCAQGCVCK         GAADKCTCCA  61 AA +4
...SCCSCCPVGCSKCAQGCVCK         GAADKCTCCA  61 AA +3
...SCCSCCPAGCEKCAKDCVCKGEEGAKAEAEKCSCCQ  68 AA +1
...SCCPCCPPGCAKCARGCICK         GGSDKCSCCP  62 AA +3
   SCCsCCP GC KCA gCvCK         g adKC CC
```

B.

	mMT-I	mMT-II	mMT-III	mMT-IV
mMT-I	0	15	19	27
mMT-II		0	24	24
mMT-III			0	26
mMT-IV				0

Figure 2. Comparison of the four murine MT proteins. A. *The amino acid sequences (in one letter code) are indicated. The bottom line shows those amino acids that are conserved in all four MTs in bold capital letters and those that are found in 3 of the MTs in lower case letters. The number of amino acids and net charge of the proteins is shown on the right.* **B.** *The number of amino acid differences in various pairwise comparisons are indicated. Each insertion is considered to be only one difference.*

One scenario for the evolution of this gene family is that there was an ancient gene duplication followed by insertion of three bases into the first exon of the 5' MT gene giving rise to a MT with 62 amino acids, then both the 5' and 3' MT genes duplicated again and one member of the 5' pair incurred an insertion of 18 bp into the third exon giving rise to the 68 amino acid MT-III protein.

MT gene expression

The MT-I and II genes are expressed at similar levels in most murine organs and they are induced coordinately by metals, glucocorticoids and lipopolysaccaride (Searle et al., 1984; Yagle and Palmiter, 1985). LPS probably induces MT by stimulating macrophages to produce interleukins, IL-I and IL-6 (De et al., 1990). The MT-I gene is also induced by hemopexin in hepatocytes, presumably because they have hemopexin receptors (Alam and Smith, 1992). These two MTs are present at high levels in day-16 to day-18 fetal liver, perhaps in response to fetal corticosterone production (Quaife et al., 1986). In the unstressed adult, the highest levels of MT-I and II are found in male germ cells and coagulation gland. After induction by metals, the highest levels of MT-I and II mRNAs are found in liver, kidney, intestine and pancreas (Durnam and Palmiter, 1981). MT-III mRNA appears to be restricted to brain and based on immunocytochemistry experiments with human tissue (Uchida et al., 1991), it is probably localized in astrocytes. The organs that express MT-IV mRNA are currently being examined.

A locus control region for the MT-I and MT-II genes

Because the MT-I and MT-II genes are regulated coordinately and have a much wider tissue distribution than MT-III and MT-IV, it is possible that these two genes are in a separate chromosomal domain from the other MT genes. Our current working hypothesis is that one chromosomal domain is separated from another by locus control regions (LCR) that bind specific factors and may attach the DNA to the nuclear matrix, thereby creating chromosomal loops. The location of LCRs may be revealed as sites of DNase I hypersensitivity due to the disruption of the regular nucleosomal packaging of DNA by specific DNA binding proteins.

MacArthur and Lieberman (1987) described the location of DNase I hypersensitive sites that flank the MT-I and MT-II genes in S49 cells. These thymus-derived cells are unusual in that they do not normally express the MT genes, because of DNA methylation (Compere and Palmiter, 1981). However, variants can be selected that are resistant to Cd and these variants invariably express MT-I, MT-II or both genes. DNase I hypersensitive sites appear in the promoter region of whichever genes are expressed. In addition, there are hypersensitive sites about 5 kb upstream of the MT-II gene and about 4 kb 3' of the MT-I gene that are present regardless of whether any of the MT genes are expressed (see Fig. 1). Based on the hypothesis that these hypersensitive sites might mark part of the MT LCR, we set out to test whether large regions of DNA that included these sites would confer upon a test gene properties similar to those ascribed to the LCR region located 5' of the β-globin cluster (Grosveld et al., 1987; Townes and Behringer, 1990).

For these experiments, we constructed a marked MT-I gene that we call MT-I*. It has a 2 bp insertion that converts the BglII site in the 5' untranslated region to an EcoRV site which allows detection of any transcripts from that gene by oligonucleotide hybridization or PCR. The MT-I* gene is centrally located within a 3.8 kb EcoRI fragment. This DNA was used to make transgenic mice. This MT-I* transgene was expressed in a manner similar to many previous constructs in which the MT-I promoter was fused to a wide variety of different structural genes. Namely, only some (about 70%) of the transgenic mice expressed the transgene and the level of expression was not related to transgene copy number. The tissue-specific pattern of expression did not resemble the endogenous MT-I gene and the gene was uninducible in the fetus or by glucocorticoid hormones, unlike the endogenous MT-I gene.

In contrast, when transgenic mice were made with a MT-I* gene that was flanked by 10 kb of DNA normally located 5' of the MT-II gene and 7 kb of DNA normally located 3' of the MT-I gene (See Fig. 1) we observed remarkably good copy-number-dependent expression and the level of transgene expression in the liver was almost equal to that of the endogenous MT-I genes on a per gene basis. The tissue-specific pattern of transgene expression resembled that of the endogenous genes and the MT-I* gene was highly inducible by heavy metals, glucocorticoids and lipopolysaccaride (Table 1). These results indicate that the 17 kb of flanking region contains sequences with properties similar to those ascribed to locus control regions. The location of essential elements within the flanking region is being investigated.

Mice expressing excess metallothionein

Because the marked MT-I* gene retains an intact open reading frame, we suspect that these mice express elevated levels of MT-I protein in all the tissues where MT-I is normally expressed. In some transgenic lines of mice, total MT-I mRNA levels are 10 to 20-fold higher than normal. These mice have been maintained on a normal diet for several generations with no obvious pathological effects. Growth, reproduction and longevity appear normal. These mice will provide a means of assessing whether excess MT production provides protection against various physiological stresses.

Table 1. Induction of MT-I* by metals, dexamethasone and endotoxin

Treatment	Control	Transgenic
	Hepatic MT-I mRNA molecules/cell	
none	920 (10)	9,450 (3)
$ZnSO_4$	2,500 (5)	37,000 (3)
$CdSO_4$	5,950 (3)	68,900 (1)
Dexamethasone	3,600 (3)	78,600 (2)
Endotoxin (LPS)	8,740 (2)	168,000 (2)

Transgenic mice from line 3928-3 (37 transgenes) were either untreated, maintained on 25 mM $ZnSO_4$ in their drinking water for 5-6 days, injected with $CdSO_4$ (1 mg/kg, intraperitoneal), injected with dexamethasone (5 mg/kg), or injected with lipopolysaccaride (3 mg/kg). Mice were killed 5-7 hr after injections and total nucleic acids were isolated from liver for determination of total MT-I mRNA concentration by solution hybridization. The numbers in parentheses indicate the number of liver samples analyzed.

Mice unable to express metallothionein

We have used homologous recombination in embryonic stem cells to inactivate the MT-I gene, the MT-II gene, or both genes (see Fig. 3). These cells are being used to generate mice that cannot express one or both of these MTs. At this time, heterozygous mice with both MT-I and II alleles disrupted have been generated and they are currently being inbred to create mice that cannot express either of these ubiquitously-expressed MT genes. These experiments will soon reveal whether MT is essential for normal mammalian development or response to various physiological stresses.

We have also used homologous recombination to inactivate the murine MT-III gene in ES cells. When mice that are homozygous for the gene disruption are made they will provide insight into whether this unique, brain-specific MT plays and important role in neural function.

Figure 3. Vectors used for MT gene disruption by homologous recombination. *The MT-III replacement vector (left) has a pol-neo gene replacing most of the endogenous MT-III gene. The MT-I and MT-II gene disruption vector has a pol-neo gene replacing about 2 kb of sequence between the two MT genes. Each of the MT genes has oligonucleotide insertions that introduce nonsense mutations near the amino terminus of the MT coding region. Depending on the site of recombination, this vector can result in cells with either or both of the endogenous MT genes replaced by the mutant versions. For both vectors, the appropriate recombination events were detected by PCR using the primers (horizontal arrows) indicated.*

REFERENCES

Alam, J. and Smith, A. (1992) Heme-Hemopexin-mediated Induction of Metallothionein Gene Expression. *J. Biol. Chem.* **267**, 16379-16384.

Compere, S.J. and Palmiter, R.D. (1981) DNA Methylation Controls the Inducibility of the MOuse Metallothionein I Gene in Lymphoid Cells. *Cell* **25**, 233-240.

Cox, D.R. and Palmiter, R.D. (1983) The Metallothionein-I Gene Maps to Mouse Chromosome 8: Implications for Human Menkes' Disease. *Hum. Genet.* **64**, 61-64.

De, S.K., McMaster, M.T. and Andrews, G.K. (1990) Endotoxin Induction of Murine Metallothionein Gene Expression. *J. Biol. Chem.* **265**, 15267-15274.

Durnam, D. and Palmiter, R.D. (1981) Transcriptional Regulation of the Mouse Metallothionein-I Gene by Heavy Metals. *J. Biol. Chem.* **256**, 5712-5716.

Durnam, D.M. and Palmiter, R.D. (1987) Analysis of the Detoxification of Heavy Metal Ions by Mouse Metallothionein. *Experientia Suppl.* **52**, 457-463.

Durnam, D.M., Perrin, F., Gannon, F. and Palmiter, R.D. (1980) Isolation and Characterization of the Mouse Metallothionein-I Gene. *Proc. Natl. Acad. Sci. U.S.A.* **77**, 6511-6515.

Grosveld, F., van Aussendelft, G.B., Greaves, D.R., and Kollias, G. (1987) Position-independent, High-level Expression of the Human β-Globin Gene in Transgenic Mice. *Cell* **51**, 975-985.

MacArthur, C.A. and Lieberman, M.W. (1987) Different Types of Hypersensitive Sites in the Mouse Metallothionein Gene Region. *J. Biol. Chem.* **262**, 2161-2164.

Palmiter, R.D., Findley, S.D., Whitmore, T.E. and Durnam D.M. (1992) MT-III, a Brain-specific Member of the Metallothionein Gene Family. *Proc. Natl. Acad. Sci. USA* **89**, 6333-6337.

Quaife, C., Hammer, R.E., Mottet, N.K. and Palmiter, R.D. (1986) Glucocorticoid Regulation of Metallothionein during Murine Development. *Devel. Biol.* **118**, 549-555.

Searle, P.F., Davison, B. L., Stuart, G.W., Wilkie, T.M., Norstedt, G. and Palmiter, R.D. (1984) Regulation, Linkage and Sequence of the Mouse Metallothionein I and II Genes. *Mol. and Cell. Biol.* **4**, 1221-1230.

Townes, T.M. and Behringer, R.R. (1990) Human Globin Locus Activation region (LAR): Role in Temporal Control. *Trends Genetics* **6**, 219-223.

Uchida, Y., Takio, K., Titani, K., Ihara, Y. and Tominaga, M. (1991) The Growth Inhibitory Factor that is Deficient in Alzheimer's Disease Brain is a 68 Amino Acid Metallothionein-like Protein. *Neuron* **7**, 337-347.

Yagle, M.K. and Palmiter, R.D. (1985) Coordinate Regulation of Mouse Metallothionein I and II Genes by Heavy Metals and Glucocorticoids. *Mol. Cell. Biol.* **5**, 291-294.

Metallothionein III
K. T. Suzuki, N. Imura and M. Kimura (eds)

METALLOTHIONEIN STRUCTURE / FUNCTION INTERFACE

P.C. Huang

College of Life Science, National Tsing Hua University
Hsin-Chu 30043, Taiwan, R.O.C.

(on leave from Johns Hopkins University, Baltimore 21205, MD, U.S.A.)

Metal sequestration function of metallothionein was examined and correlated with structural constraints which could be conveyed by the highly conserved amino acid residues in each of the two domains.

We mutagenized the coding sequences of this protein so that codons at single or multiple sites could be specifically substituted, or the distance between domains expanded. Native and engineered coding sequences were expressed with appropriate shuttle vectors in yeast, *Saccharomyces cerevisiae* and mammalian cells.

Metallothioneins from the recombinant constructs were subjected to analyses with circular dichroism, metal affinity and exchange reactions as well as ligand/solution accessibility. Structural analysis was also carried out by ^{113}Cd NMR. For functional tests, the altered sequences were subjected to transgenic experiments assaying for the ability of these engineered metallothioneins to confer resistance to metal sensitive yeast hosts. Thus cultured mammalian cells (CHO wild type and cadmium resistant mutants, and mouse 3T3 cells) were used to examine the effect of various episomal metallothionein on metal homeostasis and protection of specific cellular components towards metal toxicity. Selected metalloenzymes and a cytoskeletal metalloprotein were examined for metal exchange reactions with metallothioneins.

A possible role of metallothioenien to rescue metal intoxicated metalloproteins/enzymes is proposed.

Organisms have evolved to depend upon certain trace elements for metabolism while coping with reactive, toxic levels of the same or other metal ions. Nature is burdened with an increasing level of toxic metals. Biological function and bioavailability of metals in a cell invariably involve metal interaction with proteins/peptides. Indeed, over seventy enzymes have been recognized that require metals as a cofactor for activity (Vallee and Auld 1990), and a library of regulatory proteins has been shown to contain finger-shape motifs which bind zinc (Berg 1990). Metals may interact with a protein and dictate its conformation, resulting in normal or altered stability and function. All metalloproteins are thus potential targets for metal toxicity. How a cell discriminates and manipulates the essential versus toxic metals is the central question in understanding the detoxification mechanisms, with which normality of a cell is to be maintained. We shall focus on the possible involvement of metallothionein as a metal scavenger/donor, concentrating our efforts with one antagonist pair of metals: zinc, an essential element and cadmium, a toxic pollutant.

Mammalian metallothionein is now known to exist as four isospecies, all enriched in cysteines. On the basis of a survey of metallothionein sequences from diverse sources, either determined directly or inferred from their cDNAs, it is shown that MT1, MT2 and MT4 metallothioneins consist of 61 amino acids and MT3 has 68 (Palmiter et al. 1992 and this volume). In all cases there are 20 cysteines in each metallothionein, which binds seven equivalents of bivalent d^{10} ions such as Zn, Cd, Hg, Pb, Bi, Sn, Co, Ni, Bi and Tc. Higher metal to protein stoichiometries are observed with univalent ions, Cu(I), Ag(I) and Au(I) (see review by Kagi and Schaffer 1988). Typical apparent binding constants (M^{-1}) at neutral pH have been estimated to be 2 x 10^{-12}, 4 x 10^{-14} and 2 x 10^{-16} for Zn, Pb and Cd, respectively (cited in Kagi and Schaffer 1988). Spectroscopic and magneto-optical studies indicate that metal binding by metallothionein assumes a tetrahedral tetrathiolate coordination (Vasak et al. 1987). Hence a total of 42 intramolecular cys-metal(II)-cys crosslinks exist for the bivalent ions, 24 in one and 18 in the other metal binding cluster.

Features of the two metal binding clusters in mammalian metallothioneins, corresponding to alpha (carboxyl half) and beta (amino terminal half) domains (Winge and Miklossy 1982), have been derived from studies of spatial structure of metallothioneins from rabbit (Arseniev et al. 1988), rat (Schultze et al. 1988), and human (Messerle et al.1990, 1992) by 2-D NMR, and the metal positions of the two domains by homonuclear and heteronuclear NMR (Otvos and Armitage 1980, Armitage and Boulanger 1983). All cys-Cd bonds have been unambiguously assigned by $^{1}H^{113}Cd$ NMR correlation measurements (Frey et al. 1985) and more recently by X-ray diffraction analysis of fossilized metallothionein crystals (Robbins et al. 1991). Thus, each of the metallothionein domains is about 15-20A in diameter, with the amino acid residues forming abundant helical turns wrapped around the metal core. The two domains are linked by a conserved tripeptide hinge, lys-lys-ser.

In spite of the rich information on its structure, metallothionein is a protein in search of a function (Karin 1985). The propensity for it to bind metals with selectivity and with different affinities have prompted suggestions that its function is pleiotropic and metal specific. For cadmium, metallothionein's sole function may be detoxification, and it appears to be the only protein in the cell to exert such a role (Hamer 1986). How does cadmium exert its toxicity and what is the mechanism for metallothionein to exercise its defensive action are questions yet to be answered.

Proposed roles of metallothionein have included its serving as a metal-buffering ligand to sequester or to dispense metals, and having a specialized function in normal cellular metabolism and development (review in Kagi and Schaffer 1988). Through gene amplification (Griffith et al. 1983, Durnam and Palmiter 1987, Huang et al. 1987) and transfer (Thiele et al. 1986, Ecker et al. 1986, Chernaik and Huang 1991) experiments, the presence of metallothionein is shown to associate strongly with Cd detoxification.

Suggestion for metallothionein to act as a stress protein is proposed based on the versatile inducibility of metallothionein gene transcription by a wide range of stressful conditions and on the presence of multiple, interdigitated cis-regulatory elements upstream of metallothionein genes (Huang and Lin 1989). Such an activity can be mediated through interactions with other metalloproteins. Reactivation of apo-enzymes, presumably by donating the needed zinc ions (Udom and Brady 1980, Li et al. 1981), and dynamic fluctuation in metal binding have been shown by experiments with alkylating agent displacement (Bernhard et al. 1986), metal titration (Nielson and Winge 1984), saturation-transfer NMR spectroscopy (Otvos et al. 1987, Nettesheim et al. 1985), and circular dichroism (Stillman and Zelazowski 1989). While the exchange may be intra- as well as intermolecular, an effect of thionein on DNA binding by transcriptional metallo-transactivating factors has been demonstrated (Zeng et al. 1991a and b).

From these observations, we subscribe to a notion that, metallothionein may function in detoxification by metal sequestration, reducing the bioavailability levels of toxic metals in the cells, and by metal exchange, resurrecting the poisoned metalloenzymes/ proteins from inactivation. We hypothesize that these functional pathways are regulated, respectively, in one or the other domains of metallothionein. We tested this hypothesis by substitutional mutagenesis, coupled with structure and functional analyses.

Within the coding sequences of metallothionein, specific codons at unique, single or multiple sites could be substituted without affecting the metal binding and stability of the gene product. By cloning individual domain-exonic sequences and altering the orientation of, or distance between, we tested to see if the two domains in metallo-thionein are independent. In order to accomplish these tests, suitable prokaryotic and eukaryotic systems for metallo-thionein expression, for functionality assays and for the recovery of quantitative amounts of material for physico-chemical studies were developed.

1. Search for native variation in metallothionein sequences

The ubiquitous presence of metallothionein suggests that this protein has survived the selective pressure of evolution, and naturally occurring sequences are reflective of structural constraints imposed on function (Huang et al. 1991). Examining metallothionein species from fish (Andersen et al. 1986) to birds, we were able to demonstrate that two distinct isospecies of metallothionein exist in pigeon, differing from each other in their amino acid sequences (Lin et al. 1989). However, in other avian species, notably ducks, only one unique metallothionein isoform is present (Lin et al. 1990b). Furthermore, complete sequence determinations show that two different genera of ducks (verified by their mitochondrial genomes, Lin and Huang 1990a) carry the same sequence of metallothionein, which is identical to that of their intergeneric hybrid and of chicken (Lin et al. 1990c).

These results suggest that the onset of diversity of metallothionein coincided with the emergence of avian species and that the single metallothionein species we have examined is ancestral in nature. Its amino acid sequence may therefore reflect the early constraint placed on metallothionein structure for the proper function of the molecule.

2. Engineered metallothioneins and their characterization

We reported earlier an improved procedure with which to generate a repertoire of mutants for CHO MT2 (Pine and Huang 1987), of which about three hundred have been identified by DNA sequencing to be substitution mutations at single or multiple sites (Pine et al. 1985). These mutations, however, are limited to cysteine to tyrosine replacement, and were obtained at random sites. Since aromatic amino acids are absent in mammalian metallothionein, the effects seen with tyrosine substitutions must be interpreted with caution. Additional mutants were generated.

i. cysteine mutants

A complete set of mutant metallothionein coding sequences in which codons for each of the twenty individual cysteines were replaced by serine codons, and cloned into a yeast expression system. The recombinant metallothionein gene is placed under the control of a constitutive promoter on a multi-copy coli-yeast shuttle vector, and expressed in a metal sensitive host whose endogenous copperthionein gene had been previously deleted. Our results show that the expressed metallothioneins conferred increased metal resistance to the yeast host, as assayed by a sensitive assay in which population doubling times were monitored in rich media supplemented with a sub-lethal dose of $CdCl_2$. Earlier measurements on mutants with single cysteine replacements at 12 different positions revealed two mutant classes. One class (cys-ser at positions 5, 13, 19 and 33) had the same detoxification capacity as natural metallothionein from CHO MT2. A second class (cys-ser at positions 7, 15, 26, 29, 44, 48, 50 and 60) conferred to the host markedly less resistance to cadmium resistance. All five bridging cysteine mutants studied (at positions 7, 15, 44, 50, and 60) conferred lower cadmium resistance. In contrast, mutation of four out of seven terminal cysteines (at positions 5, 13, 19, and 33) was shown to be inconsequential. Mutations in the alpha-domain tend to be more detrimental than in the beta-domain to metallothionein in conveying cadmium resistance, suggesting that the contribution of individual cysteines to the detoxification of metallothionein is site specific (Chernaik and Huang 1991).

We have also studied the effects of replacing one or more cysteine residues with different amino acids. Metallothioneins which contain a cysteine mutation at position 50 (C50Y, C50S, C50A and C13,50Y) conferred markedly less cadmium resistance than did wild type metallothionein, or metallothionein with a single cysteine mutation at position 13 (C13Y and C13S). Thus similar effects are seen when a given cysteine is substituted with serine, tyrosine or alanine at these sites (Cismowski et al. 1992).

Wild type and three of the mutant CHO MT2s (C13Y, C50Y and C13,50Y) were purified to homogeneity from yeast grown in subtoxic levels of either $CdCl_2$ or $^{113}CdCl_2$ and analyzed for their chemical and spectroscopic properties. Wild type and mutant metallothioneins, in their cadmium bound forms, all showed characteristic ultraviolet absorption spectra with shoulders at 245-250 nm due to cadmium-thiolate binding. Upon acidification these absorption shoulders were abolished. In all cases two distinct titrations were seen, presumably corresponding to the two independent cadmium binding domains in each of the proteins. Analysis of domain structures was performed with both the sulfhydryl reagent 5,5'-dithio-bis(2-nitrobenzoic acid) and with the protease subtilisin. These studies indicated that both mutations affected domain structure by disrupting the normally tight protein clusters. Circular dichroism spectra obtained for wild type and mutant metallothioneins indicated unique structural rearrangements in mutants containing a cysteine 50 to tyrosine alteration. These results suggest that cysteines at both positions 13 and 50 are important for stable metal coordination. Cysteine 50, however,appears to be more critical to cadmium detoxification function than cysteine 13 (Cismowski and Huang 1991b).

By growing yeast expressing wild type or mutant metallothioneins in medium containing $^{113}CdCl_2$ we have prepared ^{113}Cd-metallothionein in its native form(Cismowski and Huang 1991a), in contrast to conventional preparation by re-constitution, for ^{113}CdNMR analyses. The one-dimensional ^{113}Cd-NMR spectrum of the recombinant wild type CHO MT2 was compared to the spectra of native rat and rabbit liver metallothioneins. The close correspondence between the ^{113}Cd chemical shifts in these metallothioneins is consistent with the presence of two separate metal clusters, A and B, corresponding respectively to the alpha- and beta-domains, in the recombinant metallothionein, as well as a conservation of cysteine to metal connectivities between the native and recombinant metallo-thioneins. A one-dimensional ^{113}Cd-NMR spectrum was also determined for each of the three purified mutant metallothioneins. The spectra of the two single mutants (C13Y and C50Y) were consistent with the introduction of substantial disruptions in metal coordination in the mutated domain, while the non-mutated domain showed no significant perturbations. In the double mutant C13,50Y, the ^{113}Cd resonances indicated that major alterations in the overall structure were present. These data have been incorporated into a model consistent with the *in vivo* and *in vitro* properties of these mutants (Cismowski et al. 1991b).

ii. lysine mutants

There are eight lysines in mammalian metallothioneins at invariant positions of 20, 22, 25, 30, 31, 43, 51, and 56, respectively. Each of these lysines has been systematically substituted with either a glutamate or glutamic acid. Additionally, several doubly substituted mutants (e.g. K25,56E; K51,56E) have been created along with some triple mutants (e.g. K43,51,56E); the sites of these mutations fall within or between domains. Selected mutants containing replacements at lysines 30 and 31, the interdomain hinge region, [single(K30E/Q; K31E/Q); double mutants (K30,31 E/Q)] have been characterized. The results show that at least one charged amino acid must be present in the hinge for the proper expression of metallothionein in the yeast expression system. (Cody and Huang 1992, submitted).

iii. insertion mutants

Specific peptides of varying lengths were inserted between the two metal cluster domains of CHO metallothionein, which normally are spanned by only three amino acids, lys-lys-ser. This was achieved by inserting defined oligo-nucleotides of up to four tandem repeats of dodecanucleotides into an Alu-1 endonuclease cleavage site, which separates the two exonic regions of the MT2 coding sequence. The native and altered sequences were cloned into the high expression E. coli-yeast shuttle vector, and used to transform yeast cells whose endogenous metallothionein gene was deleted. All constructs were shown to be functional in rendering the host cells resistant to 300 uM of either copper or cadmium, when grown on minimal media plates. The level of mutant protein accumulated in the cells, however, diminished with the length of the insert. Determinations for copy number of the chimeric plasmids and metallothionein mRNAs in the transformed cells showed that the replicational and transcriptional capacities of the long and short constructs were equivalent. Differences in protein level may thus be due to varied fidelity in the translational and/or post-translational processes. These results show that the stability of metallothionein depends upon the proximity of its two domains. Since as many as 12 amino acids may be inserted between the domains, it is possible to engineer metallo-thionein with a variety of functional or targeting motifs at this site (Rhee et al. 1990). We learned earlier that metallothionein is stable as a fusion protein with an additional peptide of 50 residues in length (from lambda N protein) linked to its amino terminus (unpublished data).

3. Additional biological systems for functional studies

As we aimed at using cultured cell systems to analyze native and engineered metallothioneins, we have examined constitutive metallothionein expression in CHO cells in which metallothionein genes have been preferentially amplified (Huang et al. 1987). These cells were used to demonstrate that metallothionein gene expression is enhanced under various stress conditions, including exposure to non-transition metals (Rhee and Huang 1989) and interferons (Morris and Huang 1987), and that the transcription may be auto-regulated (Morris and Huang 1989).

Exploring this regulatory mode at an organismal level, through a series of collaborative studies, we have learned that adrenalectomy down-regulates metallothionein synthesis through the control of glucocorticoid level in rats (Hidalgo et al. 1992). Chronic feeding of rats over a two year period with cadmium in drinking water ad lib, resulted in a highly significant elevation of metallothionein excreted in the urine (Shaikh et al.1989). Interestingly, metallothionein induction is ontogenically controlled; metallothionein is not induced by cadmium in new born mice (Thomas et al. 1987, 1988 a and b). These studies lay the foundation for a possible approach to probe the function of metallothionein through the use of transgenic systems.

4. Metallothionein function in eukaryotic cells : an alternative monitoring system other than yeast

The use of yeast as an expression system allows both a sensitive bioassay for metal resistance/detoxification and an efficient factory for the production of mammalian metallothioneins (Cismowski and Huang 1991a). We surmise that certain mutations which did not show an effect on metal detoxification function in yeast may be defective and would be detectable only in their native host. Functions such as targeting and compart-mentalization may be different in yeast and higher eukaryotic cells. We endeavored therefore to develop assays that would allow us to examine this possibility, taking advantage of the availability of well characterized mouse 3T3 cells, and a collection of cadmium resistant mutants generated in our laboratory earlier (Corrigan and Huang 1981, 1982, Huang et al. 1980).

Actin was chosen to represent the major cytoskeletal protein that is poisoned by cadmium (Mills and Ferm 1989, Prozialeck and Niewnhuis 1991). Actin undergoes polymerization as cells live and divide (Pollard 1990), the kinetics of which are affected by environmental variables such as pH, temperature, and types and concentration of metals (Zimmerle and Frieden 1988). Under environmental modulation, it is suspected that actin may assume a wide spectrum of conformational states, from mainly monomeric to polymers which self assemble rapidly, depending on the metals it binds (Nowak et al. 1988). Preliminary studies by us (in preparation) and others suggest that actin filaments may be one of the cellular targets for damage by cadmium (Mills and Ferm 1989, Prozialeck and Niewenhuis 1991), and that metallothionein can reverse such a toxic effect. The damaging effect was observed microscopically.

Fluorescent microscopy was used to verify the effect of cadmium on actin and the ability of metallothionein in reversing. Morphological changes of actin were observed with the aid of phallotoxin, an actin specific fluorescent dye. As actin bundles or stress fibers are well organized in normal cells,they are shortened when cells are treated with cadmium. They become granular and amorphous, or totally disappear depending upon the concentration and duration of treatment.

The ability of metallothionein in metal detoxification is dependent on its concentration in the cells. This is shown by CHO mutants with amplified metallothionein genes (Huang et al. 1980). The level of metallothionein in cells may also be regulated by introducing exogenous metallothionein coding sequences into the host. A coli-BPV shuttle vector was constructed to transfect mouse 3T3 cells (courtesy of K. Morton), which have well discernable cytoskeletal structures. The construct carries a drug resistance marker, so that copy number of the vector, hence metallothionein level in the cell, can be controlled by the level of neomycin in the medium. By comparison with the host cells which have only one pair of metallothionein isogenes, the ability of native metallothionein to render metal resistance in the transfected cells was observed. In these studies, the level of mRNA was monitored by quantitative Northern blot hybridization using metallothionein cDNA as the probe (Morris et al. 1987, 1988, Rhee and Huang 1989), and the level of metallothionein was monitored by ^{35}S cysteine labeling and electrophoresis .

It would be interesting if the wild type metallothionein sequence in the vector can be replaced with cassettes of mutants and changes in the extent of protection observed.

5. Metallothionein as metal donor/receptor - protein/protein exchange

Assays for metal exchange involving metallothionein and other metalloproteins/enzymes that are targets for cadmium toxicity were developed. In this series of experiments, we examined the ability of metallothionein to restore the activity of Zn-enzymes when zinc in the enzyme is replaced by cadmium. Earlier studies show that Zn-thionein may be a physiological donor of zinc to apometalloenzymes such as alcohol dehydrogenase and alkaline phosphatase (Udom and Brady 1980). Such a finding has been supported by results from ligand substitution reactions of various thioneins with EDTA and apo-carbonic anhydrase (Li et al. 1980).

We have concentrated on the possible role of metallothionein to resucue cadmium toxicity on actin, a major cytoskeletal protein, and carboxypeptidase A, a well characterized metalloenzyme.

Monomeric G-actin was prepared from chicken cells following the procedure of Spudich and Watt (1971) (courtesy of D. Lin). The actin was labelled with N-(1-pyrenyl) iodoacetamide and the polymerization followed spectrophoto-metrically. The effective concentration of cadmium needed to inhibit actin polymerization was determined. Blockage of the nucleation-elongation was examined at excitation and emission wavelengths 365 and 386nm, respectively (Kouyama and Mihashi 1981), using a fluorescence spectrophotometer (Perkin-Elmer 600-105). Under these conditions, we observed that actin polymerization (as reflected by increases in fluorescence in this assay) was completely stopped by the presence of 12.5 uM of $CdCl_2$, and the blockage was reversed by the addition of 1 to 4uM of Zn-metallothionein (An et al. in preparation).

While actin represents a metal binding, structural component of the cell, we reason that Zn-containing enzymes may also be targets for cadmium damage and that their defense mechanism may be shared. Zinc and cadmium are isomorphous in chemical coordination. A number of these enzymes can be readily transformed into corresponding Cd-containing species without apparent consequences. However, several Zn-enzymes have shown enhanced or reduced activities in the presence of cadmium (Flik et al. 1987, Jamall and Smith 1985).

The dual-function carboxypeptidase A is of particular interest, since its esterase activity is enhanced, but its peptidase activity inhibited when its bound Zn^{2+} is replaced by Cd^{2+} (Coleman and Vallee 1961, Auld and Holmquist 1974). It is possible that cadmium substitution induces changes in the coordination which alters the tertiary structure not only in the immediate vicinity of the metal binding sites but also in the overall conformation of this enzyme (Yamashita et al. 1990). Whether metallothionein or its mutants can reverse these effects is unknown, and was examined in this study.

Carboxypeptidase activity was measured under conditions with which the ester and peptide hydrolytic activities are differentially affected by cadmium. The model compounds, hippuryl-dl-B-phenyllactate and carbobenzoxyglycyl-L-phenyalanine, was used respectively for the assay of esterase and peptidase activities. A range of enzyme-substrate ratios was tested in the presence or absence of varying concentrations of cadmium chloride. The activities were recorded spectrophotometrically as previously described (Auld and Holmquist 1974, Coleman and Vallee 1962), and the reaction kinetics analyzed.

Zinc was replaced by cadmium in carboxypeptidase A by equilibrium dialysis against buffers containing cadmium chloride, or indirectly by first subjecting the Zn-enzyme to dialysis by 1,10 phenanthroline(in metal-free dialysis tubing) which chelates the zinc, and then adding cadmium to the apoenzyme. It was clear that cadmium inhibited peptidase activity, and the effect is reversible in the presence of Zn-metallothionein (in preparation).

Although we were able to observe the restoration of peptidase activity of the apoenzyme, whose zinc had been previously removed by dialysis against 1,10 phenanthroline, by Zn-metallothionein, the kinetics and extent of metal exchange have yet to be determined. This can be done using radiolabels, ^{109}Cd or ^{65}Zn, or by atomic absorption spectrometry to trace the separation of Zn-Carboxypeptidase A and metallothionein by electrophoresis on non-denaturing gel. Carboxypepetidase and metallothionein are also readily fractionated on Sephadex G75 column due to their distinct molecular weights (35,000 vs 6800), hence the exchange can be followed by monitoring the switch over of metals in the respective fractions. Once established, this assay will be used on different amino acid substitution mutants. We hope to be able to correlate the nature and site of the substitution with the metal exchange function.

This study may well be extended to include examinations of the interaction of metallothionein with other zinc-binding factors which recognize the trans-activating metal responsive elements (MRE) which are involved in regulating metallothionein gene expression. One such element has recently been identified the upstream regulatory sequences of one of the CHO metallothionein isogenes (Hung et al. 1991).

In summary, in our continued effort to understand the function of metallothionein, we have examined the contribution of conserved amino acids to the stability and metal sequestering activity of mammalian metallothioneins, and possible differential capacity of each of the two metallothionein domains to serve as a metal scavenger/donor.

Our working hypothesis was based on the observation that mammalian metallothionein consists of two homologous but distinguishable domains within which certain amino acids are invariable in nature. We envision that these amino acids are maintained for efficient metal binding or exchange, and that each domain fulfills a different but related function as a metal donor/receptor entity. This would be demonstrated if the function of metallothionein could tolerate only limited change in some of these highly conserved amino acids.

A set of site-specific metallothionein mutants substituted for each of the 20 cysteines and 8 lysines have been used in this study. Their metal detoxification capacity and stability in vivo (in yeast and mammalian cells) was studied with a yeast expression system and with transformed mammalian cells. The role of these highly conserved amino acids in metallothionein structure and function apparently vary with position, whether in the alpha or beta domain. We have selected among these mutants stable variants that showed altered function in metal detoxification or in metal exchange for physico-chemical characterization and correlate the change with metal binding capacity, affinity, domain accessibility, domain ellipticity and metal-thiol cluster arrangement.

We have also examined Cd/Zn exchange reactions involving metallothionein and other metalloproteins (e.g. actin) and metalloenzymes (e.g. carboxypeptidase A). While the results provide a mechanistic understanding of cadmium toxicity, the ability for metallothionein to rescue cells from metal intoxication offers a new dimension to its pleiotropic function.

Acknowledgment - Supported in part by grants from NSC 820211-B007021 and NIH GM RO1 32606. Contributions by Chris Cody, Fuying Hou, Matt Johnson, An Ying, Dianne Lin, Steve Lesko, Karen Hickey, Peter Pan, and William Chang to various phases of the work in preparation are greatly appreciated.

Literature

Andersen, R.A., Frazier. J. and Huang, P.C. (1986) Transition metal binding proteins from three Chesapeake Bay fish species. Env. Health Perspectives 65: 149-156.

Armitage, I. and Boulanger, Y. (1983) Cadmium-113 NMR. in NMR of Newly Accessible Nuclei. (Laszlo, P. ed.) Academic Press, New York.

Arseniev, A., Schultze, P., Worgotter, E., Braun, W., Wagner, G., Vasak, M., Kagi, J.H.R. and Wuthrich, K. (1988) Three dimensional structure of rabbit liver [Cd7] metallothionein-2a in aqueous solution determined by nuclear magnetic resonance. J. Mol. Biol. 201: 637-657.

Auld, D. S. and Holmquist, B. (1974) Carboxypeptidase A. Differences in the mechanisms of ester and peptide hydrolysis. Biochemistry 13: 4355-4361.

Bernhard, W.R., Vasak, M. and Kagi, J.H.R. (1986) Cadmium binding and metal cluster formation in metallothionein: a differential modification study. Biochemistry 25:1975-1980.

Berg, J. (1990) Zinc finger domains - hypothesis and current knowledge. Annu. Biophys. Biophys. Chem. 19: 405-421.

Chernaik, M.L. and Huang, P.C. (1991) Differential effect of cysteine-to-serine substitutions in metallothionein on cadmium resistance. Proc. Natl. Acad. Sci., USA 88: 3024-3028.

Cismowski, M.J. and Huang, P.C. (1991a) Purification of mammalian metallothionein from recombinant systems. Methods in Enzymology 205: 312 319.

Cismowski, M.J. and Huang, P.C. (1991b) Effect of cysteine replacements at positions 13 and 50 on metallothionein structure. Biochemistry 30: 6626-6632.

Cismowski,M.J., Narula, S.S., Armitage,I., Chernaik, M. and Huang, P.C. (1992) Mutation of invariant cysteines of Chinese hamster metallothionein alters metal binding capacity, cadmium resistance and ^{113}Cd NMR spectrum. J. Biol. Chem. 266: 24390-24397.

Coleman. J. E. and Vallee, B.(1961) Metallocarboxy-peptidase: stability constants and enzymatic characteristics. J. Biol. Chem. 236: 2244-2249.

Coleman, J. E. and Vallee, B.(1962) Apocarboxy-peptidase-substrate complexes. J.Biol.Chem. 237: 3430-3436.

Corrigan, A.C. and Huang, P.C. (1981) Cellular uptake of Cd and Zn. Biol. Trace Element Res. 3: 197-216.

Corrigan, A.C. and Huang, P.C. (1982) Cd and Zn flux in wild type and Cd resistant CHO cells. Biol. Trace Element Res. 5: 25-53.

Durnam, D.M., and Palmiter, R.D. (1987) Analysis of detoxification of heavy metal ions by mouse metallothionein. Experientia Suppl. 52: 457-463.

Ecker, D.J., Butt, T.R., Sternberg, E.J., Neeper, M.P., Debouck, C., Gorman, J.A. and Crooke, S.T. (1986) Yeast metallothionein function in metal ion detoxification. J. Biol. Chem. 261: 16895-16900.

Flik, G., van de Winkel J. G.J., Part, P., Bonga, S.E.W. and Lock, R.A.C. (1987) calmodulin-mediated cadmium inhibition of phosphodiesterase activity, in vitro. Arch. Toxicology 59: 353-359.

Frey, M.H., Wagner, G., Vasak, M., Sorensen, O.W., Neuhaus, D., Worgotter, E., Kagi, J.H.R., Ernst, R.R. and Wuthrich, K. (1985) Polypeptide-metal cluster connectivities in metallothionein 2 by novel ^{1}H-^{113}Cd heteronuclear two-dimensional NMR experiments. J. Am. Chem. Soc. 107: 6847-6851.

Griffith, B.B., Walter, R.A., Enger, M.D., Hildebrand, C.E. and Griffith, J.K. (1983) cDNA cloning and nucleotide sequence comparison of Chinese hamster metallothionein I and II mRNAs. Nucl. Acids Res. 11: 901-910.

Hamer, D. H. (1986) Metallothionein. Annu.Rev.Biochem. 55: 913-951.

Hildago, J. Rhee, S.J., Huang, P.C. and Garvey, J.S. (1992) Physiological role of glucocorticoids on the regulation of rat liver metallothionein synthesis. Hormone Metabolic Res. 24: 233-236.

Huang, P.C. and Lin, L. Y. (1989) Metallothionein as a stress protein. in Biological Active Protein and Peptides (Chiou et al. eds.) Acad. Sinica Press, pp. 69-80.

Huang, P.C. and Cody, C. (1991) Enhanced transcription and amplification of eukaryotic metallothionein genes as an adaptive and defense mechanism in response to cadmium. in Adaptive Response and Defense Mechanisms (Seu, J. ed.), Taegu, Korea pp. 53-70.

Huang, P.C., Smith, B., Bohdan, P. and Corrigan, A. J. (1980) Effect of zinc on cadmium influx and toxicity in cultured CHO cells. Biol. Trace Element Research 2: 211-220.

Huang, P.C., Morris,S., Dinman,J., Pine, R. and Smith, B. (1987) Role of metallothionein in detoxification and tolerance to transition metals. Experientia Suppl. 52: 439-446.

Huang, P.C., Cody, C. Cismowski, M. Chernaik, M. Rhee, I.K. and Lin, L.Y. (1991) Native and engineered metallo-thioneins. in Metallothionein in Medicine and Biology (C.Klaassen and K. Suzuki eds.) CRC Press, FL pp.87-101.

Hung, S., Yu, C., Chou, T., Huang, P. C. and Lin, L.Y. (1991) Molecular cloning of Chinese hamster metallothionein II gene and its 5'-flanking region. Biochim. Biophys. Acta. 1090: 255-258.

Jamall, I.S. and Smith, J.C. (1985) Effects of cadmium on glutathione peroxidase, superoxide dismutase, and lipid peroxidation in the rat heart: a possible mechanism of cadmium cardiotoxicity. Tox.Appl.Pharm. 80: 33-42.

Kagi, J.H.R. and Schaffer, A. (1988) Biochemistry of metallothionein. Biochemistry 27: 8508-8515.

Karin, M. (1985) Metallothioneins, proteins in search of function. Cell 41: 9-10.

Kouyama, T. and Mihashi, K. (1981) Fluorimetry study of N-(1-pyrenyl) iodoacetamide-labelled F-actin. Eur. J. Biochem. 114: 33-38.

Li, T-Y., Kraker, A.J., Shaw, C.F. and Petering, D.H. (1980) Ligand substitution reactions of metallothioneins with EDTA and apo-carbonic anhydrase. Proc. Natl. Acad. Sci., USA 77: 6334-6338.

Li, T.Y., Minkel, D.T., Shaw, III, C.F. and Petering, D.H. (1981) On the reactivity of metallothioneins with 5,5'-dithiobis-(2-nitrobenzoic acid). Biochem. J. 193: 441-446.

Lin, L.Y., Lin, W.C. and Huang, P.C. (1989) Pigeon metallothionein consists of two isospecies. Biochim. Biophys. Acta 1037: 248-255.

Lin, L.Y. & Huang, P.C. (1990a) Complete homology in metallothionein from two genera of ducks and their hybrids. Biochem. Biophys. Res. Comm. 168: 182-187.

Lin, L.Y., Chen, I.P., Tseng, C.S. and Huang, P.C. (1990b) Maternal transmission of mitochondrial DNA in ducks. Biochem. Biophys. Res. Comm. 168: 188-193.

Lin, L.Y., Liu, L.F., Tam, M.F., Huang, P.C., Vestling, M. and Fenselau, C. (1990c) Primary sequence of duck metallothionein. Biochim. Biophys. Acta 1041: 31-35.

Messerle, B.A., Schaffer, A., Vasak, M., Kagi, J.H.R. and Wuthrich, K. (1990) Three-dimensional structure of human [^{113}Cd7] metallothionein-2 in solution determined by nuclear magnetic resonance spectroscopy. J. Mol. Biol. 214: 765-779.

Messerle, B.A., Schaffer, A., Vasak, M., Kagi, J.H.R. and Wurthrich, K. (1992) Comparison of the solution conformations of human [Zn7]-metallthionein-2 and [Cd7]-metallthionein-2 using nuclear magnetic resonance spectroscopy. J. Mol. Biol. 225: 433-443.

Mills, J.W. and Ferm, V.H. (1989) Effect of cadmium of F-actin and microtubules of madin-Darby canine kidney cells. Tox. Appl. Pharm. 101: 245-254.

Morris, S. and Huang, P.C. (1987) Transient response of amplified metallothionein genes in CHO cells to induction by alpha-interferon. Mol. Cell Biol. 7: 600-605.

Morris, S. and Huang, P.C. (1989) Intracellular metallothionein concentration affects the rate of zinc or cadmium influx and metallothionein mRNA concentration. Expt. Cell Res. 185: 166-175.

Nettesheim, D.G., Engeseth, J.R. and Otvos, J.D. (1985) Products of metal exchange reactions of metallothionein. Biochemistry 24: 6744-6751.

Nielson, K.B. and Winge, D.R. (1984) Preferential binding of copper to the beta domain of metallothionein. J. Biol. Chem. 259: 4941-4946.

Nowak, E., Strzelecka-Golaszewska, H. and Goody, R.S. (1988) Kinetics of nucleotide and metal ion interaction with G-actin. Biochemistry 27: 1785-1792.

Otvos, J.D. and Armitage I.M. (1980) Structure of the metal clusters in rabbit liver metallothionein. Proc. Natl. Acad. Sci., USA 77: 7094-7098.

Otvos, J.D., Engeseth, H.R., Nettesheim, D.G. and Hilt, C.R. (1987) Inter-protein metal exchange reactions of metallothionein. Experientia Suppl. 52: 171-178

Pine, R. and Huang, P.C. (1987) An improved method to obtain a large number of mutants in a defined region of DNA. Methods in Enzymology 154: 415-430.

Pine, R., Cismowski, M., Liu, S. and Huang, P.C. (1985) Construction and characterization of a library of metallothionein coding sequence mutants. DNA 4: 115-126.

Palmiter, R., Findley, S.D., Whitmeore, T.E. and Durnam, D.M. (1992) MT-III, a brain-specific member of the metallo-thionein gene family. Proc. Natl. Acad. Sci.,USA 89: 6333-6337.

Pollard, T.D. (1990) Actin. Current Opinion in Cell Biology 2: 33-40.

Prozialeck, W.C. and Niewenhuis, R.J. (1991) Cadmium (Cd2+) disrupts intercellular junctions and actin filaments in LLC-PK1 cells. Toxic. Appl. Pharm. 107: 81-97.

Rhee, I.K., Lee, K.S. and Huang, P.C. (1990) Metallothionein with interdomain hinges expanded by insertion mutagenesis. Protein Engineering 3: 205-213.

Rhee, S. J. and Huang, P.C. (1989) Metallothionein accumulation in CHO cdr cells in response to lead treatment. Chemical-Biological Interactions 72: 347-361.

Robbins, A.H., McPhee, D.E., Williamson, M., Collett, S.A., Xuong, N. H., Furey, W.F., Wang, B.C. and Stout, C.D. (1991) Refined crystal structure of Cd, Zn metallothionein at 1.8A resolution. J. Mol. Biol. 221: 1269-1293

Schultze, P., Worgotter, E., Braun, W., Wagner, G., Vasak, M., Kagi, J.H.R. and Wuthrich, K. (1988) Conformation of [Cd_7]-metallothionein-2 from rat liver in aqueous solution determined by nuclear magnetic resonance spectroscopy. J. Mol. Biol. 203: 251-268.

Shaikh,Z., Harnett,K.M., Perlin, S.A. and Huang, P.C. (1989) Chronic cadmium intake results in dose-related excretion of metallothionein in urine. Experientia 45: 146-148.

Spudich, J.A. and Watt, S. (1971) The regulation of rabbit skeletal muscle contraction. J. Biol. Chem. 246: 4866-4871.

Stillman, M.J. and Zelazowski, A.J. (1989) Domain-specificity of Cd^{2+} and Zn^{2+} binding to rabbit liver metallothionein 2. Biochem. J. 262: 181-188.

Thiele, D.J., Walling, M.J. and Hamer, D.H. (1986) Mammalian metallothionein is functional in yeast. Science 231: 854-856.

Thomas, D.J., Winchurch, R.A. and Huang, P.C. (1987) Ontogenic variations in acute lethality of Cd in C57Bl/6J mice. Toxicology 47: 317-323.

Thomas, D.J., Morris, S. and Huang, P.C. (1988a) Age dependence of metallothionein inducibility in mouse liver. in Trace Elements in Man and Animals (Hurley, L. et al. eds) pp. 293-295. Alan R. Liss, Inc. NY.

Thomas, D.J., Morris, S. and Huang, P.C. (1988b) Age-dependent variation for inducibility of metallothionein genes in mouse liver by cadmium. Developmental Genet. 9: 13-22.

Udom, A.O. and Brady, F.O. (1980) Reactivation in vitro of zinc-requiring apo-enzymes by rat liver zinc-thionein. Biochem. J. 187: 329-335.

Vasak, M., Worgotter, E., Wagner, G., Kagi, J.H.R. and Wuthrich, K. (1987) Metal coordination in rat liver metallothionein-2 prepared with or without reconstitution of the metal clusters, and comparison with rabit liver metallothionein-2. J. Mol. Biol. 196: 711-719.

Vallee, B.L. and Auld, D.S. (1990) Zinc coordination, function, and structure of zinc enzymes and other proteins. Biochemistry 29:5647-5659.

Winge, D.R. and Miklossy, K-A. (1982) Domain nature of metallothionein. J. Biol. Chem. 257: 3471-3476.

Yamashita, M.M., Wesson, L., Eisenman, G. and Eisenberg, D. (1990) Where metal ions bind in proteins. Proc. Natl. Acad. Sci. ,USA 87: 5648-5652.

Zeng, J., Vallee, B.L. and Kagi J.H.R. (1991a) Zinc transfer from transcription factor IIIA fingers to thionein clusters. Proc. Natl. Acad. Sci., USA 88: 9984-9988.

Zeng, J., Heuchel,M.R., Schaffner, W. and Kagi, J.H.R. (1991b) Thionein (apometallothionein) can modulate DNA binding and transcription activation by zinc finger containing factor Sp1. FEBS Lett. 279: 310-312.

Zimmerle, C.T. and Frieden, C. (1988) pH-induced changes in G-actin conformation and metal affinity. Biochemistry 27: 7759-7765.

Metallothionein III
K. T. Suzuki, N. Imura and M. Kimura (eds)

THE RELATIONSHIP BETWEEN NUCLEAR LOCALIZATION OF METALLOTHIONEIN AND PROLIFERATION OF RAT HEPATOCYTES

K. Tsujikawa[1], K. Sagawa[1], N. Suzuki[1], T. Shimaoka[1], Y. Kohama[1], N. Otaki[2], M. Kimura[3] and T. Mimura[1]

[1]Faculty of Pharmaceutical Sciences, Osaka University, 1-6 Yamadaoka, Suita 565, Osaka, Japan, [2]National Institute of Industrial Health, 6-21-1 Nagao, Tama, Kawasaki, Kanagawa 211, Japan, and [3]Department of Molecular Biology, Keio University School of Medicine, 35 Shinanomachi, Shinjuku-ku, Tokyo 160, Japan.

SUMMARY; To clarify the functions of metallothionein (MT) in cell proliferation, its induction and the relationship between its subcellular localization and the cell cycle were examined in rat liver. MT was induced significantly in the livers of rats 24 h after CCl_4-administration and 70% partial hepatectomy. Moreover, MT induction, as well as DNA synthesis, was initiated in primary cultured adult rat hepatocytes, by the hepatocyte growth factors, EGF and insulin.

Flow cytometric analysis revealed that MT was translocated into the nuclei from the cytoplasm of hepatocytes during S to G_2M phases of the cell cycle, following stimulation of the above growth factors. The translocation of MT into nuclei as well as DNA synthesis was completely suppressed by a tyrosine kinase inhibitor in *in vitro* experiments.

These results suggested that MT was a tyrosine kinase- and cell cycle-dependent, nuclear protein. MT might function in the signal transduction of growth factors having a tyrosine kinase type receptor.

INTRODUCTION

Reported in 1956, Metallothionein (MT) was first isolated from equine kidney cortex as a low-molecular-weight, Cd, Zn, and Cu-binding, cysteine-rich, cytoplasmic protein by Margoshes, Vallee, and Kagi (Margoshes and Vallee, 1957 ; Kagi and Vallee, 1960). Since then, MT has been found throughout the animal kingdom as well as in plants, eukaryotic microorganisms, and some prokaryotes with highly conservative structures (Kagi and Kojima, 1987). Therefore, MT has been supposed to play a crucial role in some fundamental life processes. At present, biological functions of MT are known to be detoxification of cadmium and other heavy metals (Bremner, 1987), regulation of zinc and copper metabolism (Bremner, 1991), scavenging of superoxide radicals (Thornalley and Vasak, 1985) and resistance to anticancer agents (Endresen et al., 1983). However, definitive physiological functions of MT are not yet clearly understood.

Recently, we found by immunohistochemical study that MT was present in the nuclei of primary cultured gastric mucosal cells of newborn rats (Tsujikawa et al., 1992). This result suggested that MT might function as a nuclear protein as well as a cytoplasmic protein. The nuclear localization of MT was first reported in rat liver and kidney by Banerjee et al. (Banerjee et al., 1982). Further studies have demonstrated that MT is also localized in the nuclei of both fetal and newborn rat hepatocytes (Andrews et al., 1987), that its intranuclear localization decreased with age, and that at about 2 weeks MT is localized mainly in the cytoplasm, similar to the adult rat liver pattern (Panemangalore et al., 1983 ; Templeton et al., 1985 ; Clarkson et al., 1985 ; Cherian et al., 1987 ; Nishimura et al., 1989a). Moreover, MT was also found to be localized in nuclei of human fetal and neonatal liver (Fuller et al., 1990) and kidney cells (Nartey et al., 1987), human testicular embryonal carcinoma (Kontozoglou et al., 1989), human thyroid tumors (Nartey et al., 1987), human prostatic hyperplasia (Suzuki et al., 1991) and rat lung cells (Hart et al., 1985). In recent years, Nishimura et al. reported that MT was localized predominantly in the nuclei of certain, actively proliferating cells of the genital organs in male and female rats (Nishimura et al., 1989b ; Nishimura et al., 1990). Such findings suggest a close association of MT and cell proliferation. However, the relationship between nuclear localization mechanisms of MT and cell proliferation remains unclear, possibly due in part to the focus of the previous research having been on the functions of cytoplasmic MT.

In this study, we examined the increase in the level of MT and the relationship between localization of MT in cellular nuclei and the cell cycle, in proliferating hepatocytes of rats for the following reasons : 1) Hepatocytes in normal adult liver are essentially quiescent (G_0 phase in the cell cycle), yet retain the capacity for extensive proliferation as demonstrated during liver injury and regeneration (Grisham, 1962 ;Brodsky and Uryvaeva, 1977). 2) Hepatocytes in the remnant liver after 70% hepatectomy and in hepatotoxin-injured liver are stimulated by growth factors *in vivo* (Nakamura et al., 1984). 3) Rat hepatocytes can be easily cultured *in vitro* (Seglen, 1976). 4) MT exists in the nuclei as well as in the cytoplasm of regenerating rat hepatocytes (Nishimura et al., 1989a). 5) The regulatory mechanism of liver regeneration is not yet clearly understood. The development of research concerning the regenerative mechanism of hepatocytes is necessary for treatment of fulminant hepatitis, and hepatic insufficiency occurring after hepatectomy.

MATERIALS AND METHODS

Treatment of animals : For studies on the regenerating livers, 7-week-old male Wistar rats underwent approximately 70% hepatectomy following Higgins and Anderson (1931). The remnant livers were excised at different times following the operation. A sham operation was performed by freeing the liver from its surrounding connective tissues. Hepatitis was induced by oral administration of CCl_4 (2 ml/kg) to 16-week-old male Wistar rats (Tanaka, 1982), allowing subsequent extraction of both regenerating and hepatitis-bearing liver cell samples.

Primary culture of rat hepatocytes : Parenchymal hepatocytes in male Wistar rats were isolated and cultured as monolayers following Nakamura et al. (1983). The cells were cultured in Williams' medium E supplemented with 5% calf serum. Nuclei were isolated from the hepatocytes following Nakamura et al. (1983), suspended in 1 ml of Williams' E medium supplemented with 5 % calf serum, and subsequently used for flow cytometric analysis (Tsujikawa et al., 1992).

Assay of MT concentration and DNA synthesis : EGF (20 ng/ml) and insulin (0.1 μM) were added to primary cultures 22 h after plating and cultured for further 12 or 24 h. MT levels were measured by RIA, and DNA synthesis was assayed by measuring the incorporation of ^{3}H-thymidine into DNA with or without hydroxyurea (10 mM) in 2 h cultures as described by Tsujikawa et al. (1991).

Radioimmunoassay : The liver and hepatocytes cultured in 10 cm-diameter dishes were homogenized (50%, w/v) and sonicated (1 ml/dish), respectively, in 10 mM Tris-HCl, pH 8.6, containing 5 mM 2-mercaptoethanol, and centrifuged at 36,000 g for 1 h at 4 °C. MT concentration in each supernatant was determined by radioimmunoassay (RIA) (Ikei et al., 1989).

Indirect immunofluorescence : The rat hepatocytes were stained with anti-MT antiserum (1;1000 dilution with 2% FCS-PBS) followed by fluorescein-conjugated goat anti-rabbit antibody (1:100 dilution with PBS) (Tsujikawa et al., 1991) and observed with an Olympus confocal laser biological microscope LSM-GB 200.

Fluorescent flow cytometry : Isolated hepatocyte nuclei were fixed and permeabilized by exposure to methanol at -20 °C for 2 min. The fixed nuclei were washed and incubated with anti-MT antiserum (1:1000 dilution) at 37 °C for 1 h followed by fluorescein isothiocyanate (FITC)-conjugated goat anti-rabbit antibody (1:1000 dilution) at 37 °C for 1 h. The washed nuclei were suspended in propidium iodide solution (PI, 20 μg/ml) and measured with an EPICS elite flow cytometer system (Coulter). A single-parameter histogram displays fluorescence intensity (in channel units) on the x axis and cell number on the y axis. Dual-parameter contour plots display fluorescence intensities (in channel units) on the x and y axes and cell number on the z axis.

Statistics : Student's *t*-test was employed to determine the statistical significance of the data obtained in this study.

RESULTS

MT induction in hepatocytes *in vivo* and *in vitro* : MT levels in the rat liver and serum were determined 24 h after partial hepatectomy and oral administration of CCl_4 by RIA. MT was induced significantly in the CCl_4-injured liver and remnant liver after partial hepatectomy (Table I). The administration of CCl_4

Table I. MT concentrations in livers and sera of CCl_4-administered and partially hepatectomized rats

Treatment	MT concentration (μg / ml)	
	Liver	Serum
1) Control	0.82 ± 0.28	N.D.
CCl_4-administration	13.54 ± 3.22 *	0.05 ± 0.02
2) Sham operation	33.00 ± 9.07	N.D.
Hepatectomy	68.17 ± 7.65 #	N.D.

Each value represents mean ± S. E. (n = 3). *, Significantly different from the control group at $p < 0.05$. #, Significantly different from the sham operation group at $p < 0.05$. N.D. : not detectable.

Table II. MT concentrations in primary cultured rat hepatocytes stimulated with EGF, insulin, and zinc chloride

Treatment	MT concentration (ng / dish)	
	12 h a)	24 h a)
Control	103 (100) b)	136 (100) b)
EGF + Ins	172 (166)	124 (91)
$ZnCl_2$	984 (951)	1776 (1303)
EGF + Ins + $ZnCl_2$	1254 (1212)	2426 (1779)

Hepatocytes were stimulated with EGF (20 ng/ml), insulin (Ins, 10^{-7} M), and $ZnCl_2$ (50 μM) 24 h after plating. Each value represents mean for triplicate dishes. a) Hepatocytes were stimulated at the indicated times. b) % of control.

also increased the MT level in the serum, such being dependent upon the degree of injury caused to the hepatocytes by CCl_4. Table II shows MT levels in primary cultured adult rat hepatocytes stimulated with EGF and insulin, and/or $ZnCl_2$. The MT level in hepatocytes stimulated with EGF and insulin was 1.7-fold higher than that in the control hepatocytes 12 h after stimulation. However, the MT level had returned to that of the control after a further 12 h period. Zn significantly increased MT levels in normal hepatocytes in a time-dependent manner. Arithmetical and multiple increases in MT levels were recognized 12 and 24 h, respectively, after the concomitant addition of EGF, insulin, and Zn.

<u>Subcellular localization of MT</u> : Immunohistochemical studies using anti-MT antiserum revealed that MT was localized in the cytoplasm of normal adult rat hepatocytes (Fig. 1A). However,

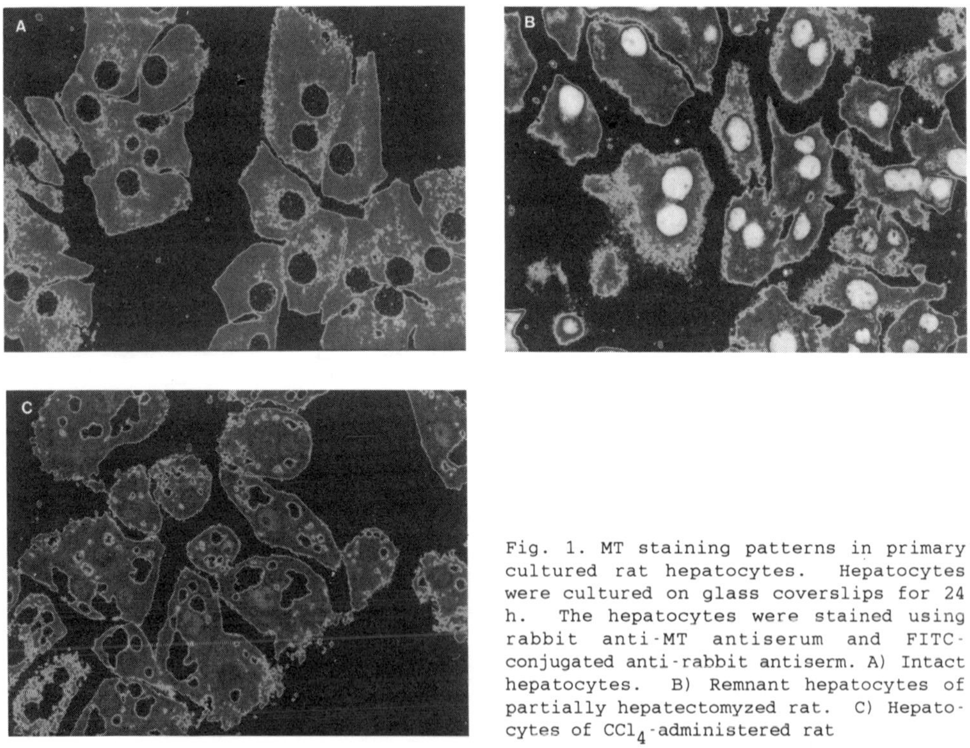

Fig. 1. MT staining patterns in primary cultured rat hepatocytes. Hepatocytes were cultured on glass coverslips for 24 h. The hepatocytes were stained using rabbit anti-MT antiserum and FITC-conjugated anti-rabbit antiserm. A) Intact hepatocytes. B) Remnant hepatocytes of partially hepatectomyzed rat. C) Hepatocytes of CCl_4-administered rat

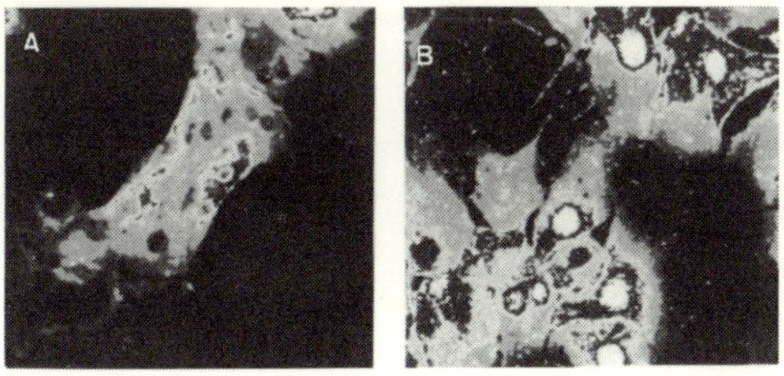

Fig. 2. MT staining patterns in primary cultured rat hepatocytes stimulated with EGF and insulin. A) Intact hepatocytes. B) EGF and insulin-stimulated hepatocytes.

Figs. 1B and 1C show that MT was localized in the nuclei as well as in the cytoplasm of regenerating and CCl_4-injured rat liver cells, respectively. MT staining pattern of primary cultured hepatocytes 12 h after stimulation with EGF and insulin is presented in Fig.2. Although non-treated hepatocytes showed strong immunostaining only in the cytoplasm (Fig. 2A), growth factor-stimulated hepatocytes were stained in the nucleus as well as in the cytoplasm (Fig. 2B).

<u>Flow cytometric analysis of MT in nuclei of hepatocytes</u> : The relationship between nuclear localization of MT and cell cycle was studied in rat liver remaining after 70% hepatectomy using a flow cytometer. In adult rat parenchymal hepatocytes, it has been documented that the nuclear population is composed of 2n C_0 and 4n-G_0 phases, and that new populations (2n-S and -G_2M, and 4n-S and -G_2M) appear in the regenerating hepatocytes. Fig. 3 shows the appearance of 4n-G_2M phase nuclei in the remaining hepatocytes 12 h after 70% hepatectomy, by PI staining (DNA level). This population was most notable 24 h after the operation with hepatocytes proliferating normally and DNA synthesis in the regenerating liver also peaking at that time. FITC staining (MT level) revealed that the level of MT in the nuclei peaked 24 h after the hepatectomy (Fig. 4). Moreover, MT-positive staining was recognized in both diploid and tetraploid nuclei in the S to G_2M phases, but not in those in the 2n-G_1 phase. In addition, there were two distinct populations of nuclei in the 4n and 2n-G_2M phases (Fig. 5).

Since MT is a low molecular weight protein (about 7 kDa), the nuclear transport of MT from the cytoplasm might occur by simple

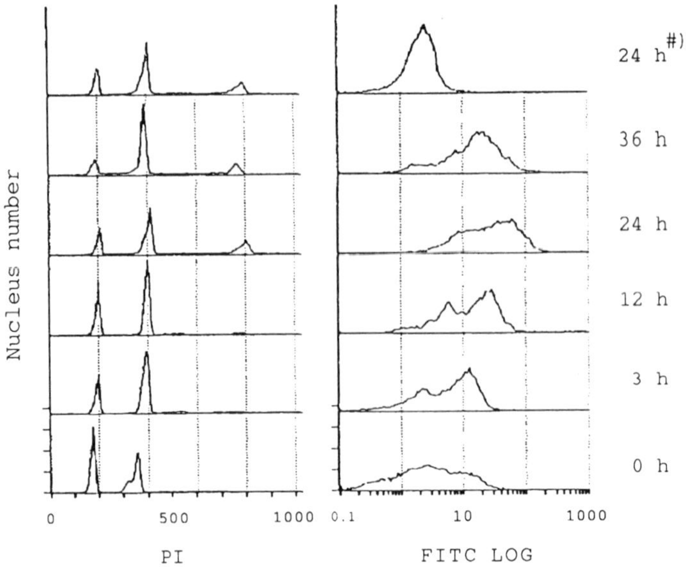

Fig. 3. Time course of fluorescence intensity in nuclei of hepatocytes remaining after partial hepatectomy. Nuclei were obtained from the parenchymal hepatocytes remaining after partial hepatectomy at the indicated times. #) The nuclei obtained from the parenchymal hepatocytes 24 h after hepatectomy were stained with normal rabbit serum. Based on 10000 nuclei.

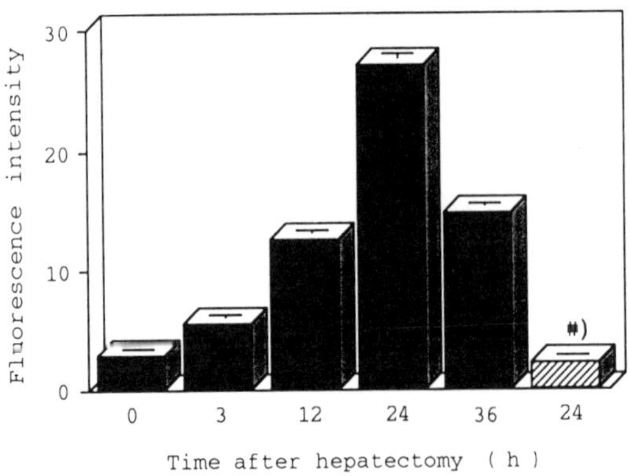

Fig. 4. Time course of fluorescence intensity of FITC in nuclei of hepatocytes remaining after partial hepatectomy. Nuclei were obtained from the parenchymal hepatocytes remaining after partial hepatectomy at the indicated times. #) The nuclei obtained from the parenchymal hepatocytes 24 h after hepatectomy were stained with normal rabbit serum. Fluorescence intensity indicates mean ± S.E. of 10000 nuclei.

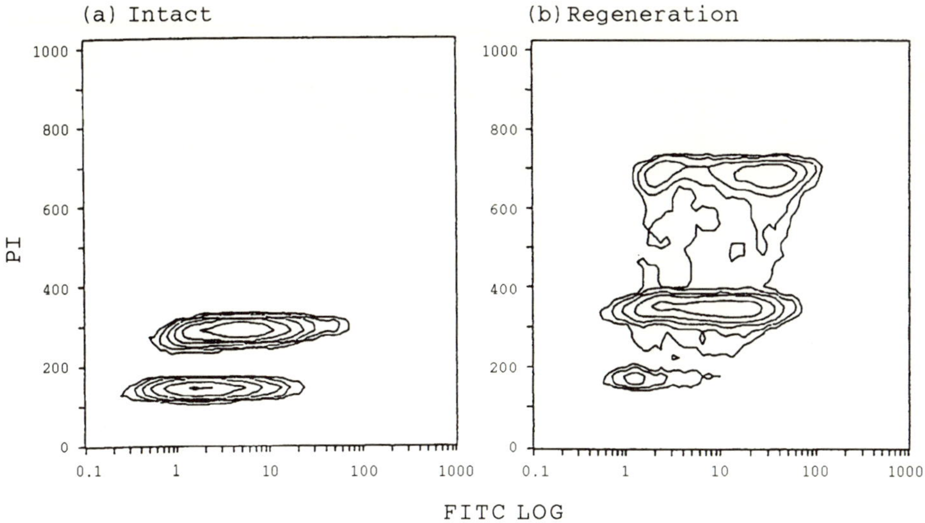

Fig. 5. Flow cytometric analysis of nuclei in hepatocytes of remnant rat liver. Each value represents mean of 10000 nuclei.

diffusion, causing an increased level of MT in the regenerating liver. Therefore, MT levels in rat hepatocytes were increased by the administration of zinc, with the degree of MT staining of the nuclei being determined by flow cytometric analysis. MT concentration had increased 50-fold in zinc-administered rat liver compared with normal rat liver. However, no marked changes of MT or DNA levels were recognized in the nuclei (Fig. 6). These results suggested that the nuclear localization of MT in hepatocytes was not dependent on simple diffusion.

Effect of a tyrosine kinase inhibitor on nuclear localization of MT : Flow cytometric analysis of FITC-staining revealed that MT levels in the nuclei of primary cultured hepatocytes had increased 12 h after stimulation by EGF and insulin. An increase in nuclear MT levels was inhibited completely by the concomitant addition of a tyrosine kinase inhibitor, genistein (Fig. 7). Genistein also

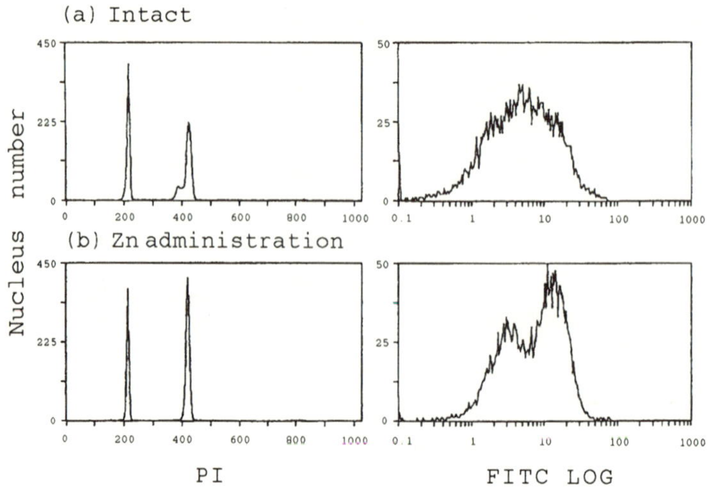

Fig. 6. Flow cytometric analysis of nuclei in hepatocytes of rat administered with zinc acetate. Zn (10 mg/kg) was subcutaneously administered daily, for 1 week. Based on 10000 nuclei.

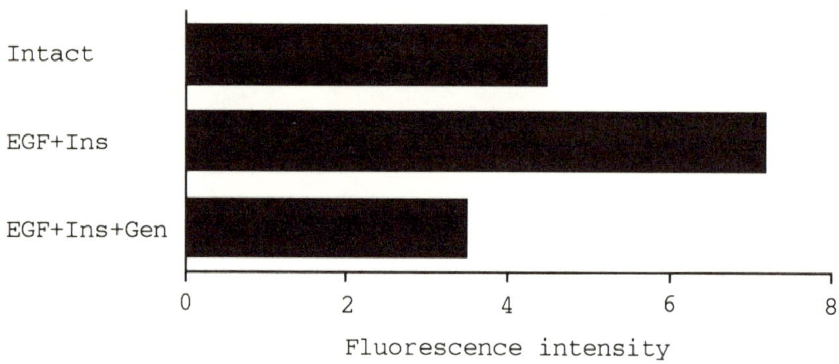

Fig. 7. Fluorescence intensity in nuclei of primary cultured rat hepatocytes treated with EGF, insulin, and genistein. Hepatocytes were treated with EGF (20 ng / ml), insulin (Ins, 10^{-7} M), and genistein (Gen, 20 μg / ml) 24 hr after plating. Each value represents mean for triplicate dishes.

inhibited DNA synthesis in the hepatocytes stimulated with EGF and insulin (Fig. 8).

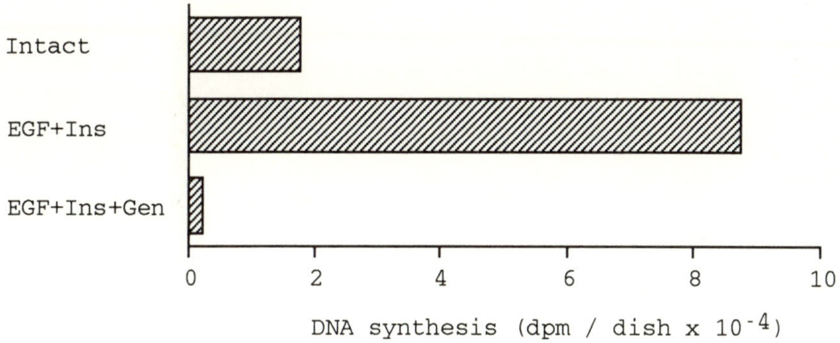

Fig. 8. DNA synthesis in primary cultured rat hepatocytes treated with EGF, insulin, and genistein. Hepatocytes were treated with EGF (20 ng / ml), insulin (Ins, 10^{-7} M), and genistein (Gen, 20 µg / ml) 24 hr after plating. Each value represents mean for triplicate dishes.

DISCUSSION

We considered the functions of MT in cell proliferation from the viewpoints of the induction, the subcellular localization, and the signal transduction.

1) Induction: The liver has a remarkable capacity to undergo compensatory growth (Van Thiel et al., 1987). In response to the tissue deficit caused by partial hepatectomy in rats, the remaining intact lobes grow rapidly and the normal liver mass is fully restored within about 10 days (Grisham, 1962). Recently, a growth factor that strongly stimulated DNA synthesis of adult rat hepatocytes in primary culture, was isolated from the sera of partially hepatectomized rats (Nakamura et al., 1984) and rat platelets (Russel et al., 1984). This hepatocyte growth factor (HGF) has also been found in the plasma of patients with fulminant hepatitis (Gohda et al., 1986) and with partial hepatectomy for the removal of a hepatoma (Selden et al., 1986). At present, HGF is thought to be a hepatotrophic factor that triggers liver regeneration after partial hepatectomy and liver injury, in both humans and rodents (Nakamura, 1991). HGF is synthesized by non-parenchymal liver cells of CCl_4-administered (Kinoshita et al.,

1989) and partially hepatectomized rats (Nakamura et al., 1984). MT is known to be greatly enhanced by a wide variety of agents and conditions, whereas it remains unclear whether or not HGF can induce MT synthesis in the liver (Kagi, 1991). This study showed that EGF and insulin, growth factors of hepatocytes in vitro, significantly increased MT levels in primary cultured hepatocytes (Francavilla et al., 1986). These growth factors were known to be MT inducers in a hepatoma cell line (Karin et al., 1987) and a mouse mammary carcinoma (Kobayashi et al., 1991). These results suggest a possible close association of MT synthesis with cellular proliferation.

2) Subcellular localization: Ohtake et al. (1978) found that Zn-MT was induced in regenerating rat livers and suggested its involvement in the regulation of normal hepatocyte growth, rather than neoplastic growth processes. However, Webb (1987) refuted this suggestion. Clearly, the involvement of MT in cell proliferation must be reconsidered. Following the discovery of the nuclear localization of MT in the liver and kidney cells by Banerjee et al. (1982), its nuclear localization has been reported in several kinds of tissue cells as well as in the hepatocytes. These papers said that MT in nuclei might closely associate with cell proliferation and differentiation, whereas there was no clear evidence.

Nuclear localization of MT was found in proliferating cells but not in quiescent ones. The premise that the initiation factor would be the growth factor for these cells, was demonstrated in this study. That is, MT was a cell cycle-dependent nuclear protein, and the initiation factor bringing about the translocation of MT comprised growth factors such as EGF and insulin. These results gave rise to a further consideration that MT might be involved in the process of transduction of the signal from these growth factors into the cell nuclei.

The nuclear pore can be considered to be freely permeable to proteins with a molecular weight of less than about 60 kDa (Bonner, 1978). Since the molecular weight of MT is about 7 kDa, it ought to diffuse very quickly through the nuclear pore, forming an equilbrium state between the cytoplasm and nucleus. However, MT is localized mainly in the cytoplasm of normal adult hepatocytes. This contradiction can not be explained at present. Two mechanisms have been suggested for nuclear localization of certain nuclear proteins. Resident proteins such as histone and SV40 large T antigen are translocated into nuclei soon after synthesis (Dingwall and Laskey, 1986). Non-resident proteins are translocated into nuclei by stimulation (Picard and Yamamoto, 1987). Several kinds of NLS (nuclear localization signal) have been identified for both types of proteins (Garcia-Bustos et al., 1991). These consist of a short peptide that is rich in positively-charged amino acids, such as lysine and arginine, and usually contain proline. Whilst NLS has never been discovered in MT molecules, it is significant that MT contains 8 lysine and 2 proline residues in its 61 amino acids (Kagi and Kojima, 1987), suggesting the possible existence of a new NLS. There are at least two possibilities that may explain the differential nuclear distribution of MT in S and G_2M phases of the cell cycle : 1) MT might be bound to a macromolecule (molecular weight > 60 kDa) in normal adult hepatocytes. The stimulation of growth factors such as EGF, insulin, and HGF could dissociate these molecules, resulting in the nuclear localization of MT. 2) MT might be localized in hepatocyte nuclei with a macromolecule(s) with which it is associated. In this case, two further possibilities exist : i) The macromolecule(s) may be a cell cycle-dependent nuclear protein. ii) An alternative mechanism of nuclear entry of MT might be the masking of NLS by the macromolecule(s). Demasking may be achieved by the growth factors

3) Signal transduction: HGF, EGF, and insulin, which have receptors with tyrosine kinase activity (Willingham and Pastan, 1982 ; Ullrich and Schlessinger, 1990 ;.Bottaro et al., 1991), initiate their signaling by interacting with each receptor, and finally cause DNA synthesis (Carpenter and Cohen, 1990 ; Nakamura, 1991). The signal transduction mechanisms of these factors are thought to be similar, although precise mechanisms remain unclear. Genistein, a tyrosine kinase inhibitor, is known to inhibit

tyrosine phosphorylation and DNA synthesis by HGF and EGF (Nakamura, 1991 ; Arakaki et al., 1992). It is highly significant that the nuclear localization of MT by EGF and insulin was inhibited by genistein, because MT might be a factor which transduces growth factor signals to nuclei.

Acknowledgements
We are indebted to Professor Nobuyoshi Shimizu, Keio University School of Medicine, for helpful suggestions during this study. We also thank Messrs. Naohiro Yasuda, Takumi Imai, Makoto Kakutani, and Misses. Kaori Suzuki and Mariko Hara for expert technical assistance.

REFERENCES

Andrews, G. K., Gallant, K. R., and Cherian, M. G. (1987) Eur. J. Biochem. 166, 527-531.

Arakaki, N., Hirono, S., Kawakami, S., Tsubouchi, H., Ishii, T., Hara, H., and Daikuhara, Y. (1992) Biochem. Biophy. Res. Commun. 185, 22-28.

Banerjee, D., Onosaka, S., and Cherian, M. G. (1982) Toxicology 24, 95-105.

Bonner, W. M. (1978) in Protein Migration and Accumulation in Nuclei, Busch, H. Ed, Academic press, New York, Vol. 6, pp 97-148.

Bottaro, D. P., Rubin, J. S., Faletto, D. L., Chan, A. M., Kmiecik, T. E., Vande, W. G., and Aaronson, S. A. (1991) Science 251, 802-804.

Bremner, I. (1987) Experientia Suppl. 52, 81-107.

Bremner, I. (1991) Methods Enzymol. 205, 25-35.

Brodsky, W., and Uryvaeva, I. V. (1977) Int. Rev. Cytol. 50, 275-332.

Carpenter, G., and Cohen, S. (1990) J. Biol. Chem. 265, 7709-7712.

Cherian, M. G., Templeton, D. M., Gallant, K. R., and Banerjee, D. (1987) Experientia Suppl. 52, 499-505.

Clarkson, J. P., Elmes, M. E., Jasani, B., and Webb, M. (1985) Histochem. J. 17, 345-352.

Dingwall, C., and Laskey, R. A. (1986) Ann. Rev. Cell. Biol. 2, 367-390.

Endresen, L., Bakka, A., and Rugstad, H. E. (1983) Cancer Res. 43, 2918-2926.

Francavilla, A., Ove, p., Polimeno, L., Sciascia, C., Coetzee, M. L., and Starzl, T. E. (1986) Cancer Res. 46, 1318-1323.

Fuller, C. E., Elmes, M. E., and Jasani, B. (1990) J. Pathol. 161, 167-172.

Garcia-Bustos, J., Heitman, J., and Hall, M. N. (1991) Biochim. Biophys. Acta 1071, 83-101.

Gohda, E., Tsubouchi, H., Nakamura, H., Hirono, S., Takahashi, K., Koura, M., Hashimoto, S., and Daikuhara, T. (1986) Exp. Cell. Res. 166, 139-150.

Grisham, J. W. (1962) Cancer Res. 22, 842-849.

Hart, B. A., Cherian, M. G., and Angel, A. (1985) Toxicology 37, 171-179.

Higgins, G. M., and Anderson, R. M. (1931) Arch. Pathol. 12, 186-202.

Ikei, H., Kodaira, T., Shimizu, F., Nakajima, K., Tohyama, C., Saitoh, K., Kimura, M., and Otaki, N. (1989) Rinsyokennsa 33, 215-219.

Kagi, J. H. R., and Vallee, B. L. (1960) J. Biol. Chem. 235, 3460-3465.

Kagi, J. H., and Kojima, Y. (1987) Experientia Suppl. 52, 25-61.

Kagi, J. H. (1991) Methods Enzymol. 205, 613-626.

Karin, M., Haslinger, A., Heguy, A., Dietlin, T., and Cooke, T. (1987) Mol. Cell. Biol. 7, 606-13.

Kinoshita, T., Tashiro, K., and Nakamura, T. (1989) Biochem. Biophys. Res. Commun. 165, 1229-1234.

Kobayashi, S., Suzuki, J. S., and Tohyama, C. (1991) in Metallothionein in Biology and Medicine, Klaassen, C.D., and Suzuki, K.T. Eds., CRC Press, Boston, pp 237-248.

Kontozoglou, T. E., Banerjee, D., and Cherian, M. G. (1989) Virchows Arch. A Pathol. Anat. Histopathol. 415, 545-549.

Margoshes, M., and Vallee, B. L. (1957) J. Am. Chem. Soc. 79, 4813-4814.

Nakamura, T., Tomita, Y., and Ichihara, A. (1983) J. Biochem. 94, 1029-1035.

Nakamura, T., Nawa, K., and Ichihara, A. (1984) Biochem. Biophy. Res. Commun. 122, 1450-1459.

Nakamura, T. (1991) Prog. Growth Factor Res. 3, 67-85.

Nartey, N. O., Banerjee, D., and Cherian, M. G. (1987a) Pathology 19, 233-238.

Nartey, N. O., Frei, J. V., and Cherian, M. G. (1987b) Lab. Invest. 57, 397-401.

Nishimura, H., Nishimura, N., and Tohyama, C. (1989a) J. Histochem. Cytochem. 37, 715-22.

Nishimura, N., Nishimura, H., and Tohyama, C. (1989b) J. Histochem. Cytochem. 37, 1601-1607.

Nishimura, H., Nishimura, N., and Tohyama, C. (1990) J. Histochem. Cytochem. 38, 927-933.

Ohtake, H., Hasegawa, K., and Koga, M. (1978) Biochem. J. 174, 999-1005.

Panemangalore, M., Banerjee, D., Onosaka, S., and Cherian, M. G. (1983) Dev. Biol. 97, 95-102.

Picard, D., and Yamamoto, K. (1987) EMBO J. 6, 3333-3340.

Russel, W. E., McGowan, J. A., and Bucher, N. L. R. (1984) J. Cell. Physiol. 119, 183-192.

Seglen, P. O. (1976) Methods Cell. Biol. 13, 29-83.

Selden, C., Johnstone, R., Darby, H., Gupta, C., and Hodgson, H. J. F. (1986) Biochem. Biophy. Res. Commun. 139, 316-366.

Suzuki, T., Umeyama, T., Ohma, C., Yamanaka, H., Suzuki, K., Nakajima, K., and Kimura, M. (1991) Prostate 19, 35-42.

Tanaka, K. (1982) Dev. Toxicol. Environ. Sci. 9, 237-49.

Templeton, D. M., Banerjee, D., and Cherian, M. G. (1985) Can. J. Biochem. Cell. Biol. 63, 16-22.

Thornalley, P. J., and Vasak, M. (1985) Biochem. Biophys. Acta 827, 36-44.

Tsujikawa, K., Imai, T., Kakutani, M., Kayamori, Y., Mimura, T., Otaki, N., Kimura, M., Fukuyama, R., and Shimizu, N. (1991) FEBS Lett. 283, 239-42.

Tsujikawa, K., Suzuki, N., Sagawa, K., Itoh, M., Sugiyama, T., Shimaoka, Y., Kohama, Y., Mimura, T., Otaki, N., and Kimura, M. (1992) submitted.

Ullrich, A., and Schlessinger, J. (1990) Cell 61, 203-212.

Van Thiel, D. H., Gavaler, J. S., Kam, I., Francavilla, A., Polimeno, L., Schade, R. R., Smith, J., Diven, W., Penkrot, R. J., and Starzl, T. E. (1987) Gastroenterology 7, 362-366.

Webb, M. (1987) Experientia Suppl. 52, 483-98.

Willingham, M. C., and Pastan, I. (1982) J. Cell. Biol. 94, 207-212.

Metallothionein III
K. T. Suzuki, N. Imura and M. Kimura (eds)

REGULATION OF METALLOTHIONEIN BIOSYNTHESIS IN GENITAL ORGANS OF MALE RATS

Chiharu Tohyama[1], Junko S. Suzuki[1,2], Shino T. Homma[1], Noriko Nishimura and Hisao Nishimura

[1]Environmental Health Sciences Division, National Institute for Environmental Studies, Tsukuba 305, Japan, [2]Department of Biology, Kyoritsu College of Pharmacy, Shiba-koen, Tokyo 105, Japan, and Department of Hygiene, Aichi Medical University, Nagakute-cho, Aichi 480-11, Japan

SUMMARY: It has been reported that testis and prostate lack metallothionein (MT), and for this reason these organs have been considered to be susceptible to toxicity and carcinogenicity by cadmium. In contrast to this notion, recent studies by immunohistochemical and biochemical analyses have revealed the presence of MT and its mRNA in the rat and mouse testes as well as other auxiliary genital organs. Since the detection and determination of MT and MT mRNA not only in these tissues has been carried out by the use of various methods in many laboratories, it is essential to solve the controversy raised by the data which support the presence or absence of MT in the male genital organs.

Metallothioneins (MTs) are classified as a family of stress protein and the expression MT genes are known to be regulated not only by heavy metals such as cadmium (Cd), zinc (Zn) and copper, but also by various endogenous factors such as cytokines and steroid hormones (see reviews, Hamor, 1986; Bremner, 1987).

Waalkes and associates have reported that the male genital organs such as testis and prostate lack MT (Waalkes et al., 1984a; 1984b, 1989), and speculated that the severe toxicity found in the testis after Cd treatment as well as injection site tumors in the prostate are caused partly due to the absence of MT (Waalkes and Perantoni, 1986; Waalkes et al., 1989), a role of which is known to lessen metal toxicity (see reviews by Shaikh, 1982). Although steroid hormones, glucocorticoid, estrogen and progesterone and vitamin D_3 were shown to be more or less involved in the regulation processes (Bremner, 1987; Karasawa et al., 1987), how MT gene expression is regulated in the testis and other male genital organs are not clear. Here, we review earlier papers which describe the presence or absence of MT in male genital organs of experimental animals and humans, and discuss the underlying problems and future perspectives in research. In this review, we try to make clearer the intention of research objectives based on experimental evidence at the time each research was performed.

The testis is extremely susceptible to Cd; for example, Cd is retained in this tissue much less (less than 0.2% of the dose) than in other tissues such as liver (45% of the dose) after a single injection (Singh et al., 1974), and spermatogenic cells are degenerated and necrotized even by such small amounts of Cd. The knowledge of toxic effects of Cd has been extended to the carcinogenic potential not only in the testis but also the prostate from experimental animals (Waalkes and Perantoni, 1989; Waalkes et al., 1992).

Since MT was shown to have a protective role against Cd insult, several investigators have attempted to determine whether MT is present in testes in rodents. Nordberg (1971) observed elevated amounts of Cd in MT fractions of testes from mice treated with repeated injections of Cd. Other studies also demonstrated Cd-binding protein, similar in relative molecular mass to MT, is present in testes of rodents by the use of ^{109}Cd as a tracer (Singh et al., 1974; Chen et al., 1974). Later, Brady and Webb (1981) estimated the amounts of MT by Zn content and reported that most of the MT isoforms in the testis is MT-2. Since MT consists of a large amount of cysteine residues but lacks leucine residues, they examined incorporation of radio-labeled cystine and leucine to the MT fraction. Their observation that a significant amount of ^{35}S-cystine but not of ^{3}H-leucine is incorporated into this fraction supports the notion that the protein is actually MT. Chellman et al. (1985) reported that when mouse strains susceptible (129/J) or resistant (A/J) to Cd-induced testicular necrosis were administered Cd so as to accumulate a similar Cd concentration in the testis of both strains, Cd is bound three times more to an MT-like protein fraction obtained from A/J testes compared to that from 129/J testes. No induction of MT was observed in the testis of both strains.

The amounts of MT or MT-like proteins were quantified by Cd-binding method (Onosaka et al., 1978) in various tissues including testes from rodents (Onosaka and Cherian, 1981; Waalkes and Klaassen, 1985). Even the testis from untreated rats was found to have relatively high level of MT compared to other tissues, and minimal induction was only found in the testis by Zn. In addition, MT mRNA level was also found highly expressed in the testis (Durnam and Palmiter, 1981). Interestingly, however, MT was hardly induced in the testicular cells. These observations have raised a significant question in terms of toxicological involvement of MT with a special reference to the susceptibility of the testis to Cd.

A possible drawback to the supposition that MT exists in the testis may be explained by the fact that the above analytical methods to detect MT cannot be specific to MT. For example, Cd-binding method was established by the use of liver tissue as standard but not by the testis or other tissues. There could be other factors which may affect the assay specificity. Quantification of MT by Zn amounts in the fractions, which are eluted from a gel filtration column, similar in relative molecular weight to MT may have limitations in terms of non-specificity and low resolution. Waalkes and Perantoni (1986) argued against the work of Brady and Webb (1981) that a relatively large pool of leucine as well as the low specific activity of this amino acid compared to ^{35}S-cystine may have resulted in a misleading conclusion.

Research groups of Waalkes (Waalkes et al., 1984a, 1984b; Waalkes and Perantoni, 1986) and of Whanger (Deagan and Whanger, 1985) purified and characterized Zn-binding proteins from cytosolic fractions of testes of rats. Although there was a small difference in the profile of gel filtration chromatography, they could find three Cd-binding protein peaks for the supernatant from the testes of rats administered with Cd *in vivo* (Deagan and Whanger, 1985) or for the supernatant treated with Cd *in vitro* prior to the isolation (Waalkes et al., 1984a). The protein fractions or groups of proteins were designated as testis metal-binding protein 1, 2, and 3 by order of elution. Both research groups used the third peak from the gel filtration column, the relative molecular mass of the protein is similar to MT. When the protein fraction was applied to anion-exchange column chromatography, two Cd-binding protein peaks were found to elute at the same conductance as rat liver MT-1 and 2. The significant difference between the results from the two groups was shown in the behavior of the testis Cd-binding protein on native gel electrophoresis. Whereas Deagan and Whanger (1985) found an apparent difference in a relative mobility between the testis proteins and the two isoforms of liver MTs, Waalkes et al. (1984a) found the same mobility for testis protein form 1 and hepatic MT-1, and for testis protein form 2 and hepatic MT-2. Deagan and Whanger (1985) described that the difference may be due to *in vivo* incorporation of Cd in their study and *in vitro* labeling with Cd in the study by Waalkes and coworkers, respectively. This speculation could be appropriate since metal replacement causes different order of metal incorporation into the protein moiety, which may change the tertiary structure of MT (Nielson and Winge, 1983). In contrast to the native gel electrophoresis, a significant difference between the testis proteins and rat liver MTs was observed for the behavior on SDS-polyacrylamide gel electrophoresis (Waalkes et al., 1984a), which suggests the apparent molecular weight could be different between them.

The most marked difference between the testis protein and liver MTs was found in amino acid composition as summarized in Table I. The major amino acids of the former were found to be glutamate and glycine with a low cysteine (less than 3%) content, whereas those of MT-1 and 2 are cysteine (approximately 30%), serine and lysine with a lack of aromatic amino acids and histidine. Similar observations on amino acid composion were obtained for the testis of monkeys and man (Waalkes et al., 1988a; Kaur et al., 1993), prostate of rats (Waalkes and Perantoni, 1989) and testes of rats treated with Cd and selenium (Ohta et al., 1988).

Waalkes and Perantoni (1986) adopted a new purification protocol in their later study and demonstrated the amino acid composition of two isoforms of the testis metal-binding protein (Table I). In brief, they utilized heat treatment and sequential acetone precipitation, immediately followed by a reverse phase column high-performance-liquid-chromatography using a trifluoroacetate and acetonitrile solvent gradient. Although they used rat hepatic MT-1 and -2 as markers for eluted fractions of the testis metal-binding protein from the column, a pitfall of this strategy would be that metal-binding capacity of the protein could not be retained for the fractions eluted from the column. Apparently, they presumed that the protein peaks that were eluted at similar but not identical retention times with MT-1 and -2 are the corresponding testis metal-binding protein isoforms. Amino acid analysis of the testis protein isoforms showed a marked variation in terms of numbers of glutamate, glycine, cysteine and lysine residues among their own studies (Waalkes and Perantoni, 1986; Waalkes et al., 1984a) and the one by Deagan and Whanger (1985), and Waalkes and Perantoni (1986) described possible explanations such as difference in purification procedures and diversity in strains. However, it is less likely that the difference in purification procedure affects amino acid analysis data unless the protein is degraded during isolation processes and a fragment of the protein is used for the analysis, or protein specimens are contaminated with other peptides.

It was reported that the testis metal-binding protein 3 was a oxidized and/or degraded product of the testis metal-binding protein 2 (Waalkes et al., 1984b). This interpretation was drawn from the experimental evidence that amino acid composition of the testis metal-binding protein 2 is strikingly similar to that of the protein 3, that amounts of testis protein 2 was decreased and those of the protein 3 was elevated after storage of the testis cytosol at -20° for 2 weeks and that this phenomenon was suppressed by the addition of dithiothreitol or N-ethylmaleimide to the cytosol prior to the storage. Since the relative molecular mass of the protein 2 is approximately 25,000 daltons, it is concluded that the protein 2 comprises of four monomeric forms of the former. This

Table I. Summary of amino acid composition analysis for rat testis zinc-binding protein

Amino acid	Testis[1] I	Testis[1] II	Testis[2] I	Testis[2] II	Testis[3] I	Testis[3] II	Liver[2] I	Liver[2] II
	Amino acid composition (mol/100 mol)							
Asp	6.8	9.4	10.5	11.3	10.6	10.9	6.8	6.8
Thr	5.5	4.6	5.2	5.4	7.0	11.3	6.7	2.8
Ser	8.0	5.6	9.0	11.0	9.2	7.2	15.2	18.5
Glu	10.4	11.5	10.2	11.2	22.8	24.0	2.1	4.6
Pro	4.3	6.8	6.5	3.2	5.5	5.3	3.2	3.9
Gly	11.9	11.2	16.0	16.7	3.2	3.6	9.1	6.3
Ala	8.6	6.2	8.6	9.4	5.5	7.9	5.1	7.6
Cys	7.2	5.4	2.9	2.6	1.5	1.3	34.5	32.0
Val	5.9	7.1	5.3	3.9	0.0	0.0	3.3	1.6
Met	2.5	1.4	1.9	0.0	1.7	1.7	1.4	1.7
Ile	4.6	2.6	4.3	2.2	4.5	6.4	0.2	1.4
Leu	6.7	9.0	5.2	3.3	4.7	6.4	0.3	0.4
Tyr	1.5	2.3	5.8	4.0	0.0	0.0	0.0	0.3
Phe	2.3	2.7	0.0	2.8	2.5	2.6	0.0	0.0
Lys	7.5	7.7	5.0	7.0	18.9	19.1	12.0	11.9
His	2.5	2.7	0.0	4.3	0.0	0.0	0.1	0.2
Arg	3.9	3.7	3.6	1.7	0.0	0.0	0.0	0.0

References for each data on amino acid analysis : 1. Waalkes et al., 1984a; 2. Deagan and Whanger, 1985; 3. Waalkes and Perantoni, 1986.

characteristics could be very unique to this testis metal-binding protein. How the testis metal-binding protein 3 behaves for the above new procedure remains to be characterized to clearly elucidate the relationship between the two protein forms.

The notion that MT exists at least in the testis are supported by experimental results, such as (1) immunohistochemical localization and immunochemical reactivity, (2) expression of MT mRNA, and (3) incorporation of specific amino acids.

The presence of MTs in the testes from Cd-treated rats was shown by immunohistochemical method (Danielson et al., 1982; Nolan and Shaikh, 1986): in these studies, MT was found mainly in Sertoli cells and Leydig cells but not in spermatogenic cells. Under physiological conditions, we have found that MT was localized mainly in spermatogenic cells, spermatozoa and Sertoli cells of adult rats, but not in the interstitial cells (Fig. 1; Nishimura et al., 1990). These results indicate that the spermatogenic cells, which constitutively express MT gene, are so susceptible to Cd toxicity that degenerative changes occur in these cells by Cd, the level of which can induce MT in less vulnerable cells like Leydig and Sertoli cells.

Interestingly, MT staining pattern was remarkably different from each other in terms of different sections of the seminiferous tubules (Fig. 1A). In the seminiferous tubules where spermatozoa are adhered to the adluminal space, MT was found mainly in the head of the spermatozoa and nearly no MT staining was found in the spermatogenic cells (Fig. 1B). On the other hand, in the seminiferous tubules where spermatozoa were already transferred to the epididymis for maturation, MT was found in the spermatogonia, spermatocytes and spermatids (Fig.1B).

The presence of MT in the prostate and other genital tissues has been extensively studied (Timms and Hagen, 1989; Umeyama et al., 1987; Nishimura et al., 1990). Umeyama and coworkers (1987) described that MT was localized in ventral, lateral, and dorsal lobes of the prostate as well as coagulating gland of the rat treated with Cd. They also described that the lateral lobe of saline-treated rats harbored a high level of MT. We clearly demonstrated that the staining intensity was higher in dorsolateral lobes than ventral lobe of untreated rats (Fig. 1C, 1D; Nishimura et al., 1990). The amounts of MT and Zn in these lobes appear to be correlated each other (Nishimura et al., 1990). Recently, we have investigated the expression of MT gene in the prostate, i.e., ventral and dorsolateral lobes, seminal vesicle, and coagulating gland of male Wistar rats (Tohyama et al., unpublished). MT mRNA and MT were analyzed by Northern blot

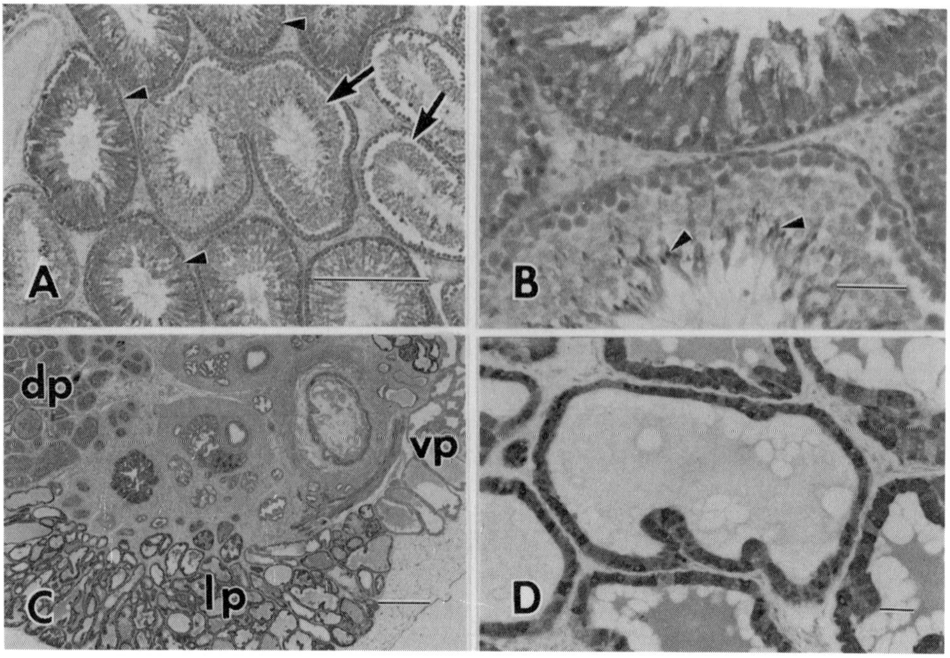

Fig. 1. Metallothionein localization in testis (A,B), and prostate (C,D). Tissues were obtained from adult male rats (8 weeks old). Seminiferous tubules of testis showed different degrees of intensity of MT immunostaining depending on the stage of spermatogenesis. (A) Spermatogenic cells and Sertoli cells showed strong MT immunostaining in seminiferous tubules which were in the process of spermatozoa production (arrowhead). In contrast, weak staining was observed in seminiferous tubules that contained spermatozoa (arrow). (B) Higher magnification of Figure 1A. MT was clearly shown in the head of spermatozoa (arrowhead). (C) Strong MT staining was found in lateral lobe (lp) and dorsal lobe (dp), but not in ventral lobe (vp). (D) Higher magnification of Figure 1C. MT was found in the cytoplasm and nucleus of the epithelial cells of the lateral lobe. Bars: (A) 10 μm, (B) 50 μm, (C) 50 μm and (D) 10 μm.

hybridization and radioimmunoassay, respectively, for quantification, and by *in situ* hybridization and avidin-biotin peroxidase complex method for histochemical localization. In intact rats, both MT mRNA and MT were present, in a cell-specific manner, in dorsolateral lobes, seminal vesicle and coagulating gland but not in ventral lobe. When rats were subjected to orchidectomy and maintained for subsequent 2 weeks, the plasma testosterone concentration was found 10% of the one observed in the intact rats, and severe atrophy of the organs as well as significant decrease in levels of both MT mRNA and MT was found. An injection of testosterone propionate to orchidectomized rats resulted in a rapid rise of plasma testosterone level and remarkably restored not only size of these organs, but also MT mRNA and MT, particularly in dorsolateral lobes and coagulating gland by day 3. In dorsolateral lobes, tissue Zn concentration did not precede the increase in MT, suggesting that MT gene expression is dependent on stimuli caused by testosterone. No change was observed in the liver tissue from intact, orchidectomized and testosterone-treated rats. How MT induction is regulated in the dorsolateral lobes remains to be studied since there are possibilities that testosterone may modulate MT gene expression through glucocorticoid responsive element present in the promoter of MT gene or through other signal transduction pathways. Nevertheless, taken together, the results obtained so far demonstrate that the male auxiliary genital organs, except ventral lobe, contain MT and that the biosynthesis of MT in the dorsolateral lobe is regulated by testosterone. It should also be pointed out that MT is hardly inducible by metal ions like Cd in the prostate as well as in the testis .

Immunochemical evaluation of the testis Zn-binding protein revealed that the protein fractions obtained from testis supernatant or eluates from gel filtration column chromatography have proteins that cross-react, in a completely competitive manner, with the sheep antiserum specific to rat MT-1 (Mehra and Bremner, 1987). Our data obtained by the use of rabbit antiserum specific to both MT-1 and 2 support this observation and showed only two distinct bands on immunoblot analysis following Western blotting (Tohyama et al., unpublished). These data indicate that the testis contains both MT-1 and 2, the observations of which are different from Brady and Webb (1981) who reported that the MT-2 is a predominant isoform of MT in the rat testis. Recently, Kaur et al. (1993) reported that they purified and isolated a Zn-binding protein whose amino acid composition is distinct from MT and similar to that of the Zn-binding testis protein from rodents and monkeys. The antiserum that was raised against this protein failed to cross-react with MT from human liver and kidney.

MT mRNA was expressed highly in a constitutive manner in the mouse testis (Durnam and Palmiter, 1981; Shiraishi et al., 1989). Recently, Andrews and coworkers (De et al., 1991) showed the presence of MT mRNA in the process of spermatogenesis in mice by using *in situ* hybridization and Northern blot analysis. Their finding was very similar to our finding on MT (Nishimura et al., 1990), in that MT mRNA was localized mainly in spermatocytes and spermatids. In isolated Sertoli cells very low level of MT was found. They also described only a basal level of expression of MT in interstitial, spermatogonial and mature sperm cells.

Since the testis protein reported by Waalkes and associates (Waalkes et al., 1984a; Waalkes and Perantoni, 1986), does not contain cysteine residues but does contain significant amounts of glutamate residues, Shaikh and Tewari (1990) evaluated the incorporation of these amino acids into MT fraction by the use of radio-labeled compounds. They found a marked incorporation of ^{35}S-cystine in the MT fraction (or the testis protein 3 fraction), which is similar to the observation of Brady and Webb (1981), and negligible incorporation of ^{3}H-glutamate into the fraction. Shaikh and Tewari (1990) argued the possibility of different pool size of amino acids, but claimed that the ^{3}H peak derived from glutamate should have been observed if the testis protein has high percentage of glutamate and negligible contents of cysteine residues even if a pool size of glutamate would be large enough to minimize the effect of incorporation of ^{3}H-glutamate.

It should be considered that DNA methylation controls MT-I gene expression in murine lymphoid cells (Compere and Palmiter,

1981) and that hypermethylation status of MT gene in the rat testis may be related to the uninducibility of the protein in this tissue (Bhave et al., 1988). If this is the case, it is not clear why amounts of MT (Nishimura et al., 1990) as well as MT mRNA (Compere and Palmiter, 1981; Shiraishi et al., 1989; De et al., 1991) are expressed highly in the testis under physiological conditions. If MT gene is expressed in the testis, it is quite conceivable that the transgenic mice carrying the interferon gene under the control of the MT promotor express high levels of interferon mRNA in the testes (Iwakura et al., 1988).

As described above, controversy whether MT exists in the testis and other male genital organs remains to be clarified. If the Zn- or Cd-binding protein indeed differs from MT, questions with aspects of not only analytical techniques but also biological significance will arise as described below.

(1) How specific are the Cd-binding or related methods? Since many workers have been utilizing these methods to estimate MT amounts in various tissues, it would be imperative to evaluate the specificity of the methods. Are there any other factors, besides coexisting metal ions such as Cu, that may affect Cd binding to the protein?

(2) How can one explain epitopic similarity between MT and the testis and prostate protein(s) if there are such distinctly different amino acid compositions between the testis protein and MT. Since several antisera from different laboratories have been utilized to determine or localize MT in the testis and other male genital organs, antigenic determinants should be fairly evaluated for the testis protein (Danielson et al., 1982; Nolan and Shaikh, 1986; Mehra and Bremner, 1987; Umeyama et al., 1987; Nishimura et al., 1990). The recent report by Kaur et al. (1993) describes that the antiserum produced against a Zn-binding protein from the human testis lacks cross-reactivity with MT, suggesting the possiblity that MT and the Zn-binding protein could be different entities and thus have different epitopic characteristics.

(3) If MT does exist, reasons for both vulnerability of the

testis to Cd insult and carcinogenicity of Cd in the testis and prostate should be searched for other mechanisms other than lack of detoxification by MT. Since the ventral prostate does express a minimal level of MT gene compared to the dorsolateral prostate (Umeyama et al., 1987; Nishimura et al., 1990), this lobe-specific MT gene expression could partly account for the occurrence of tumors in the ventral lobe of prostate (Waalkes et al., 1992).

We have found that the Zn-binding protein is extremely labile to oxidation and Zn ions are easily detached from the protein moiety during purification processes (Tohyama et al., unpublished), suggesting the possibility that the protein may be slightly different from the MT isoforms so far reported. The demonstration of the primary structure of the protein will afford critical insights into possible physiological functions of MT in male genital organs.

Data of amino acid composition on the Zn-binding protein purified from testis and prostate from rodents and primates show a remarkable distinction from MT (Deagan and Whanger, 1985; Kaur et al., 1992, Ohta et al., 1888; Waalkes and Perantoni, 1986; 1989; Waalkes et al., 1984a; 1988a; 1988b). Furthermore, the ovary of Syrian hamsters was reported to lack MT and instead to harbor a protein apparently similar to the testis Cd-binding protein (Waalkes et al., 1988c). In contrast, the presence of MT was found in female reproductive tissues including the ovary of the mice and rats (De et al., 1989; Hazelhoff-Roelfzema et al., 1989; Nishimura et al., 1989). It should also be pointed out that the data on amino acid composition show marked variation in terms of numbers of glutamate, glycine, cysteine and lysine residues (Table 1).

All the experimental evidence including the amino acid analysis so far obtained has essentially failed to settle the argument about the presence or absence of MT in male genital organs. Therefore, the clarification of amino acid sequence of the testis metal-binding protein merits for a further study and is currently under way in our laboratory.

Acknowledgements

S.T.H. was supported by a postgraduate fellowship from Science and Technology Agency. Authors wish to thank Professor Tsuguyoshi Suzuki at NIES for his encouragement throughout this study.

REFERENCES

Bhave, M. R., Wilson, M. J. and Waalkes, M. P. (1988) Toxicology, 50, 231-245.

Brady, F., and Webb, M. (1981) J. Biol. Chem. 256, 3931-3935.

Bremner, I. (1987) Experientia [Suppl] 52, 81-108.

Chellman, G. J., Shaikh, Z. A., Baggs, R. B. and Diamond, G. L. (1985) Toxicol. Appl. Pharmacol. 79, 511-523.

Chen, R. W., Whanger, P. D., Hoekstra, W. G. and Ganther, H. E. (1974) J. Reprod. Fertil. 38, 293-306.

Compere, S. J. and Palmiter, R. D. (1981) Cell, 25, 233-240.

Danielson, K. G., S. Ohi, and P. C. Huang. (1982) Proc. Natl. Acad. Sci. USA. 79, 2301-2304.

De, S. K., McMaster, M. T., Dey, S. K., and Andrews, G. K. (1989) Development 107, 611-621.

De, S. K., Enders, G. C., and Andrews, G. K. (1991) Mol. Endocrinol. 5, 628-636.

Deagan, J. T. and Whanger, P. D. (1985) Biochem. J. 231, 279-283.

Durnam, D. M. and Palmiter, R. D. (1981) J. Biol. Chem. 256, 5712-5716.

Hamer, D. (1986) Ann. Rev. Biochem. 55, 913-951.

Hazelhoff-Roelfzema, W., Tohyama, C., Nishimura, H., Nishimura, N. and Morselt, A. F. W. (1989) Histochemistry 90, 363-369.

Iwakura, Y., Asano, M., Nishimure, Y. and Kawada, Y. (1988) EMBO J. 7, 3757-3762.

Karasawa, M., Hosoi, J., Hashiba, H., Nose, K., Tohyama, C., Abe, E., Suda, T. and Kuroki, T. (1987) Proc. Natl. Acad. Sci. USA 84, 8810-8813.

Kaur, G., Nath, R. and Gupta, G. S. (1993) J. Trace Elements Exper. Med. 6, 1-13.

Mehra, R. K. and Bremner, I. (1987) Experientia [Suppl] 52, 565-572.

Nielson, K. B. and Winge, D. R. (1983) J. Biol. Chem. 258, 13063-13069.

Nishimura, N., H. Nishimura, and C. Tohyama. (1989) J. Histochem. Cytochem. 37, 1601-1607.

Nishimura, H., Nishimura, N., and Tohyama, C. (1990) J. Histochem. Cytochem. 38, 927-933.

Nolan, C. N. and Shaikh, Z. A. (1986) Toxicol. Appl. Pharmacol. 85, 135-144.

Nordberg, G. F. (1971) Environ. Physiol. 1, 171-181.

Ohta, H., Seki, Y., and Imamiya, S. (1988) Bull. Environ. Contam. Toxicol. 41, 195-200.

Onosaka, S., Tanaka, K., Doi, M., and Okahara, K. (1978) *Eisei Kagagku* 24, 128-131.
Onosaka, S. and Cherian, M.G. (1981) Toxicology 22, 91-101.
Shaikh, Z. A. (1982) *In* Biological Roles of Metallothionein. E. C. Foulkes (ed.), Elsevier North-Holland, New York. 69-98.
Shaikh, Z. A. and Tewari, P. C. (1990) Experientia 46, 694-696.
Shiraishi, N., Hayashi, H., Hirai, Y., Aono, K., Itano, Y., Kosaka, F, Noji, S. and Taniguchi, S. (1989) Toxicol. Appl. Pharmacol. 98, 501-506.
Singh, K., Nath, R. and Chakravarti, R. N. (1974) J. Reprod. Fertil. 36, 257-265.
Timms, B. G. and Hagen, J. A. (1989) Prostate 14, 367-382.
Umeyama, T., Saruki, K., Imai, K., Yamanaka, H., Suzuki, K., Ikei, N., Kodaira, T., Nakajima, K., Saitoh, H. and Kimura, M. (1987) Prostate 10, 257-264.
Waalkes, M. P. and Klaassen, C. D. (1985) Fund. Appl. Toxicol. 5, 473-477
Waalkes, M. P. and Perantoni, A. (1986) J. Biol. Chem. 261, 13097-13103.
Waalkes, M. P. and Perantoni, A. (1989) Toxicol. Appl. Pharmacol. 101, 83-94.
Waalkes, M. P., Chernoff, S. B. and Klaassen, C. D. (1984a) Biochem. J. 220, 811-818.
Waalkes, M. P., Chernoff, S. B. and Klaassen, C. D. (1984b) Biochem. J. 220, 819-824.
Waalkes, M. P., Perantoni, A. and Palmer, A. E. (1988a) Biochem. J. 256, 131-137.
Waalkes, M. P., Perantoni, A., Bhave, M.R. and Rehm, S. (1988b) Toxicol. Appl. Pharmacol. 93, 47-61.
Waalkes, M. P., Rehm, S., and Perantoni, A. (1988c) Biol. Reproduction 39, 953-961.
Waalkes, M. P., Perantoni, A. and Rehm, S. (1989) Biol. Trace Element Res. 21, 483-490.
Waalkes, M. P., Rehm, S., Perantoni, A. O. and Coogan, T. P. (1992) *In* Cadmium in the Human Environment. G. F. Nordberg, R. F. M. Herber and L. Alessio (eds.), IARC, Lyon. 391-400.
Webb, M. (1987) Experientia [Suppl] 52, 483-498.

Metallothionein III
K. T. Suzuki, N. Imura and M. Kimura (eds)

FACTORS INVOLVED IN THE TRANSCRIPTIONAL REGULATION OF METALLOTHIONEIN GENES

Shinji Koizumi[1] and Fuminori Otsuka[2]

[1]Department of Experimental Toxicology, National Institute of Industrial Health, 6-21-1, Nagao, Tama-ku, Kawasaki 214 and [2]Department of Environmental Toxicology, Faculty of Pharmaceutical Sciences, Teikyo University, Sagamiko, Kanagawa 199-01, Japan

SUMMARY: Synthesis of eukaryotic metallothioneins (MTs) is controlled mainly at the transcriptional level. Recent studies have shown that several *trans*-acting factors are involved in this regulation. In the yeast *Saccharomyces cerevisiae*, activation of the MT gene by copper(I) is mediated by the ACE1 protein, which has a copper-regulatory/DNA-binding domain that is structurally similar to the MT protein itself. By binding copper, the ACE1 protein acquires the ability to bind to the specific sequences upstream of the MT gene and activate transcription. In mammalian cells, the regulation appears to be more complicated. A number of known transcription factors, including Sp1, AP-1, AP-2, AP-4, glucocorticoid receptor and MLTF, activate transcription of the MT genes by binding to their specific upstream recognition sites. For the response to heavy metals, *cis*-elements called the metal-responsive elements (MREs) commonly found in the upstream regions of the MT genes are essential. Several MRE-binding factors with different properties have been reported so far, some of these binding to DNA only in the presence of heavy metals. How these factors function in the regulation of the MT genes is being investigated.

Metallothionein (MT) genes are transcriptionally controlled by a number of extracellular signals including heavy metals. The regulatory mechanisms have been studied mainly in the yeast and mammalian cell systems, multiple cellular *trans*-acting factors demonstrated or assumed to act via *cis*-DNA elements present upstream of the MT genes. Here we describe the factors that have been reported to interact with the regulatory sequences, particularly focusing on those responsive to heavy metals, and also the relevant factors under investigation in our laboratory.

1. Regulatory factors of yeast

The MT encoded by the *CUP1* gene of *Saccharomyces cerevisiae* (Fogel and Welch, 1982) is functionally similar to the mammalian MTs, being induced by heavy metals and suppressing the metal toxicity (Hamer *et al.*, 1985). However, MTs of both organisms are not homologous in their primary structures, also quite different in the spectrum of heavy metal response. The yeast MT is induced only by Cu and Ag, whereas the mammalian MTs are responsive to various metals including Zn, Cd, Cu, Hg and Ag (Butt *et al.*, 1984; Karin *et al.*, 1984b; Fürst *et al.*, 1988). The *cis*-element required for the heavy metal-dependent transcriptional activation of the *CUP1* gene is located in the region between -105 and -180 upstream from the transcription start site (Thiele and Hamer, 1986). This region is composed of two imperfect repeats, one of which designated as UASc (-148 to -105) is critical for the Cu-induction of the gene (Fürst *et al.*, 1988). There is no homology between the UASc and the mammalian metal responsive elements.

Cu-sensitive yeast mutants which lack the Cu-dependent expression of *CUP1* were isolated, and subsequently the genes complementing this genetic defect were cloned in two laboratories (*ACE1*: Thiele, 1988; *CUP2*: Welch *et al.*, 1989). These two genes were found to be identical (Buchman *et al.*, 1989), encoding a protein with 225 amino acids (Fürst *et al.*, 1988; Szczypka and Thiele, 1989). The yeast mutants defective in *ACE1/CUP2* lack the proteins binding to UASc, suggesting that its gene product is a transcriptional activator (Huibregtse *et al.*, 1989; Welch *et al.*, 1989). Indeed, the ACE1 proteins synthesized *in vitro* or in *E. coli* specifically bind to UASc in a Cu-dependent manner (Fürst *et al.*, 1988; Buchman *et al.*, 1989), and activate transcription from a downstream promoter *in vitro* (Culotta *et al.*, 1989). The ability of ACE1 to bind to DNA and activate transcription is dependent only on Cu(I) and Ag (I), not affected by the divalent heavy metals that induce mammalian MTs. ACE1 functions in the presence of general transcription factors including TFIID, and is suggested to be required for both the formation and maintenance of the transcriptional initiation complex (Culotta *et al.*, 1989; Kambadur *et al.*, 1990).

DNA sequencing analysis of the *ACE1* gene suggested that the ACE1 protein has two structural domains; the amino-terminal domain contains 12 cysteine residues and is rich in basic amino acids, whereas the carboxyl-terminal half lacks cysteine and is rich in acidic amino acids (Fürst *et al.*, 1988; Buchman *et al.*, 1989; Szczypka and Thiele, 1989). In fact, these are likely to be also the functional domains. The amino-terminal domain by itself can bind to the UASc in a Cu-dependent manner, suggested to be the DNA-binding/metal-regulatory domain. An analysis of a hybrid molecule between ACE1 and a heterologous yeast transcription factor indicated that the carboxyl-terminal half of the ACE1 protein acts as a transcriptional activation domain (Hu *et al.*, 1990)

The cysteine residues in the amino-terminal domain are arranged in the configuration Cys-X-Cys or Cys-X-X-Cys, which is also seen in the heavy metal-binding structure of the MT molecule , and the addition of Cu makes the ACE1 protein resistant to proteolysis. These findings suggested that a Cu-cysteine cluster analogous to that of the MT protein is formed, the resulting conformational change conferring the DNA-binding ability on the ACE1 molecule (Fürst *et al.*, 1988; Szczypka and Thiele, 1989). This hypothesis is supported by several spectroscopic analyses (Casas *et al.*, 1991; Dameron *et al.*, 1991; Nakagawa *et al.*, 1991). The mutation at any one of eleven cysteine residues within the DNA-binding domain affects the Cu-induced structural change, and the residues critical for Cu-binding and DNA-binding are interdigitated within the same domain (Hu *et al.*, 1990; Buchman *et al.*, 1990). The structural alteration of the ACE1 protein occurs as a result of highly cooperative binding of Cu ions. Based on this property, a large effect on transcription is expected by only a small change in the intracellular Cu concentration (Fürst and Hamer, 1989).

An *ACE1*-related gene of a pathogenic yeast, *Candida glabrata*, has recently been cloned. The product of this gene, AMT1, has a high homology with ACE1, particularly in the cysteine cluster in the amino-terminal region (Zhou and Thiele, 1991; Zhou *et al.*, 1992). The same group also cloned a gene coding for a protein, ACE2, that appears to have a role in the basal level transcription of the *CUP1* gene (Butler and Thiele, 1991).

2. Regulatory factors of mammalian cells

The transcriptional regulation of mammalian MT genes appears to be much more complicated than that of yeast. In the upstream regions of mammalian MT genes, there are a number of regulatory elements which serve as the recognition sites of specific transcriptional activators, including Sp1 (Lee *et al.*, 1987a; Andersen *et al.*, 1987; Mueller *et al.*, 1988), AP-1 (Lee *et al.*, 1987a; Lee *et al.*, 1987b; Angel *et al.*, 1987), AP-2 (Mitchell *et al*, 1987; Imagawa *et al.*, 1987), AP-4 (Mermod *et al.*, 1988), MLTF (Carthew *et al.*, 1987; Mueller *et al.*, 1988) and glucocorticoid receptor (Karin *et al.*, 1984a). These elements are also present in the upstream regions of other genes, but are not present in all of the MT genes.

In addition to these, the MT genes have common regulatory elements that are essential for the heavy metal-dependent transcriptional activation. These elements, the metal responsive (or regulatory) elements (MREs), were identified within the 5'-flanking regions by the analysis of deletion mutants (Stuart *et al.*, 1984; Carter *et al.*, 1984; Karin *et al.*, 1984a). The consensus sequence for the MRE is TGCRCNCGGCC (R=purine; N=any nucleotide), and usually multiple imperfect copies are present upstream of the MT genes. The TGCRCNC region (MRE core) is most conserved and particularly important in the heavy metal-dependent transcriptional activation (Stuart *et al.*, 1985; Culotta and Hamer, 1989). The synthetic MRE sequence by itself can confer metal responsiveness on heterologous promoters, and generally multiple copies are required for higher activity (Stuart *et al.*, 1985; Searle *et al.*, 1985; Westin and Schaffner,

1988; Koizumi *et al.*, 1992b). MREd of mouse MT-I (mMT-I) gene (for convenience, referred to as mMREd) was shown to have an inducible enhancer activity (Westin and Schaffner, 1988). The GC-rich sequence of mMREd adjacent to the MRE core is almost identical to the consensus sequence of the Sp1 binding site, and actually appears to bind Sp1 (Mueller *et al.*, 1988; Westin and Schaffner, 1988).

MRE was expected to be the target site of a specific transcription factor(s), like other regulatory DNA elements. In fact, MREs have been shown to be protected in heavy metal-treated cells by *in vivo* footprinting experiments (Andersen *et al.*, 1987; Mueller *et al.*, 1988). Nuclear factors binding specifically to the MRE sequences have been identified in many laboratories, using binding assays *in vitro* such as mobility shift assay, footprinting, Southwestern blotting and UV-crosslinking (summarized in Table I). Some of these factors require Cd or Zn ions for binding to MRE, being the candidates for transcriptional activators. Interestingly, no factor that responds to multiple metal species has been reported. There are also reports for factors irresponsive to exogenous heavy metals or a factor whose DNA binding is suppressed by metals. Thus, it seems that multiple factors interact with the MRE sequence and regulate MT gene transcription, although not conclusive at present. The properties of individual MRE-binding factors reported so far will be described below. For convenience, these factors are classified according to the response to exogenously added heavy metals in the DNA-binding reaction *in vitro*.

i) Factors responsive to Cd

Séguin and Hamer (1987) identified a MRE-binding factor in mouse L cell nuclear extract, using a mobility shift assay with oligonucleotide probes containing mMREd, which is most powerful in activating transcription among the MREs of the mMT-I gene (Stuart *et al.*, 1985). This factor already exists in the cells untreated with heavy metals. Competition experiments with oligonucleotides containing the mutated MRE sequences showed

Table I. Mammalian MRE-binding factors.

Factor	Origin[a]	Assay[b]	Metal Response[c]	Target Sequence[d]	Reference
	m L	MS	Cd ↑	mMREd	Seguin & Hamer (1987)
	r Fao	MS	Cd ↑	rMT-I	Andersen *et al.* (1990)
MTF-1	h HeLa	MS	Zn ↑	mMREd	Westin & Schaffner (1988)
ZAP	r Liver	MS	Zn ↑	mMREa	Searle (1990)
ZRF	h HeLa	MS	Zn ↑	hMREa	Koizumi *et al.* (1992b)
MEP-1	m Cd^r-L	SW	-	mMREd	Labbé *et al.* (1991)
MBF-I	m L	UV/FP	-	tMREa	Imbert *et al.* (1989)
MRE-BF1 MRE-BF2	h HepG2 etc.	MS/SW	-	hMREa*	Czupryn *et al.* (1992)
MREBP	h HeLa	MS	Cd,Zn,Cu ↓	hMREa	Koizumi *et al.* (1992a)

[a]The names of cell lines are indicated except liver: m, mouse; r, rat; h, human; Cd^r, Cd-resistant. [b]Main assays used for the detection of MRE-binding activities: MS, mobility shift assay; SW, Southwestern blotting; UV, UV-crosslinking; FP, DNase I footprinting. [c]Effects of exogenous heavy metals in DNA-binding assays *in vitro*: ↑, stimulation; ↓, inhibition; -, no effect. [d]Target sequences used as probes in the DNA-binding assays: mMREa and mMREd, MREa and MREd of mouse MT-I gene, respectively; hMREa, MREa of human MT-II$_A$ gene; hMREa*, MREa of human MT-I genes; tMREa, MREa of trout MT-B gene; rMT-I, upstream sequences of rat MT-I gene.

that the MRE core is essential for the factor-DNA interaction. In an assay with the extracts of cells cultured without heavy metals, Cd promoted the complex formation in a dose-dependent manner (maximum at 2.5-80µM), whereas Zn and Cu showed no effect. Formation of this complex is sensitive to EDTA.

Andersen *et al.* (1990) analyzed the extract of rat hepatoma Fao cells using a mobility shift assay with electrophoresis in a tris-glycine buffer system, and found a factor interacting with the upstream sequences of the rat MT-I gene. Competition experiments showed that this factor recognizes at least the MREc region. In an assay with the extract of cells unexposed to heavy metals, the DNA-binding activity of this factor was enhanced by 5μM Cd. In UV-crosslinking experiments, a 39kDa protein was shown to bind to an oligonucleotide containing MREc and MREd.

ii) Factors responsive to Zn

Westin and Schaffner (1988) reported a nuclear factor of HeLa cells (derived from human cervical carcinoma), named MTF-1, that binds specifically to mMREd. The DNA-binding ability of MTF-1 estimated by a mobility shift assay is activated by exogenously added Zn (maximum at 100μM), while not by Cd. As mentioned above, mMREd has an almost perfect match with the consensus of the Sp1 binding site, and its properties *in vivo* and *in vitro* appear to be more complicated than other MREs. Transfection experiments indicated that both the MRE core and the GC-rich sequences are essential for the full activity of mMREd in the heavy metal-dependent transcriptional activation *in vivo*. In the mobility shift assay, an oligonucleotide containing the mMREd sequence binds either Sp1 or MTF-1. Sp1 can bind to an oligomer with mutations in the MRE core, whereas MTF-1 requires both the MRE core and the GC-rich sequences for binding. These results prefers to the hypothesis that the interaction of mMREd with MTF-1, rather than with Sp1, is essential for the metal-inducible transcription. Methylation interference experiments indicated that the binding sites of these two factors are similar, but not identical.

Searle (1990) also identified a similar MRE-binding protein, ZAP (zinc activated protein), in the nuclear extract of rat liver, using a mobility shift assay. ZAP binds to oligomers containing the MREa of mMT-I gene (referred to as mMREa) and its variant, mMREa* (the fourth G in the core sequence is converted to A). The DNA binding of ZAP is dependent on Zn (maximum at 60-100μM), but not on Cd and Cu. Binding and competition experiments using the oligomers with single base changes in the mMREa* sequence indicated that each base in the TGCRCNCG sequence (particularly bases other than the second G and the ambiguous bases at the fourth and the sixth positions) is important for the binding of ZAP. The affinity of ZAP to each of the MREs of the mMT-I gene predicted from the above results is essentially related to their activity in the transcriptional activation *in vivo*, except that mMREd is stronger than expected.

In our laboratory, a HeLa cell nuclear factor similar to the two factors described above was identified. The characteristics of this protein, ZRF, and the relationship with MTF-1 and ZAP will be described later.

iii) Factors irresponsive to heavy metals

Factors described below do not depend on exogenously added heavy metals in binding to MRE *in vitro*. This may reflect their true characteristics *in vivo*; for example, heavy metals might affect their properties other than DNA-binding. However, it remains possible that the irresponsiveness is due to saturation of the factors with endogenous heavy metals, as noted for the yeast ACE1 protein (Fürst *et al.*, 1988).

Imbert *et al.* (1989) detected a nuclear factor, MBF-I, in mouse L cells not exposed to heavy metals, using assays of DNase I footprinting with a probe containing the upstream sequences of trout MT-B (tMT-B) gene and UV-crosslinking with an oligomer probe containing mMREd. MBF-I was purified about 300-fold using chromatography through several columns, including an affinity column with MREa of tMT-B gene (referred to as tMREa) as a ligand, and shown to be a 74kDa protein. Competition experiments in both assays using mutant mMREd oligomers showed that the MRE core is particularly important for the DNA-binding of MBF-I. In footprinting, MBF-I protects tMREa, which has been shown to be functional in L cells. However, the authors noted that in footprinting experiments using a probe with the mMT-I upstream sequences, MBF-I binds only to MREe, that lacks activity in the metal-inducible transcription (Stuart *et al.*, 1985). MBF-I increases several times the transcription from the tMT-B gene promoter *in vitro* in the presence of Zn, although not in the presence of Cd or Cu. In contrast, Zn does not increase the MRE-binding activity of MBF-I in both of the *in vitro* binding assays described above, and it is possible that Zn affects the function of MBF-I in a manner other than enhancement of DNA-binding activity.

An activity protecting the sequences around mMREd in an Exonuclease III (ExoIII) footprinting assay was identified in the nuclear extracts of mouse L cells as well as Cd-resistant (Cd^r) L cells (Séguin and Hamer, 1987; Séguin and Prévost, 1988). Competition experiments with mutant oligomers showed that bases in both the MRE core and the GC-rich sequences are required for the protection. This activity, already existing in L cells unexposed to heavy metals, is not enhanced by exogenously added Cd, but sensitive to chelators such as EDTA and 1,10-phenanthroline. The inhibition of MRE-binding by chelators can be recovered by the addition of Zn, but not by the addition of other heavy metals (Séguin, 1991). Competition experiments with various oligomers indicated that this factor also binds to other MREs of the mMT-I gene as well as those of the human MT-IIA (hMT-IIA) and tMT-B genes (Labbé et al., 1991).

Using Southwestern blotting, the same group identified a 108kDa protein, MEP-1, that binds to an oligomer containing mMREd, in the nuclear extract of Cd^r-L cells (Séguin and Prévost, 1988; Labbé *et al.*, 1991). A homologue is also present in human HeLa cells. Experiments with various probes and competitors showed that the binding of MEP-1 to mMREd requires both the MRE core and the GC-rich sequences. However, MEP-1 is likely to be distinct from Sp1, since MEP-1 does not bind to an oligomer that has the Sp1 binding site but not the MRE core. MEP-1 binds to various MREs as the factor detected by ExoIII footprinting. The similarity in the recognition sequence specificity suggests that these two activities possibly represent the same protein. It is unclear whether the Cd-responsive MRE-binding factor reported by the same group (Séguin and Hamer, 1987) is identical to MEP-1 or not. By UV-crosslinking experiments, the authors detected a 115kDa MRE-binding protein which may possibly represent MEP-1.

Czupryn *et al.* (1992) detected two nuclear factors in several human cell lines including a hepatocellular carcinoma HepG2, using a mobility shift assay with an oligomer probe containing MREa of the human MT-I genes. These factors are not affected by heavy metals *in vitro*, but their levels change after induction by heavy metals such as Zn, Cd and Cu *in vivo*. One of these, MRE-BF1, is present in uninduced cells, but disappears after incubation of the cells in the presence of heavy metals. In contrast, the other factor, MRE-BF2, appears or increases after induction. Southwestern blotting experiments and an analysis of the protein-DNA complexes electroeluted from gel suggested that MRE-BF1 and MRE-BF2 are the proteins with molecular masses of 86kDa and 28kDa, respectively. Formation of both complexes is inhibited by the addition of chelators. Although no data about the recognition sequence specificity of these factors were presented, the authors noted that these complexes are competed out by an excess of the cold MREa oligomer, but not by an oligomer containing mMREd.

iv) A factor suppressed by heavy metals

In our laboratory, a nuclear factor whose binding to MRE is suppressed by several heavy metals was identified in the HeLa cell nuclei. This will be described in the next section.

3. Nuclear factors binding to MREs of the human MT-IIA gene

Recently, we detected two nuclear factors that recognize MREs of the hMT-IIA gene in HeLa cells. These were named MREBP (metal responsive element binding protein) and ZRF (zinc regulatory factor). The properties and possible functions of these factors are described in this section.

MREBP, which we found first, appears to interact weakly with MRE, detectable only by a modified mobility shift assay in which a non-specific competitor poly(dI-dC) is added after mixing of the probe and the nuclear extract (Koizumi *et al.*, 1991). MREBP binds to MREa of the human MT-IIA gene (referred to as hMREa) in the absence of heavy metals (Koizumi *et al.*, 1992a). Competition experiments indicated that the MRE core is essential for the

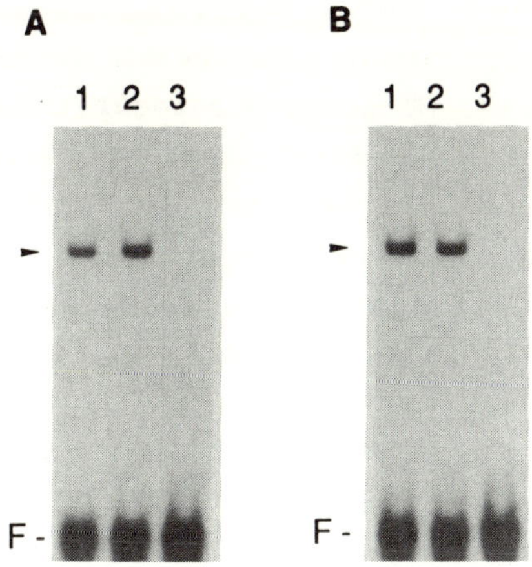

Fig. 1. Effects of heavy metals on the MREBP-MRE complex. HeLa cell nuclear proteins were incubated with a ^{32}P-hMREa oligomer probe in the presence of Cd (A) or Zn (B) and electrophoresed in 5% polyacrylamide gel as described (Koizumi *et al.*, 1991). (A) Lanes 1-3, 0, 10 and 50μM $CdSO_4$, respectively. (B) Lanes 1-3, 0, 50 and 100μM $ZnSO_4$, respectively. The arrowhead indicates the MREBP-DNA complex. F indicates the free probe.

specific binding of MREBP. Heavy metals that induce MTs, including Cd, Zn and Cu, inhibit the complex formation in high concentration ranges (>100μM for Zn, >25μM for Cd and >80μM for Cu)(Fig. 1; Koizumi *et al.*, 1992a; Koizumi and Otsuka, unpublished results). This characteristic implies that MREBP may act as a negative regulator of MT gene transcription. DNase I footprinting experiments showed that the binding affinity of MREBP differs between different MREs of the hMT-II$_A$ gene. MREBP was partially purified by chromatography through ion exchange, gel filtration and hMREa-affinity columns. Southwestern blotting experiments with a hMREa probe suggested that the MREBP activity resides in a 112kDa protein. The possible function of MREBP will be discussed later.

Competition experiments *in vivo* suggested the presence of a heavy metal-dependent, positive regulatory factor(s) (Séguin *et al.*, 1984), and *in vivo* footprinting experiments showed that MREs are protected in heavy metal-exposed cells (Andersen *et al.*, 1987; Mueller *et al*, 1988). According to these findings, it seems unlikely that the MT genes are regulated only by a negative regulator such as MREBP. Therefore, we further analyzed HeLa cell nuclear proteins for the human MRE-binding activity using mobility shift assay under various conditions. Consequently, we found a factor distinct from MREBP, which was named ZRF (Koizumi *et al.*, 1992b). This factor was detected in the nuclei of both heavy metal-exposed and unexposed cells, using a hMREa probe. ZRF is detectable with a 28bp probe that contains hMREa as well as the adjacent sequences, but not with the shorter probe used for the MREBP assay. In fact, it has been confirmed that for the binding of ZRF, additional bases are necessary besides those required for MREBP binding (Koizumi *et al.*, 1992b; Otsuka, Hamer and Koizumi, manuscript in preparation). The MRE-binding

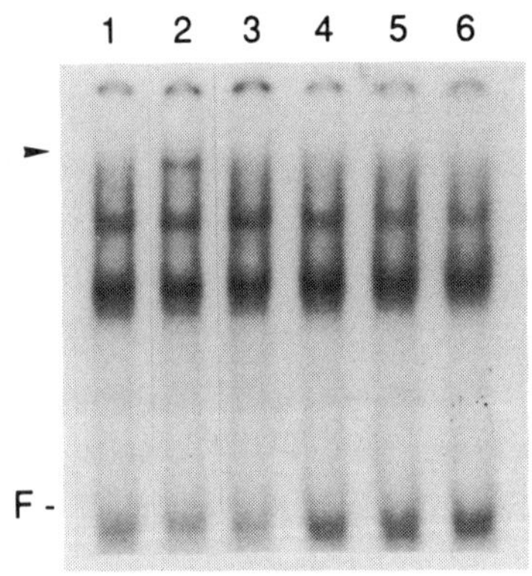

Fig. 2. Effects of heavy metals on the ZRF-MRE complex formation. HeLa cell nuclear proteins were incubated with a 28bp ^{32}P-hMREa oligomer probe with or without heavy metals and electrophoresed in 5% polyacrylamide gel as described (Koizumi *et al.*, 1992b). Lanes: 1, no addition; 2, 100μM $ZnSO_4$; 3 and 4, 10 and 100μM $CuSO_4$, respectively; 5 and 6, 10 and 100μM $HgCl_2$, respectively. The arrowhead indicates the ZRF-DNA complex. F indicates the free probe.

activity of ZRF is activated by exogenously added Zn (maximum at 100μM), but not by other heavy metals such as Cd, Cu and Hg (Fig. 2; Koizumi *et al.*, 1992b). Furthermore, ZRF strongly requires Mg and dithiothreitol for its binding to DNA. The latter requirement may possibly indicate that the cysteine residues of the ZRF protein are involved in the metal-regulated DNA-binding, analogous to the interaction of the MT protein and heavy metals via cysteine residues.

It was found difficult to purify ZRF by conventional chromatographic procedures including an affinity chromatography with hMREa-containing DNA fragments as a ligand, because of the loss of activity and the adsorption of nonspecific proteins to matrixes. However, ZRF could be purified approximately 16,000-fold by repeating a biotin-avidin purification method, in which ZRF bound to a biotinylated hMREa probe is isolated using streptoavidin beads (Otsuka, Hamer and Koizumi, manuscript in preparation). The purified protein has a molecular mass of 116kDa. In UV-crosslinking experiments as well as in the biotin-avidin system, recovery of both the 116kDa protein and the ZRF activity was highly specific to the probe DNA sequence and entirely dependent on the presence of Zn, confirming that the 116kDa protein is the entity of ZRF. In DNase I footprinting, the purified protein specifically protects MREa and MREb in the hMT-IIA upstream region (Fig. 3). A MRE-like sequence overlapping the cap site is also weakly protected. The ZRF-MRE complex preformed in the presence of Zn is easily dissociated by the addition of 20mM EDTA or 2mM o-phenanthroline, suggesting the direct and essential role of Zn ions.

The property of ZRF is similar to those of MTF-1 (Westin and Schaffner, 1988) and ZAP (Searle, 1990) described in the previous section; all of these are activated only by Zn in binding to MRE *in vitro*. However, there are minor differences in their recognition sequences. For the specific binding to MRE, MTF-1 requires the GC-rich sequence in addition to the MRE core, whereas ZRF does not require the former. Within the MRE sequence, the requirement for bases is very similar between ZRF

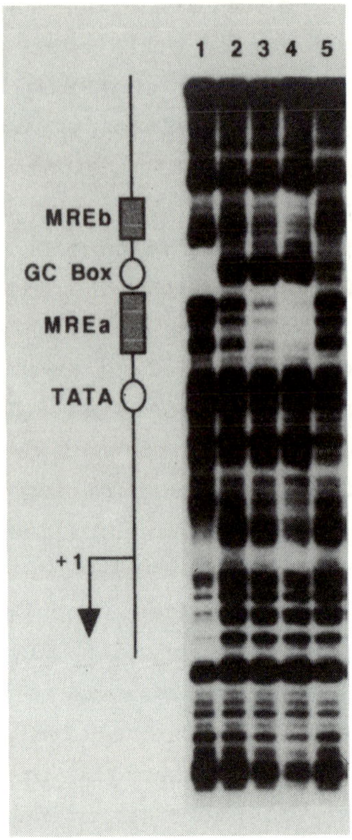

Fig. 3. DNase I-footprints of ZRF. A ZRF preparation purified by the biotin-avidin method was incubated with a ^{32}P-DNA probe containing the upstream sequences of the hMTII$_A$ gene (-174 to +71; the 5'-terminal of non-coding strand was labeled). Lanes: 1, recombinant Sp1 (1 footprint unit); 2-4, ZRF (7.5, 15 and 30μl, respectively); 5, no protein. The positions of regulatory elements are shown on the left. The cap site is indicated by +1.

and ZAP. However, the sequence just upstream of the MRE core is also important for ZRF binding, but less important for ZAP binding. At present, informations for the comparison of these factors are not sufficient, since they were assayed under different experimental conditions, for example, using the probes with different MRE sequences. The identity or homology of these factors should be concluded by the structural analysis of the genes coding for them.

In order to know the functions of MREBP and ZRF, we examined the activity of their recognition sequences *in vivo* (Koizumi *et al.*, 1992b). Four tandem copies of either the 14bp or 28bp MREa oligomers (used for the binding assays of MREBP and ZRF, respectively) were inserted upstream of the herpes simplex virus thymidine kinase (TK) gene promoter fused to the chloramphenicol acetyl transferase (CAT) reporter gene. After transfection of plasmids containing these gens into HeLa cells, the expression of CAT in the presence or absence of Zn was estimated. The 28bp oligomer could mediate the Zn-dependent expression of CAT, whereas the 14bp oligomer could not. These results strongly suggest that the binding of ZRF to MRE is important in the Zn-dependent transcriptional activation. Together with the results of the binding experiments *in vitro*, ZRF is likely to be a factor that activates transcription by binding to MRE in the presence of Zn. In contrast, MREBP does not appear to have a direct role in the Zn-inducible transcription. It seems likely that the function of MREBP is to shut off unnecessary transcription when the intracellular concentration of heavy metals is low, to prevent other transcription factors that recognize related sequences from binding to MRE, or to select particular MREs available for transcriptional activators by binding to each MRE with different affinities.

In the *in vitro* binding reaction, the ability of ZRF to bind to hMREa is activated specifically by Zn. It appeared possible that this specificity is due to the property of the MREa sequence; that is, each of the multiple MREs present within the upstream region of the hMT-II$_A$ gene may have its own response to a particular heavy metal species. However, cells transfected with the plasmid containing the 28bp hMREa-TK-CAT fusion gene expressed CAT in response not only to Zn but also to Cd and Cu (Koizumi *et al.*, 1992b). The Zn-specific DNA binding of ZRF thus does not result from the property of MREa, but is likely to be due to the property of the ZRF protein. It is known that mMREd is also responsive to multiple heavy metal species (Culotta and Hamer, 1989). It remains unclear whether the transcriptional activation of the hMT-IIA gene by heavy metals other than Zn is mediated by factors distinct from ZRF, or whether the conditions in the binding reaction *in vitro* are inappropriate for reproducing the function of ZRF *in vivo*.

Purification and gene cloning of the MRE-binding factors are now in progress in several laboratories. In the near future, more accurate informations about the mechanism of the MT gene expression will be obtained, through the studies of primary structure and expression of the regulatory protein genes, as well as the characterization of purified or recombinant proteins.

Acknowledgement

We thank Mrs. K. Suzuki for her assistance in preparing this manuscript.

REFERENCES

Andersen, R. D., Taplitz, S. J., Oberbauer, A. M., Calame, K. L. and Herschman, H. R. (1990) Nucleic Acids Res. 18, 6049-6055.

Andersen, R. D., Taplitz, S. J., Wong, S., Bristol, G., Larkin, B. and Herschman, H. R. (1987) Mol. Cell. Biol. 7, 3574-3581.

Angel, P., Imagawa, M., Chiu, R., Stein, B., Imbra, R. J., Rahmsdorf, H. J., Jonat, C., Herrlich, P. and Karin, M. (1987) Cell 49, 729-739.

Buchman, C., Skroch, P., Dixon, W., Tullius, T. D. and Karin, M. (1990) Mol. Cell. Biol. 10, 4778-4787.

Buchman, C., Skroch, P., Welch, J., Fogel, S. and Karin, M. (1989) Mol. Cell. Biol. 9, 4091-4095.

Butler, G. and Thiele, D. J. (1991) Mol. Cell. Biol. 11, 476-485.

Butt, T. R., Sternberg, E. J., Gorman, J. A., Clark, P., Hamer, D., Rosenberg, M. and Crooke, S. T. (1984) Proc. Natl. Acad. Sci. USA 81, 3332-3336.

Carter, A. D., Felber, B. K., Walling, M., Jubier, M.-F., Schmidt, C. J. and Hamer, D. H. (1984) Proc. Natl. Acad. Sci. USA 81, 7392-7396.

Carthew, R. W., Chodosh, L. A. and Sharp, P. A. (1987) Genes Dev. 1, 973-980.

Casas, F. J., Hu, S., Hamer, D. and Karpel, R. L. (1991) FEBS Lett. 281, 205-208.

Culotta, V. C. and Hamer, D. H. (1989) Mol. Cell. Biol. 9, 1376-1380.

Culotta, V. C., Hsu, T., Hu, S., Furst, P. and Hamer, D.(1989) Proc. Natl. Acad. Sci. USA 86, 8377-8381.

Czupryn, M., Brown, W. E. and Vallee, B. L. (1992) Proc. Natl. Acad. Sci. USA 89, 10395-10399.

Dameron, C. T., Winge, D. R., George, G. N., Sansone, M., Hu, S. and Hamer, D. (1991) Proc. Natl. Acad. Sci. USA 88, 6127-6131.

Fogel, S. and Welch, J. (1982) Proc. Natl. Acad. Sci. USA 79, 5342-5346.
Fürst, P. and Hamer, D. (1989) Proc. Natl. Acad. Sci. USA 86, 5267-5271.
Fürst, P., Hu, S., Hackett, R. and Hamer, D. (1988) Cell 55, 705-717.
Hamer, D. H., Thiele, D. J. and Lemontt, J. E. (1985) Science 228, 685-690.
Hu, S., Fürst, P. and Hamer, D. (1990) New Biol. 2, 544-555.
Huibregtse, J. M., Engelke, D. R. and Thiele, D. J. (1989) Proc. Natl. Acad. Sci. USA 86, 65-69.
Imagawa, M., Chiu, R. and Karin, M. (1987) Cell 51, 251-260.
Imbert, J., Zafarullah, M., Culotta, V. C., Gedamu, L. and Hamer, D. (1989) Mol. Cell. Biol. 9, 5315-5323.
Kambadur, R., Culotta, V. and Hamer, D. (1990) Proc. Natl. Acad. Sci. USA 87, 9168-9172.
Karin, M., Haslinger, A., Holtgreve, H., Richards, R. I., Krauter, P., Westphal, H. M. and Beato, M. (1984a) Nature 308, 513-519.
Karin, M., Najarian, R., Haslinger, A., Valenzuela, P., Welch, J. and Fogel, S. (1984b) Proc. Natl. Acad. Sci. USA 81, 337-341.
Koizumi, S., Otsuka, F. and Yamada, H. (1991) Chem.-Biol. Interact. 80, 145-157.
Koizumi, S., Suzuki, K. and Otsuka, F. (1992a) J. Biol. Chem. 267, 18659-18664.
Koizumi, S., Yamada, H., Suzuki, K. and Otsuka, F. (1992b) Eur. J. Biochem. 210, 555-560.
Labbé, S., Prévost, J., Remondelli, P., Leone, A. and Séguin, C. (1991) Nucleic Acids Res. 19, 4225-4231.
Lee, W., Haslinger, A., Karin, M. and Tjian, R. (1987a) Nature 325, 368-372.
Lee, W., Mitchell, P. and Tjian, R. (1987b) Cell 49, 741-752.
Mermod, N., Williams, T. J. and Tjian, R. (1988) Nature 332, 557-561.
Mitchell, P. J., Wang, C. and Tjian, R. (1987) Cell 50, 847-861.
Mueller, P. R., Salser, S. J. and Wold, B. (1988) Genes Dev. 2, 412-427.
Nakagawa, K. H., Inouye, C., Hedman, B., Karin, M., Tullius, T. D. and Hodgson, K. O. (1991) J. Am. Chem. Soc. 113, 3621-3623.
Searle, P. F. (1990) Nucleic Acids Res. 18, 4683-4690.
Searle, P. F., Stuart, G. W. and Palmiter, R. D. (1985) Mol. Cell. Biol. 5, 1480-1489.
Séguin, C. (1991) Gene 97, 295-300.
Séguin, C., Felber, B. K., Carter, A. D. and Hamer, D. H. (1984) Nature 312, 781-785.
Séguin, C. and Hamer, D. H. (1987) Science 235, 1383-1387.
Séguin, C. and Prévost, J. (1988) Nucleic Acids Res. 16, 10547-10560.
Stuart, G. W., Searle, P. F., Chen, H. Y., Brinster, R. L. and Palmiter, R. D. (1984) Proc. Natl. Acad. Sci. USA 81, 7318-7322.
Stuart, G. W., Searle, P. F. and Palmiter, R. D. (1985) Nature 317, 828-831.

Szczypka, M. S. and Thiele, D. J. (1989) Mol. Cell. Biol. 9, 421-429.
Thiele, D. J. (1988) Mol. Cell. Biol. 8, 2745-2752.
Thiele, D. J. and Hamer, D. H. (1986) Mol. Cell. Biol. 6, 1158-1163.
Welch, J., Fogel, S., Buchman, C. and Karin, M. (1989) EMBO J. 8, 255-260.
Westin, G. and Schaffner, W. (1988) EMBO J. 7, 3763-3770.
Zhou, P., Szczypka, M. S., Sosinowski, T. and Thiele, D. J. (1992) Mol. Cell. Biol. 12, 3766-3775.
Zhou, P. and Thiele, D. J. (1991) Proc. Natl. Acad. Sci. USA 88, 6112-6116.

Subject Index

Acknowledgements

The support of the following people and the financial help of the following companies is gratefully acknowledged:

Jeremias H.R. Kägi (Honorary President)
Masami Kimura (President)
Takashi Sugimura, Bert L. Vallee, Michael Webb (Advisory Board)

Organizing Committee:
Nobumasa Imura (Secretary)
Kazuo T. Suzuki (Treasurer)
Masao Akimoto, M. George Cherian, Dean H. Hamer, Curtis D. Klaassen, Akira Naganuma, Gunnar F. Nordberg, Masao Sato, Chiharu Tohyama, Ian Bremner, Bruce A. Fowler, P. C. Huang, Yutaka Kojima, Katsuyuki Nakajima, Nagahiro Saijo, Keiichi Tanaka, Dennis R. Winge

Societies:
Japan Society for Biomedical Research on Trace Elements
The Japanese Society of Toxicological Sciences
The Research Foundation for Pharmaceutical Sciences
Japanese Cancer Association
Japanese Society of Agricultural Chemistry
The Japanese Biochemical Society
The Japanese Society for Hygiene
The Molecular Biology Society of Japan
The Pharmaceutical Society of Japan

Sponsors:
Asahi Chemical Industry Co., Ltd.
Banyu Pharmaceutical Co., Ltd.
Bayer Yakuhin, Ltd.
Biotec Lab Inc Co, Ltd.
Bristol-Myers Squibb Company
Chugai Pharmaceutical Co., Ltd.
Ciba-Geigy (Japan) for the Promotion of Science
Daiichi Pharmaceutical Co., Ltd.
Dainippon Pharmaceutical Co., Ltd.
Eisai Co., Ltd.
Fujisawa Pharmaceutical Co., Ltd.
Fuso Pharmaceutical Industries, Ltd.
Horiuchi Itaro & Co., Ltd.
Igaku-Shoin Ltd.
Inoue Foundation for Science
Japan Immunoresearch Laboratories Co., Ltd.
Kanebo Ltd.
Kirin Brewery Co., Ltd.
Kodama Ltd.
Kosekihideo Shoten Co., Ltd.
Kyorin Pharmaceutical Co., Ltd.
Meiji Milk Products Co., Ltd.
Meiji Seika Kaisha, Ltd.
Mitsubishi Heavy Industries, Ltd.
Mitsui Pharmaceutical, Ltd.
Nagai, Yasushi Ph.D.
Nippon Instruments Corporation
Nippon Kayaku Co., Ltd.
Nippon Roche K.K.
Otsuka Pharmaceutical Co., Ltd.
Sankyo Co., Ltd.
Seikagaku Corporation
Shibayagi Co., Ltd.
Smith Kline Beecham
SRL, Inc.
Sumitomo Chemical Co., Ltd.
Sumitomo Pharmaceutical Co., Ltd.
Taiho Pharmaceutical Co., Ltd.
Takara Shuzo Co., Ltd. Biomedical Group
Takeda Chemical Industries, Ltd.
Teikoku Hormone MFG., Co., Ltd.
The Asahi Glass Foundation
The Japan Racing Association
The Kitasato Institute
The Tokyo Biochemical Research
The Tsukuba EXPO'85 Memorial Foundation
Toa Eiyo Ltd.
Toyama Chemical Co., Ltd.
Tsumura & Co.